Martha Hobbs.

West Newton.

April. 21st 1882.

THE PUBLISHERS BEG LEAVE TO CALL THE ATTENTION OF THE PUBLIC TO THE FOLLOWING UNSOLICITED NOTICES OF THE ELEMENTARY GEOLOGY, FROM GENTLEMEN EMINENTLY QUALIFIED TO JUDGE OF ITS MERITS.

FROM GIDEON A. MANTELL, LL.D. F.R.S. F.G.S., &c. &c. LONDON, AUTHOR OF THE WONDERS OF GEOLOGY, &c.

"I have obtained a copy of your Treatise on Geology. It is an admirable work. It has been my carriage companion for some time."

FROM PROF. B. SILLIMAN, LL.D. OF YALE COLLEGE.

"I am greatly in fault in not having answered your kind letter of Aug. 20th. with a copy of your valuable work on Geology. I took the work with me to the west in the expectation of looking it over: and although I failed to read it satisfactorily, I glanced at it enough to convince me of its high value, and shall recommend it in my Lectures."

FROM PROF. J. W. WEBSTER, OF HARVARD UNIVERSITY.

"I have just received a copy of your 'Elementary Geology,' for which I beg you to accept many thanks. I am thankful that you have found time to present us with so excellent a view of the science, and shall recommend the work warmly to the class attending my lectures.

FROM PROF. C. DEWEY, OF ROCHESTER, N. Y.

"I introduced your Geology into our Academy. Part of it is hard reasoning for minds not pretty well matured. Still it is so vastly better than any thing in the English language with which I am acquainted, that I boast over it. It is admirable for the College course."

FROM PROF. HENRY D. ROGERS, OF THE UNIVERSITY OF PENNSYLVANIA.

"I thank you sincerely for a copy of your work, and yet more for presenting us with an Elementary Treatise on Geology in a form so well adapted to the wants of instructors. Having for several years past felt the want of just such a book for my Class in the University, I hailed its appearance with real satisfaction."

FROM PROF. W. W. MATHER, GEOLOGIST TO ONE OF THE DISTRICTS OF N. YORK, AND TO THE STATE OF OHIO.

"I have examined your little work on Geology with much interest and satisfaction. It presents a large mass of matter in a small compass; is lucid, concise, and its materials are arranged in the most convenient form for the student. It seems to form a happy medium between the more elementary books for schools, and those for the more advanced students of geology. Its copious references to various works on geology, will be a great advantage to those who choose to go to the original sources and dive deeper into the various subjects discussed."

FROM PROF. J. W. BAILEY, OF THE MILITARY ACADEMY, WEST POINT.

"I have recently perused with much pleasure, your Elementary Geology, and consider it a most valuable contribution to science, and highly creditable to yourself and our country. I am glad we have such a work to which to refer students. If I had know of your publication sooner, I should have adopted it as our text book: but the Class had already provided themselves with Lyell's work. I shall recommend its adoption next year, if as is almost certain, I meet with no work in the mean time better suited to our *peculiar* wants at this Institution."

From Prof. C. B. Adams, of Middlebury College.

"Your elementary book on geology has afforded me great pleasure; and I have, since our Catalogue was printed, adopted it as a text book."

The following notices of the work, from among the many that have appeared, have been selected from some of the leading periodicals of the country.

From the American Journal of Science and Arts, for October, 1840.

"The readers of this Journal and those who know the progress of American Geology, are well aware of the important services Prof. Hitchcock has rendered to this branch of science, through a period of many years, both by his laborious explorations and his written works. In the present instance, he has attempted to prepare a work which shall fill a vacancy long felt by the instructors of geology in this country, a work which, while it gives a good view of the progress of the science in other countries, draws its illustrations mainly from American facts. From the rapid glance which we have been able to bestow upon this performance, we should think that Prof. Hitchcock had succeeding in imparting this feature to his book."

From the American Biblical Repository, for October, 1840.

"The appearance of this volume from the pen of Prof. Hitchcock, will be peculiarly gratifying to many in the community. It is designed to be used as a Text Book for classes in geology, in Colleges and other Seminaries of learning, and also, to supply the wants of the general reader, who has not the leisure to study the numerous and extended treatises that have been written on different heads of this subject. The plan of it, we think, is admirably adapted to the first of these uses, and nearly or quite as well suited to the second."

From the North American Review, for January, 1841.

"Professor Hitchcock has been too long and favorably known to scientific men, both of the new world and of the old, to make it necessary for us to say, with what ample qualifications he undertakes the task before him. His work is no 'secondary formation,' based on the published works of European writers, but in every part bears the impress of acute and original observation, and happy tact in presenting the immense variety of subjects treated in the following Sections into which the book is divided.

"The fifth Section is devoted to Organic Remains. It occupies one fourth of the whole work, and is illustrated with the best cuts in the book. We venture to say that there is not in our language so neat and compressed, yet so clear and correct an account of the 'Wonders of Geology.'"

ELEMENTARY GEOLOGY.

BY

EDWARD HITCHCOCK, D.D., LL.D.,

PRESIDENT OF AMHERST COLLEGE, AND PROFESSOR OF NATURAL THEOLOGY AND GEOLOGY.

A NEW EDITION,

REVISED, ENLARGED, AND ADAPTED TO THE PRESENT ADVANCED STATE OF THE SCIENCE.

WITH AN INTRODUCTORY NOTICE.

BY JOHN PYE SMITH, D.D., F.R.S., & F.G.S.

DIVINITY TUTOR IN THE COLLEGE AT HOMERTON, LONDON.

EIGHTH EDITION.

NEW YORK:
PUBLISHED BY NEWMAN AND IVISON,
199 Broadway.
CINCINNATI: MOORE & ANDERSON. AUBURN: J. C. IVISON & CO.
CHICAGO: S. C. GRIGGS & CO.

1852.

STEREOTYPED BY THOMAS B. SMITH,
216 WILLIAM STREET, NEW YORK.

J. D. BEDFORD, PRINTER,
138 FULTON STREET

PREFACE.

In preparing this work, three objects have been kept principally in view. The first was to prepare a Text Book for my Classes in Geology: the second, to bring together the materials for a Synopsis of Geology, to be appended to my Final Report on the Geology of Massachusetts, now in the press: And the third was, to present to the public a condensed view of the present state of geological facts, theories, and hypotheses; especially to those who have not the leisure to study very extended works on this subject. In its execution, the work differs from any with which I am acquainted, in the following particulars. 1. It is arranged in the form of distinct Propositions or Principles, with Definitions and Proofs: and the Inferences follow those principles on which they are mainly dependent. This method was adopted, as it long has been in most other sciences, for the convenience of teaching: but it also enables one to condense the matter very much. 2. An attempt has been made to present the whole subject in its proper proportions; viz. its facts, theories, and hypotheses, with their historical and religious relations, and a sketch of the geology of all the countries of the globe that have been explored. All geological works with which I am acquainted, either omit some of these subjects, or dwell very disproportionably upon some of them. 3. It is made more American than republications from European writers, by introducing a greater amount of our geology. 4. It contains copious references to writers, where the different points here briefly discussed, may be found amply treated. 5. It contains a *Palæontological Chart*, whose object is to bring under a glance of the eye, the leading facts respecting organic remains. Whether these peculiarities of the present work will be regarded by the public as improvements, important enough to deserve their patronage, time only can show.

Type of two sizes is employed in this work. The most important principles, facts, and proofs, are in larger type, to call the special attention of the student or reader: while many of the details and remarks are in smaller type. The subject is subdivided into the following heads; whose abbreviations will not need explanation: viz. *Definition: Principle: Description Inference: Remark: Proof: Details: Illustration.* Where an inference depends upon several principles, I have added a synopsis of all the proofs on which it rests.

In European countries especially, and to a good degree in our own country, geology has become a popular and even fashionable study. In most of our higher Seminaries of learning, it is explained by at least a course of lectures. But in Institutions of a lower grade, it receives far less attention than its merits deserve. Why should not a science, whose facts possess a thrilling interest; whose reasonings are admirably adapted for mental discipline, and often severely task the strongest powers; and whose results are many of them as grand and ennobling as those of Astronomy itself; (such Astronomers as Herschel and Whewell being judges,*) why should not such a science be thought as essential in education as the kindred branches of Chemistry and Astronomy? That all the parts of this Science are not yet as well settled as those of Astronomy and Chemistry, is no objection to making it a branch of education, so long as every intelligent man must admit that its fundamental facts and principles are well established.

Amherst College,
Aug. 1, 1840.

THIRD EDITION.

In presenting thus early a third edition of this work to the public, I will only say, that I have done all in my power to introduce into it all the important discoveries and improvements which have been recently made in the science.

To enforce still more strongly the remarks of Dr. Smith, in the unsolicited Introductory Notice of this work which follows, respecting the importance of an acquaintance with Geology to the minister and the missionary, I will quote a few sentences from the letters of two esteemed missionary friends, now in active service in distant lands. Rev. Justin Perkins, American Missionary in Ooroomiah in Persia, under date of Oct. 1, 1839, thus writes.

"Did not my missionary work press upon me so constantly, and with such mountain weight, I should feel strongly tempted to study geology, (of which I know very little,) so wonderfully

---

* "Geology, in the magnitude and sublimity of the objects of which it treats, undoubtedly ranks, in the scale of sciences, next to astronomy."—*Sir John Herschel.*

interesting, in a geological point of view, does the face of Persia appear to me. Indeed, I often feel that this interesting and important science has peculiar claims on American Missionaries. Visiting, as they do, all portions of the world, they enjoy opportunities of contributing to it, with almost no sacrifice of time or effort, which are possessed by no other class of American citizens. I know not that I can better atone for my own deficiency in this respect, than by requesting you, in my behalf, to urge upon the missionary students in College, the high importance of their obtaining a good practical knowledge of geology and mineralogy, while attending your lectures, as they would enhance their usefulness, in future life. It is the combined light of ALL TRUTH, *scientific*, as well as *religious*, which is to render so perfect and glorious the splendor of millenial day!"

Rev. Ebenezer Burgess, Missionary at Ahmednuggur in India, writes under date of Nov. 1, 1841, as follows:

"Did I possess an intimate acquaintance with mineralogy, it would be of great use to me in going over the country. When we go out to evangelize, it is very pleasant to be able to *geologize* and *botanize*: and such abilities render our trips far more subservient to the preservation of health. I think students err very much in not devoting to your department (in College) the full amount of time which is allotted to it in the course of studies. Perhaps they do differently in these days of reform. Every individual, who has a moderate ability for such studies, could with a little system and resolution, and without detriment to his other studies, acquire that knowledge of these branches which would be of great value to him in after life, especially if he intends to be a missionary. There are times when an additional source of relaxation or amusement is worth just about a man's life. There is now an individual connected with this mission, who has been raised from a state of great debility and weakness, by turning his attention to botany and mineralogy: at least, such appear to be the means which God has used. Of all situations, the missionary in heathen lands most needs pleasant sources of relaxation. When his mind sinks to a certain point under discouragement, or the weight of responsibility, if he has nothing with which to divert it, it begins to prey upon itself; and then there is no remedy but to leave the field."

Amherst College,
Apl. 1842.

CONTENTS.

INTRODUCTORY NOTICE.

BY DR. J. PYE SMITH, OF LONDON.

In a manner unexpected and remarkable, the opportunity has been presented to me of bearing a public testimony to the value of Dr. Hitchcock's volume, ELEMENTARY GEOLOGY. This is gratifying, not only because I feel it an honor to myself, but much more as it excites the hope that, by this recommendation, theological students, many of my younger brethren in the evangelical ministry, and serious christians in general, who feel the duty of seeking the cultivation of their own minds, may be induced to study this book. For them it is peculiarly adapted, as it presents a comprehensive digest of geological facts and the theoretical truths deduced from them, disposed in a method admirably perspicuous, so that inquiring persons may, without any discouraging labor, and by employing the diligence which will bring its own reward, acquire such a knowledge of this science as cannot fail of being eminently beneficial. It is no exaggeration to affirm that Geology has close relations to every branch of Natural History and to all the physical sciences, so that no district of that vast domain can be cultivated without awakening trains of thought leading to geological questions; and, conversely, the prosecution of knowledge in this department, cannot fail to excite the desire and to disclose the methods of making valuable acquisitions to the benefit of human life. In our day, through every degree of extensiveness, from the perambulation of a parish to the exploring of an empire, TRAVELLING has become a "universal passion," and action too. Within a very few years, the interior of every continent of the earth has been surveyed with an intelligence and accuracy beyond all example. Who can reflect, for instance, upon the activity now so vigorously put forth, for introducing European civilization, the arts of peace, the enjoyment of security, and the influence of the most benign religion, into the long sealed territories of Central Asia, and not be filled with astonishment and delightful anticipation? Similar labors are in progress upon points and in directions innumerable, reaching to the heart of all the other vast regions of the globe: and the men to whom we owe so much and from whom so much more is justly expected, are geologists, as well as transcendent naturalists in the other departments. Whoever would run the same career must possess the same qualifications. Even upon the smallest scale of provincial travelling for health, business, or beneficence, acquaintance with natural objects opens a thousand means of enjoyment and usefulness.

The spirit of these reflections bears a peculiar application to the ministers of the gospel. To the pastors of rural congregations, no means of recreating and preserving health are comparable to these and their allied pursuits; and thus, also, in many temporal respects, they may become benefactors to their neighbours. In large towns the establishment of libraries, lyceums, botanic gardens, and scientific associations, is rapidly diffusing a taste for these kinds of knowledge. It would be a perilous state for the interests of religion, that "precious jewel" whose essential characters are "wisdom, knowledge, and joy," if its professional teachers should be, in this respect, inferior to the young and inquiring members of their congregations. For those excellent men who give their lives to the noblest of labors, a work which would honor angels, "preaching among the heathen the unsearchable riches of Christ;" a competent acquaintance with natural objects, is of signal importance, for both safety and usefulness. They should be able to distinguish mineral and vegetable products, so as to guard against the pernicious and determine the salubrious; and very often geological knowledge will be found of the first utility in fixing upon the best localities for missionary stations; nor can they be insensible to the benefits of which they may be the agents, by communicating discoveries to Europe or the United States of America.

To answer these purposes, and especially in the hands of the intelligent and studious ministers of CHRIST, this work of Professor Hitchcock appears to me especially suited. Though I flatter myself that I have studied with advantage the best English treatise on Geology, and find ever new improvement and pleasure from them; and have also paid some attention to French and German books of this class; I think it no disparagement to them to profess my conviction that, with the views just mentioned, this is the book which I long to see brought into extensive use. The plan on which it is composed, is different from that of any other, so far as I know, in such a manner and to such a degree, that it is not an opponent or rival to any of them. Yet, in this arrangement of the matter, there is no affectation: all is plain, consecutive and luminous. It is more comprehensive with regard to the various relations and aspects of the science, than any one book with which I am acquainted; and yet, though within so moderate limits, it does not disappoint by unsatisfactory brevity or evasive generalities. Such is the impression made upon me by the first edition of the "Elementary Geology," and I cannot entertain a doubt but that the ample knowledge and untiring industry of the author will confer every practicable improvement upon his proposed new edition.

I received a deep conviction of the Professor's extraordinary merits, from his "Report" upon the Geology, Botany and Natural History generally, of the Province of Massachusetts. made by the command of the state government; a large volume, published in 1833, and the second edition in 1835: and from his papers in the "American Biblical Repository," which were of great service to me in composing a book on "The Relation between the Holy Scriptures and some parts of Geological

Science." But I did not till recently know that he was a "faithful brother and fellow-laborer in the gospel of Christ." An edifying manifestation of this, it has been my privilege to receive in Dr. Hitchcock's "Essay and Sermon on the lessons taught by sickness," prefixed to "A Wreath for the Tomb, or Extracts from Eminent Writers on Death and Eternity." It is my earnest prayer that great blessings from the God of all grace may attend the labors of my honored friend.

J. PYE SMITH.

Homerton College, near London,
March 16, 1841.

EIGHTH EDITION.

THE main thing attempted in this edition, has been to introduce all the important discoveries and new theoretical views which have appeared since the publication of the third edition. This has required an alteration of a large part of the stereotype plates, so rapidly is the science advancing: and yet these discoveries scarcely affect any important principles.

I had intended to append to this edition a simplified copy of a new Geological Map of the entire globe, which has lately been published by M. Boue, under the auspices of the Geological Society of France, and also a Geological Map of the United States. But finding that they would swell the size and the price of the work too much, they will be published in a separate form, with a few pages of description, under the title of "*An Outline of the Geology of the Globe, and of the United States in particular; with two Geological Maps.*" This little work, which Mr. Newman will shortly bring out, ought to be in the hands of all who study the Elementary Geology, as it will give at a glance the great outlines of all that is known, either by observation or analogy, of the geological structure of the globe.

Amherst College, January, 1847.

ELEMENTARY GEOLOGY.

SECTION I.

GENERAL ACCOUNT OF THE CONSTITUTION AND STRUCTURE OF THE EARTH, AND OF THE PRINCIPLES ON WHICH ROCKS ARE CLASSIFIED.

Definition. GEOLOGY is the history of the mineral masses that compose the earth, and of the organic remains which they contain.

Remark 1. Some writers divide Geology into two branches; 1. *Geognosy*, or Positive Geology, which embraces only the known facts of the science. 2. *Geogony*, or Speculative Geology, which attempts to point out the causes of those facts, and the inferences that result from them. See *Traité Elémentaire de Geologie par M. Rozet: Discours Préliminaire, p. V. Also, Dict. Classique d' Hist. Naturelle: Art. Geologie.* Others make three divisions. 1. Physical Geography. 2. Geognosy. 3. Geogony. See *Elements de Geologie, par J. G. Omalius d' Halloy, p.* 1. Others embrace all legitimate theory under Geogony, and confine the hypothetical part of the subject to Geology. See *Tableau des Terrains, par Alexandre Brongniart, p.* 2. But these distinctions are of little importance, and not much used by English and American writers.

Rem. 2. A very philosophical and excellent division of Geology is proposed by Professor Whewell. 1. *Descriptive, or Phenomenal Geology,* embraces the facts. 2. *Geological Dynamics* gives an exposition of the general principles by which such phenomena can be produced. 3. *Physical Geology* states the doctrines as to what have been the causes of the existing state of things. *History of the Inductive Sciences, Vol.* 3. *p.* 488, *London,* 1837.

Rem. 3. A division of geology of some practical value is the following.

1. *Economical Geology*, or an account of rocks with reference to their pecuniary value, or immediate application to the wants of society.

2. *Scenographical Geology*, or an account of rocks as they exhibit themselves to the eye in their general outlines: in other words, an account of natural scenery.

Observation. With the exception of vegetation, natural scenery is all produced by the rocks that constitute the surface: and the geologist can often determine the nature of rocks by the peculiarities of their great outlines.

3. *Scientific Geology*, or the history of rocks in their relation to science, or philosophy.

Def. Every part of the globe, which is not animal or vegetable, including water and air, is regarded as *mineral.*

Def. The term *rock*, in its popular acceptation, embraces only the solid parts of the globe: but in geological language, it includes also the loose materials, the soils, clays, and gravels,—that cover the solid parts.

Principle. The form of the earth is that of a sphere, flattened at the poles: technically, an oblate spheroid. The polar diameter is about 26 miles shorter than the equatorial.

Proof 1. Measurement of a degree of the meridian in different latitudes. 2. Astronomical phenomena; particularly the precession of the equinoxes.

Inference. Hence it is inferred that the earth must have been once in a fluid state; since it has precisely the form which a fluid globe, revolving on its axis with the same volocity as the earth, would assume.

Prin. Taken as a whole, the earth is about five times heavier than water; or 2.5 times heavier than common rocks.

Proof 1. Careful observations upon the relative attracting power of particular mountains and the whole globe, with a zenith sector. 2. The disturbing effect of the earth upon the heavenly bodies.

Inf. We hence learn that the density of the earth increases from the surface to the centre: but it does not follow that the nature of the internal parts is different from its crust. For in consequence of condensation by pressure, water at the depth of 362 miles, would be as heavy as quicksilver; and air as heavy as water at 34 miles in depth: while at the centre, steel would be compressed into one fourth, and stone into one eighth of its bulk at the surface. *Mrs. Somerville's Connexion of the Physical Sciences, p.* 91. *Fifth London Edition.*

Description. The surface of the earth, as well beneath the ocean as on the dry land, is elevated into ridges and insulated peaks, with intervening vallies and plains. *Rozet, p.* 86.

Descr. The highest mountains are about 28,000 feet above the ocean level, and the mean height of the dry land does not exceed two miles. *De la Beche's Manual of Geology, p.* 2. *Third Edition.*

Details. The height of a few of the most elevated mountains on the globe is as follows: *See Encyclopædia of Geography, Vol.* 3. *p.* 605.

Himmalayah Mountains, E. Indies,	27,677 feet.
Rocky Mountains,	12,000 "
Mouna Kea,	18,000 "
Chimborazo, Andes,	21,000 "
Ararat,	17,700 "
Mount Blanc, Europe,	15,668 "

Descr. The mean depth of the ocean is probably between two and three miles. *De la Beche's Man. Geo. p.* 2. It has been calculated from the phenomena of tides that the Atlantic in its middle part is above nine miles deep. *Phillips's Geology, p.* 23. *Edinburgh,* 1838.

Inf. Hence it appears that the present dry land might be spread over the bottom of the ocean, so that the globe should be entirely covered with water. For nearly three fourths of the surface is at present submerged.

Obs. While the great lakes of North America are elevated some of them even 600 feet above the ocean, some of the inland seas of Asia are sunk below its level. It seems to be at last settled that the Caspian Sea is 84 feet below the Black Sea. *Murchison's Geology of Europe, Vol.* 1. *p.* 322, *London,* 1845. *Lyell's Principles of Geology, Vol.* 1. *p.* 267, *Second American, from the Sixth London Edition, Boston,* 1842. The Dead Sea in Palestine is sunk 1337 feet below the Mediterranean. *Biblical Researches in Palestine, &c., by Messrs. Robinson and Smith, Vol.* 2. *p.* 222, *Boston,* 1841: *Also American Jour. Science, Vol.* 42. *Jan.* 1842, *p.* 215.

Stratification.

Def. The rocks that compose the globe are divided into two great classes, the STRATIFIED and UNSTRATIFIED.

Def. Stratification consists of the division of a rock into regular masses, by nearly parallel planes, occasioned by a peculiar mode of deposition. Strata vary in thickness from that of paper to many yards.

Obs. Strata are often very tortuous and sometimes quite wedge shaped. Nevertheless, the fundamental idea of stratification is that of parallelism in the layers. *Macculloch's Classification of Rocks, p.* 100: *also his System of Geology, Vol.* 1. *p.* 67: *also Greenough's Geology, chap.* 1.

Def. The term *stratum* is sometimes employed to designate the whole mass of a rock, while its parallel subdivisions are called *beds* or layers. The term bed is also employed to designate a layer, whose shape may be more or less lenticular, or wedge shaped, included between the layers of a more extended rock; as a bed of gypsum, a bed of coal, a bed of iron, &c. In this case the bed is sometimes said to be subordinate.

Def. When beds of different rocks alternate, they are said to be interstratified.

Def. A *seam* is a thin layer of rock that separates the beds or strata of another rock: ex. gr. a seam of coal, of limestone, &c.

Descr. A bed or stratum is often divided into thin laminæ, which bear the same relation to a single bed, as that does to the whole series of beds. This division is called the *lamination* of the bed; and always results from a mechanical mode of deposition. The appearance of fissility which it gives to a rock, is often deceptive; since the layers separate with great difficulty. This is especially true in gneiss.

Descr. The lamination is sometimes parallel to the planes of

* See note A. Addenda.

stratification; sometimes they are much inclined to each other; and often it is undulating and tortuous.

Fig. 1., shows the different kinds of lamination.

Fig. 1.

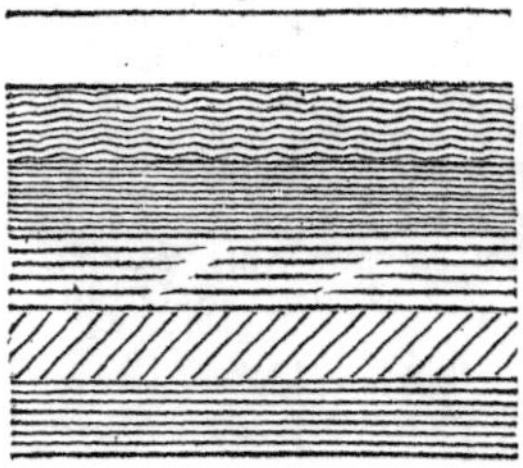

Without Laminæ.
With waved Laminæ.
Finely Laminated.
Coarsely Laminated.
Obliquely Laminated.
Parallel Laminæ.

Fig. 2, is a case of very contorted lamination, in a stratum of gneiss two or three feet thick, copied from a loose block in Colebrook, in Connecticut.

Fig. 2.

Contorted Laminæ of Gneiss: Colebrook, Ct.

Fig. 3, represents a bowlder of mica slate, 5 feet high and 6 feet broad, lying on the east side of the river stage road in Lyme, New Hampshire, north of the centre of the town. It will be seen, that in the upper part, where the laminæ have been bent most, there is a crack. In the lower part are two large tuberculous masses of quartz.

Fig. 3.

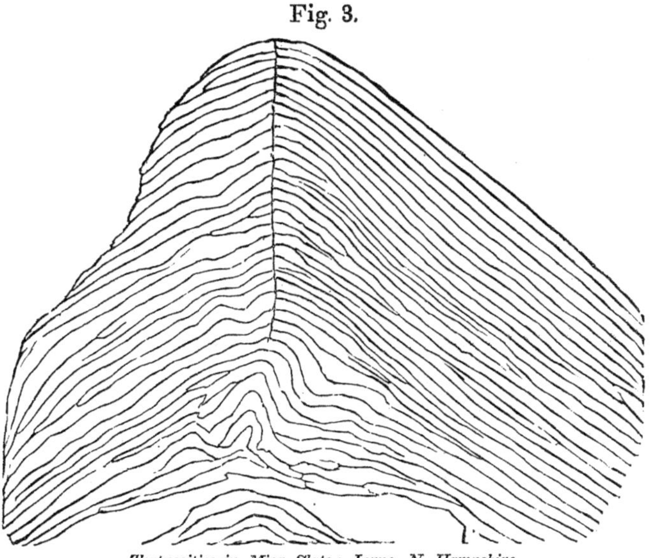

Tortuosities in Mica Slate: Lyme, N. Hampshire.

Inf. The laminæ of beds could not have been deposited originally in the curved position shown in the preceding figures: hence the flexure must have been the result of some subsequent operation.

Inf. Hence the layers at some period after their deposition, must have been in a state so plastic that they could be bent without breaking.

Origin of the varieties of Lamination.

Causes. All the lamination of stratified rocks was undoubtedly produced originally by deposition in water, and the varieties have resulted from modifying circumstances. 1. The parallel laminæ are the result of quiet deposition upon a level surface. 2. The waved lamination, in many instances, is nothing but *ripple marks;* such as are seen constantly upon the sand and mud at the bottom of rivers, lakes, and the ocean. In the secondary rocks this is too manifest to be mistaken. 3. The oblique lamination has generally been the result of deposition upon a steep shore, where the materials are driven over the edge of an inclined plane. 4. Highly contorted lamination has often resulted from lateral and vertical pressure, as illustrated by Fig. 4.

Fig. 4.

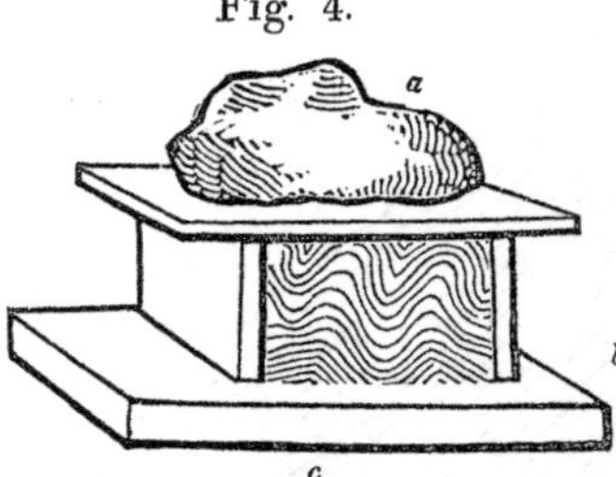

Illustration. If pieces of cloth of different colors be placed upon a table *c*, and covered by a weight, *a*, and then lateral forces *b*, *b*, be applied; while the weight will be somewhat raised, the cloth will be folded and contorted precisely like the laminæ of many rocks; as is shown in the figure.

Prin. The agency of water and heat is sufficient to bring rocks, in nearly every known case, into that plastic state which is necessary to make them bend without breaking.

Proof. 1. Water alone renders clay eminently plastic; and it imparts a degree of plasticity to nearly every variety of unconsolidated strata; so that this, without heat, may have prepared all such strata for the flexures which they exhibit. Water also renders some solid rocks quite flexible; as limestones and sandstones; and it penetrates very deep into the solid crust of the globe; so that some of the flexures, even in the solid rocks, may have been produced by forces acting upon them when saturated with water.

2. Flexures are the most abundant and extensive in the vicinity of rocks that have been melted. And it is admitted that all the older stratified rocks have been exposed more or less to the influence of heat from the unstratified; and it is also true, that rocks may be heated almost to the melting point without destroying their stratified and laminated structure. So that in heat, we have a cause sufficient to prepare rocks for any possible degree of contortion.

Prin. Volcanic forces have operated from beneath upon most of the older rocks, whereby they have been bent upward. The weight of the ocean, of drift, &c., has bent them downward; gravity and other agencies more local, have produced a lateral pressure; especially when the strata were highly inclined: and these various agencies will account for nearly every case of flexure, not only of the laminæ, but of the beds also. But more on this point further on.

Descr. In clay beds containing disseminated carbonate of lime, we frequently find nodules of argillo-calcareous matter, sometimes spherical, but more usually flattened. These are generally called *claystones*, and the common impression is, that they were rounded by water. But they are unquestionably the result of molecular attraction. The slaty divisions of the clay extend through the concretions: and on splitting them open, a leaf, or a fish, or some other organic relic, is frequently found. In New England, however, both the slaty cleavage and the organic nucleus are usually wanting.

Rem. In England these claystones are sometimes supposed to have been turned in a lathe, and to have been used for money. In this country they are usually thought to be the work of water, or of the aborigines. I find that those of New England have at least six predominant forms. *Final Report on the Geology of Massachusetts, Vol.* 2. *p.* 410.

Fig. 5.

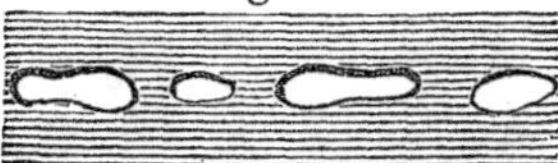

Fig. 5, will convey an idea of the manner in which these concretions are situated in the clay.

Descr. Similar concretions abound in argillaceous iron ore, which is often disseminated in clay beds or shale. These nodules, are usually made up of concentric coats of the ore: but sometimes the slaty structure of the rock containing them, extends through them, and organic relics are found to form their nucleus.

Fig. 6. represents a concretion of hydrate of iron from the clay cliffs of Gay Head in Massachusetts. The axis consists of a piece of lignite, and its resemblance to a pear is very striking: or rather to a large garden squash: for its diameter is more than seven inches.

Fig. 6.

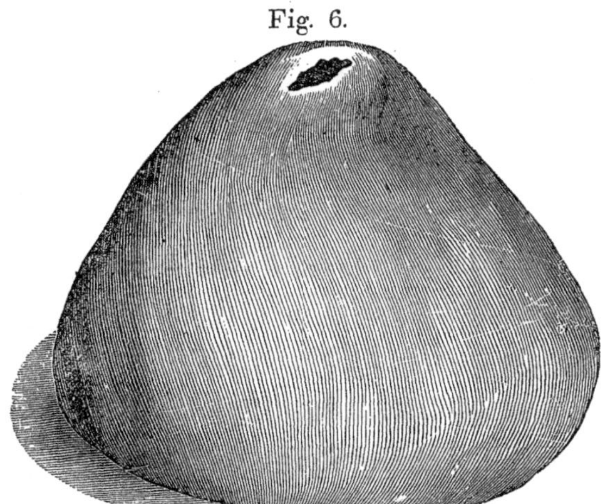

Concretion of Iron Ore: Gay Head.

Descr. The internal parts of these concretions of limestone and hydrate of iron, often exhibit numerous cracks, which sometimes divide the matter into columnar masses, but more frequently into irregular shapes. When these cracks are filled with calcareous spar, as is often the case in calcareous concre-

Fig. 7.

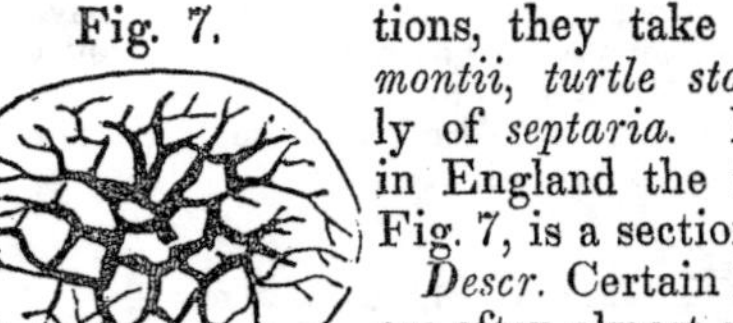

tions, they take the name of *ludus helmontii*, *turtle stones*, or more frequently of *septaria*. From these is prepared in England the famous Roman cement. Fig. 7, is a section of one of these.

Descr. Certain limestones called oolites, are often almost entirely composed of concretions made up of concentric layers: but the spheres are rarely so large as a pea.

Descr. The concretionary structure, however, often exists in limestone on a very large scale, forming spheroidal masses not only many feet, but many yards in diameter.

Divisional Structures.

Def. Both the stratified and unstratified rocks are traversed by divisional planes, called *joints;* which divide the mass into determinate shapes, which are different from beds and their laminæ. Those only which occur in the stratified rocks will be here noticed.

Descr. The most important of these joints, called *master-joints*, are more or less parallel, and so extended as to imply some general cause of production.

Descr. When these joints cross the beds obliquely, as they usually do, and there are two sets of them, they divide the rock into rhomboidal masses of considerable regularity; though wanting in that perfect equality in the corresponding angles of the prisms which is found in crystals of a simple mineral.

Figs. 8 and 9, are examples: the first from the unconsolidated clay beds of West Springfield in Massachusetts, and the latter from the gneiss of Monson in the same State.

Fig. 8. Fig. 9.

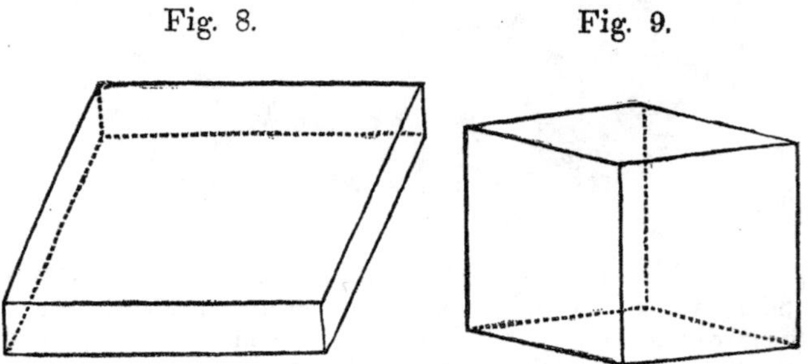

Descr. Other divisional planes separate the rock into irregular fragments: and sometimes the fissures are filled with calcareous spar, or other mineral substance.

Cleavage.

Fig. 10.

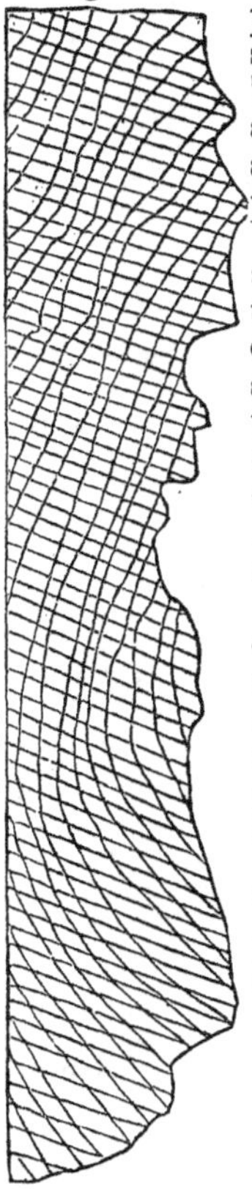

Cleavage Planes intersecting curved Strata

Descr. Some rocks are divided by a set of parallel planes, coincident neither with the stratification, the lamination, nor the joints. These are called *cleavage planes*, because they are supposed always to result from a crystalline arrangement of the particles of the rock, superinduced after its original deposition.

Descr. This cleavage is most common in argillaceous slate, and in many cases constitutes its slaty structure. But in many instances this structure is the result of original deposition, and corresponds to, or rather constitutes, the lamination. This is particularly true of the finer slates, both argillaceous, micaceous, talcose and chloritic, in this country. *Lyell's Elements of Geology, Vol.* 2. *p.* 390. *Second Amer. Edition, Boston,* 1842.

Descr. The cleavage planes are often inclined to the planes of stratification and lamination at an angle of 30° or 40°, and sometimes the the two planes dip in opposite directions. The cleavage planes are remarkable for their almost perfect parallelism, while the strata and their laminæ are usually contorted.

Fig. 10, exhibits a set of cleavage planes crossing the curved strata in the slate rocks of Wales.

In Fig. 11, are exhibited the planes of stratification, BB, B B; the joints, A A, A A; and the slaty cleavage, d, d.

Fig. 11.

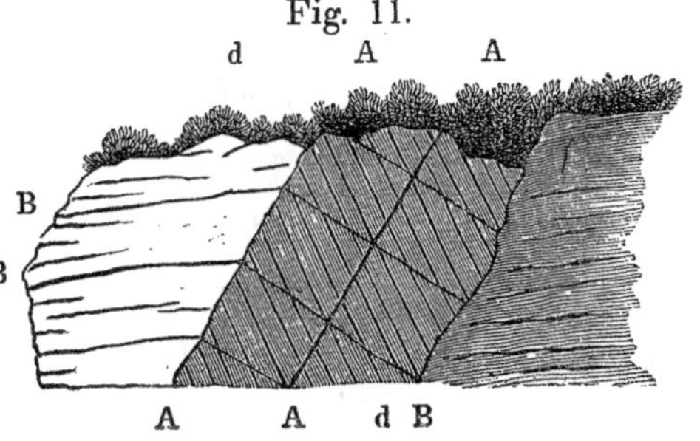

Descr. Strata and laminæ may be distinguished from joints and cleavage, 1. By the alternation of different materials in the former. 2. By a difference of organic remains in the successive layers. 3. By ripple marks and tortuosities. 4. By a difference in color of successive portions of the rock.

Descr. Joints may be distinguished from cleavage planes chiefly by two marks. 1. A jointed structure rarely extends through large masses of rocks; at least not without more interruption than is found in cleavage. 2. The portion of rock included between two joints is not capable of a subdivision by parallel planes; but in cleavage the subdivision may be carried on to an extreme degree of fineness.

Origin of the different Structures in the Stratified Rocks.

First Cause. Original deposition from water. This will explain all the phenomena of stratification and lamination.

Second Cause. Desiccation. By contracting the mass of the rock, it is compelled to separate into fragments: but this can explain only some of the more irregular divisional structures, such as those of septaria and iron stone.

Third Cause. A mechanical force acting beneath, by which fissures are produced, either parallel or radiating from a centre, or without symmetry in their direction. This may explain some varieties of joints.

Fourth Cause. Heat. This must be supposed intense enough to give so much of mobility to the particles that they can obey molecular attraction, and assume a partially crystalline form. In this way, probably nearly every case of cleavage was produced, and many cases of a jointed structure.

Fifth Cause. Water. If by this agent the particles can be made to move among one another,—(as they will do even by partial diffusion,)—they can then assume a crystalline arrangement, as in the case of heat. And the example of distinct joints, which I have given in Fig. 8, occurring in unconsolidated horizontal clay beds, seems to require such a cause as this for its explanation; although I have seen no similar case described by writers on geology. But in West Springfield and Deerfield in Massachusetts, these joints are very numerous and distinct; occurring, however, in only a few of the layers of clay, while those above and below are unaffected. This clay has certainly never been subjected to any great degree of heat, being of very recent origin.

Sixth Cause. Galvanic Electricity. The recent experiments of Mr. Robert Weare Fox of Great Britain, show that clay, subjected to a long voltaic action becomes laminated, so as to resemble clay slate in its structure. Very probably an electric agency is essential in those cases where heat and water seem to produce the effect; and that these latter causes operate chiefly by exalting the electricities and giving mobility to the particles.

Rem. In the difficult subject of the structure of rocks, I have followed chiefly the original views of Professor Sedgwick, in his paper on the slate of England and Wales, in the *Geological Transactions, Vol.* 3, *p.* 461, *Second Series.* Also Prof. John Phillips, in his *Treatise on Geology,* 2 *Vols. London,* 1837 *and* 1839; also Mr. Lyell, in his *Elements of Geology.* The subject will probably need still further attention before it is perfected. Yet much has been done.

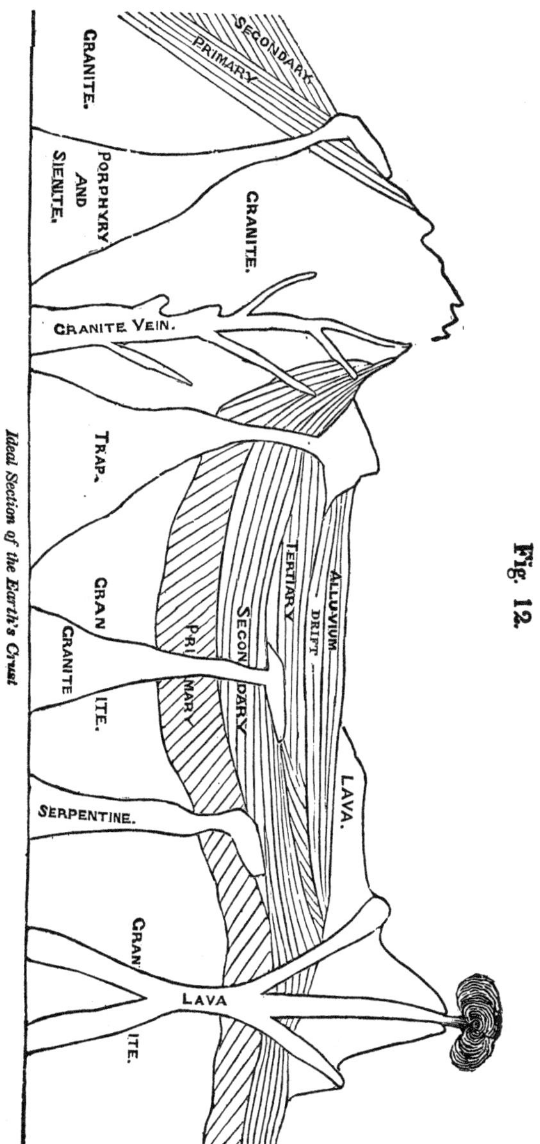

Fig. 12.

Ideal Section of the Earth's Crust

Descr. The unstratified rocks occur in four modes. 1. As irregular masses beneath the stratified rocks. 2. As veins crossing both the stratified and unstratified rocks. 3. As beds of irregular masses thrust in between the strata. 4. As overlying masses. All these modes are shown in Fig. 12.

Rem. The phenomena of veins, being very important, require a more detailed explanation.

Def. Veins are of two kinds. 1. Those of segregation. 2. Those of injection. The former appear to have been separated from the general mass of the rock by elective affinity, when it was in a fluid state; and consequently they are of the same age as the rock. Hence they are often called contemporaneous veins.

Illus. Fig. 13, represents a bowlder of granitic gneiss, in Lowell, about five feet long, traversed by several veins of segregation, whose composition differs not greatly from that of the rock, except in being harder and more distinctly granitic. Where veins of this description cross one another, they coalesce so that one does not cut off the other. The surface of the gneiss near Merrimack river in Middlesex County is often covered with reticulations produced by veins of this description.

Fig. 13.

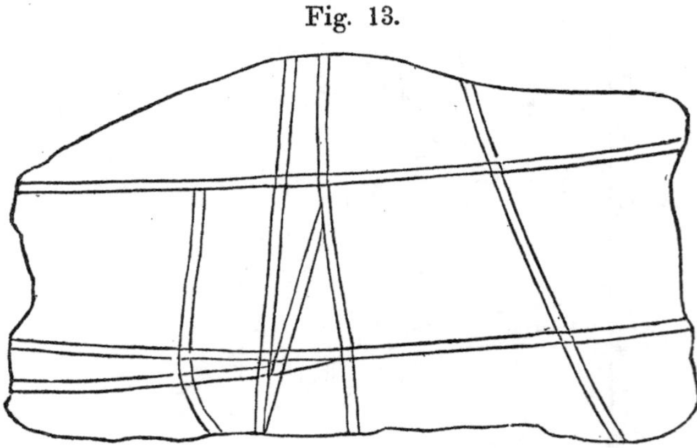

Veins of Segregation in Gneiss: Lowell.

Obs. Quartz veins are usually veins of segregation. But I have seen some of this description in Massachusetts a foot or more in width, perfectly straight and continuous, for a great distance through talcose slate, that are with difficulty referable to such an origin.

Def. The second class were once open fissures, which at a subsequent period, were filled by injected matter.

Descr. Veins of segregation are frequently insulated in the containing rock; they pass at their edges by insensible gradations into that rock; and are sometimes tubercular or even nodular.

Descr. Injected veins can often be traced to a large mass of similar rock, from which, as they proceed, they often ramify and become exceedingly fine, until they are lost. Usually, especially in the oldest rocks, they are chemically united to the walls of the containing rock; but large trap veins have often very little adhesion to the sides.

Fig. 14, exhibits granite veins protruding from a large mass of granite into hornblende slate in Cornwall.

Fig. 14.

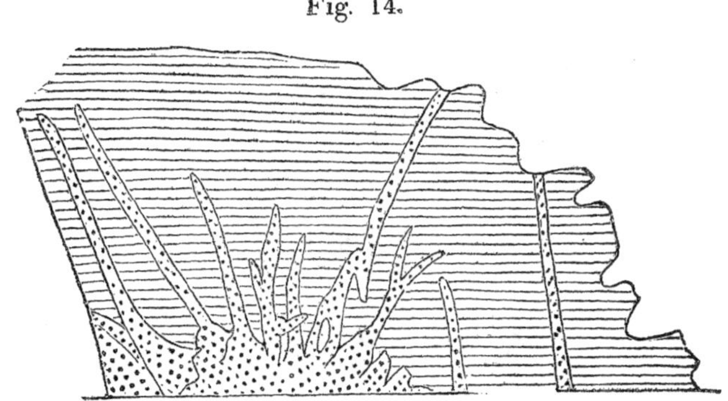

Granite Veins in Hornblende Slate: Cornwall, Eng.

Def. The large veins that are filled with trap rock or recent lava, are usually called *dykes*. These differ from true veins, also, in rarely sending off branches.

Inf. It is hence inferred, either that the matter of dykes was less fluid, because less hot, than granite, or that the rocks through which veins pass, had a higher temperature than those into which dykes were injected: so that the latter sooner cooled the fluid matter than the former. Hence, when the small lateral fissures were produced, as they probably usually were by the heat of the injected matter, the granite flowed into them, while the trap had become too hard

Descr. Trap dykes are sometimes several yards wide, and extend 60 or 70 miles; as in England and Ireland.

Descr. Dykes and veins frequently cross one another; and in such a case, the one that is cut off, is regarded as the oldest.

Rem. Undoubtedly this rule, in general, can be depended on for determining the relative age of injected veins. Yet it is easy to conceive how a vein of considerable size might be filled with matter not very fluid, without filling all the lateral fissures; and should these be subsequently filled with more perfectly fluid matter, they would appear to be cut off by the larger vein; and hence, by the rule, be regarded as the oldest, although in fact more recent.

Prin. By this rule it may be shown that granite has been erupted at no less than four different epochs.

Rem. I do not find any European writer describing more than three epochs of the eruption of granite. *Macculloch's System of Geology, Vol.* 1. *p.* 137.

Illus. Fig. 15, represents a bowlder of granite in Westhampton, Mass., whose base was the product of the earliest epoch of eruption. This is traversed by the granite vein, *a, a, a,* which was injected at a second epoch; *b,* is a granite vein cutting *a,* and therefore produced at a third epoch; while *b,* as well as *a,* are cut off by the granite veins *c,* and *d,* of a fourth epoch.

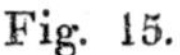
Fig. 15.

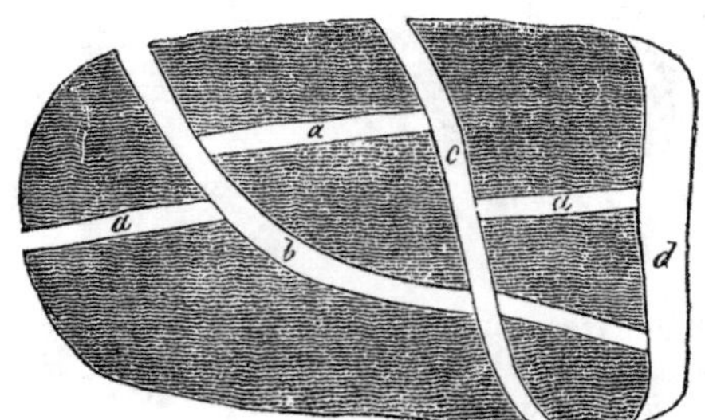

Granite Veins in Granite: Westhampton.

Prin. By the same rule can be proved successive eruptions of the trap rocks.

Illus. Fig. 16, shows several interesting veins in syenite in the north part of Cohasset, Mass. No. 1, is the basis of syenite. No. 2, a dyke of porphyry, 10 feet wide. No. 3, dykes of common greenstone, the largest 20 feet wide. No. 4, dykes of common greenstone, 5, 6, and 8 feet wide. Here then we have four successive epochs of eruption.

Fig. 16.

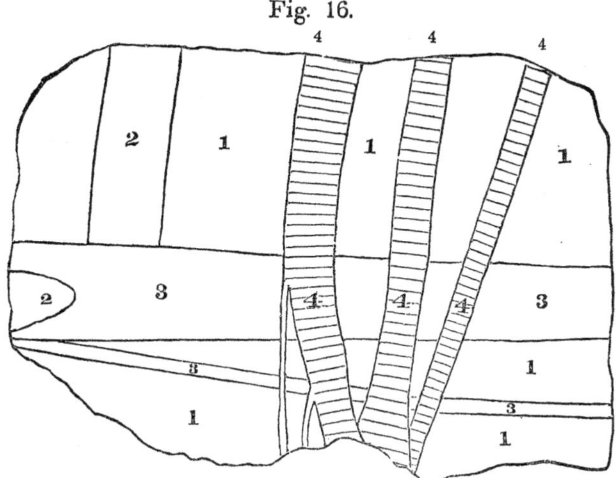

Dykes in Syenite: Cohasset, Mass.

Fig. 17, exhibits veins in metamorphic gneiss on the sea shore in Beverly, Mass. The gneiss, is almost converted into syenite. No. 1, is a vein of granite. No. 2, dykes of common greenstone. No. 3, dykes of porphyritic greenstone. These last are obviously the oldest; and one of them is very much displaced.

Fig. 17.

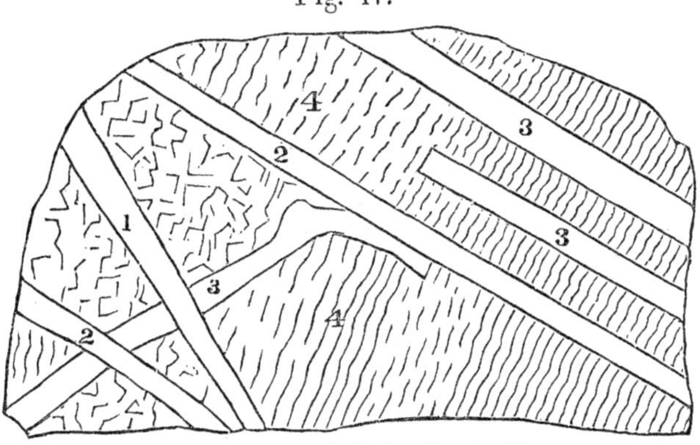

Dykes in Metamorphic Gneiss: Beverly, Mass.

Descr. In one remarkable example of veins of different kinds, I have been able to trace *eleven epochs* of the eruption of unstratified rocks. (*See* ***Plate*** 2.)

Illus. This case is in the city of Salem, Mass., near the entrance of the bridge leading to Beverly on the west side. The basis rock, No. 1, is syenitic greenstone. The oldest vein, or dyke, (2) is greenstone, a few inches wide. Nos. 3 and 4, are veins of reddish granite,—nearly all felspar, which cut across the greenstone dyke, No. 2. These are very numerous; much more so than is shown upon the drawing; and of very irregular width,—often branching out into strings a mere line in breadth. They belong to at least two epochs of eruption: for some of them are cut off by the others, and probably still more eruptive epochs might be traced among them: but they are so complicated that I have not been able to do it. No. 5, is a dyke of greenstone, which cuts off 3 and 4. No. 6, which is 40 inches wide, is porphyritic greenstone, and cuts off No. 5. A small dyke and nearly parallel, a little to the left, appears to be of the same age. No. 7, is porphyritic greenstone cutting off No. 6. No. 8, (of which there are two running nearly parallel,) intersects Nos. 5 and 7, and is granite or felspar. No. 9, consists of two large dykes of greenstone, which cut off all the others that have been described, except No. 8; and perhaps this also: but the intersection is covered by soil. No. 10, of which there are small veins near the bottom of the sketch, and near the top, and is of the same kind of granite as Nos. 3 and 4, intersects nearly all the preceding veins. Finally, No. 11, consists of the same kind of granite veins, a mere line in width, running diagonally across the sketch. The whole space represented, is 36 by 27 feet, and the lower part of it is covered by the ocean at high tide, and the upper surface by soil.

I have spent a good deal of time in examining this complicated and very interesting net work of veins and dykes; and I cannot see why we have not evidence here, of the extraordinary fact—unique so far as I know—of *eleven* successive eruptions of granite and trap rock. Or if we regard the basis rock as metamorphic;—that is, formed by the fusion of gneiss;—and it may be so,—still, we have ten subsequent injections!

Descr. Veins and dykes usually cross the strata at various angles. But not unfrequently for a part of their course they have been intruded between the strata; and hence have been mistaken for beds, and have given rise to the inquiry whether granite is not stratified.

Descr. Dykes are usually nearly straight; but granite veins are sometimes very tortuous.

Illus. Fig. 19, shows two small but very distinct granite veins in homogeneous micaceous limestone in Colrain, Mass. If these are injected veins, as they appear to be, it is extremely difficult to account for their tortuosity.

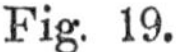

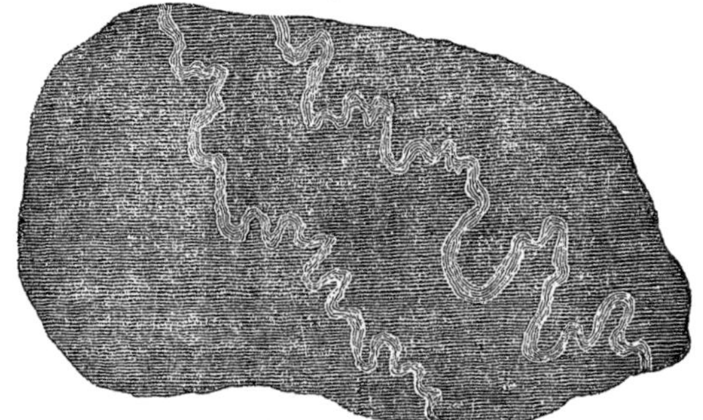

Granite Veins in Micaceous Limestone: Colrain, Mass.

Fig. 20.

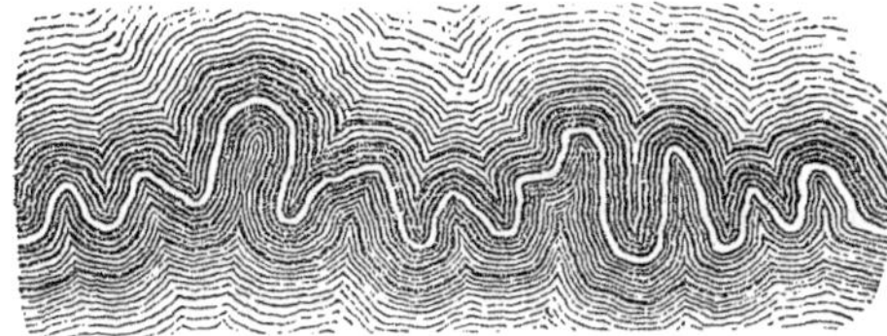

Fig. 20, represents a granite vein only one eighth of an inch thick, conforming to the flexures of mica slate in Conway, Ms. Perhaps this may be a segregated vein.

Fig. 21, (see next page) is a tortuous vein of granite in talcose slate in Chester, Mass., which does not conform to the flexures of the slate. It is possible, however, that it may correspond to the original curves of deposition, and that the present slaty structure is superinduced; though I have strong reasons to think that this is not the case.

Descr. In modern volcanos the lava is ejected from circular vents, called craters. But the older unstratified rocks, although evidently of volcanic origin, appear to have been protruded along extended fissures, either across the strata, or in the same direction as their strike. It is possible, however, that craters did once exist, but have been swept away by powerful denudation of the surface.

Example. A range of greenstone commences at New Haven in Connecticut, and runs diagonally across the valley of the Connecticut, until it terminates in Franklin County in Massachusetts. It must be 70 or 80 miles in length, and one or two in breadth, and it conforms generally to the strike of the sandstone. Similar ranges exist in New Jersey.

Descr. The unstratified rocks, especially when exposed to the weather, are usually divided into irregular fragments by fissures in various directions.

Descr. Sometimes, however, these rocks have a concretionary structure on a large scale; that is, they are composed of concreted layers whose curvature is sometimes so slight, that they are mistaken for strata.

Fig. 21.

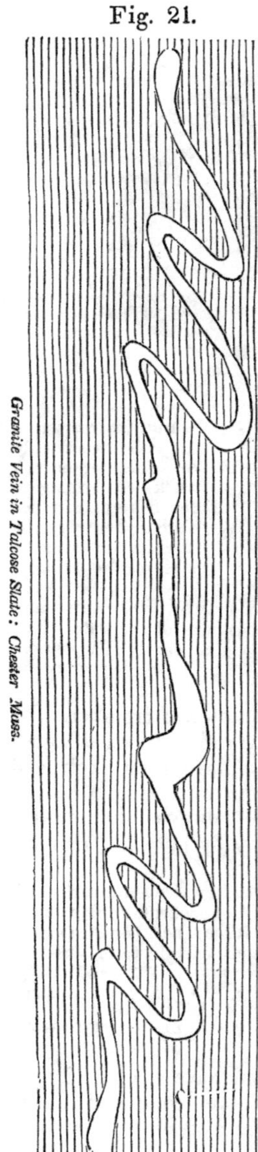

Granite Vein in Talcose Slate: Chester Mass.

Obs. Cases of this sort can be distinguished from stratification, first by the concreted divisions not extending through the whole rock; secondly, by the want of a laminar or slaty structure in the parallel masses.

Exam. A fine example of this concreted structure occurs at one of the quarries in syenite near Sandy Bay on Cape Ann. Another occurs in granite, at the granite quarry in the hill east of the village of Worcester. Another in Fitchburg. Another in the trap rock at Nahant.

Def. An interesting variety of jointed structure in some of the unstratified rocks, is the prismatic or columnar, by which large masses of rocks are divided into regular forms, from a few inches to several feet in diameter; but with no spaces between them.

Obs. This curious phenomenon will be more particularly described in a subsequent section.

Descr. The layers of the stratified rocks are sometimes horizontal; but more frequently they are tilted up so as to dip beneath the horizon at every possible angle.

Def. The angle which the surface of a stratum makes with the plane of the horizon is called its *inclination*, or *dip;* and the direction of its upturned edge is called its *strike* or *bearing.*

Obs. Of course horizontal strata have neither strike nor dip. The exposure of a stratum at the surface is called in the language of miners, its *out-crop* or *basseting.*

Descr. As a general fact, the newer or higher rocks are less inclined than those below. The highest are usually horizontal; while the oldest are often perpendicular.

Obs. This admits of too many exceptions to be employed as a means for determining the age of rocks. Thus, a considerable part of the gneiss rocks of New England (usually the oldest of the stratified rocks,) has a less dip than the sandstones, or even than some of the tertiary rocks of the same region.

Descr The instrument employed for ascertaining the dip of a stratum, is called a *clinometer*. Every geologist, however, ought to be able to determine the dip with sufficient accuracy for most purposes by the eye. A good pocket compass will answer for finding the strike.

Descr. Unstratified rocks do not probably occupy one tenth part of the earth's surface.

Descr. In Great Britain, says Dr. Macculloch, "they do not cover a thousandth part of the superficies of the island." In Massachusetts, they occupy nearly a quarter of the surface.

Prin. These rocks, however, we have reason to suppose, occupy the internal parts of the earth to a great depth, if not to the centre; over which the stratified rocks are spread with very unequal thickness, and in many places are entirely wanting.

Explanation. Fig. 12, will convey a better idea than language, of the relative situation of the two classes of rocks. The different groups of stratified rocks, viz. Alluvium, Drift, Tertiary, Secondary and Primary, are here shown resting upon one another, and upon granite beneath. This granite, also, is shown protruding to the surface; and upon its sides lie the stratified rocks highly inclined: veins of syenite, porphyry, trap, serpentine and lava, are also shown protruding through the granite, and coming from beneath it; as they must do, because they have been erupted since the granite. Veins of syenitic granite of a posterior date are likewise shown, penetrating the stratified rocks to the top of the secondary strata, which is the most recent granite yet discovered. Syenite and porphyry rise no higher than the top of the secondary: but the trap rises to the top of the tertiary; and finally, modern lava overspreads alluvium. The stratified rocks are represented as inclined at different angles; the lowest being the most tilted up. Although, therefore, this is not a section of any particular portion of the earth's crust, yet it will give a correct idea of the relative situations of the groups both of stratified and unstratified rocks. For a much larger and more detailed section of this sort, see *Buckland's Bridgewater Treatise*, *Plate* 1.

Formations.

Rem. It is not possible in geology as in other departments of natural history, to describe species with definite and invariable characters; because each rock is found to be made up of varieties, often very numerous, which insensibly graduate into one another; as do also the rocks themselves in many instances. In botany, mineralogy and zoology, species are separated by definite lines, and never do thus pass into one another. If, therefore, we employ in geology the same exactness of specific description as in these sciences, we shall impose on nature a logical precision which will not fit her in this department of her works. Another method must, therefore, be adopted. See *Macculloch's Geological Classification of Rocks*, *Chap.* 1.

Def. Each rock, in its most extended sense, consists of several varieties, agreeing together in certain general characters, and occupying such a relative situation with respect to one another, as to show that all of them were formed under similar circumstances, and during the same geological period. Such a group constitutes a *formation.* Ex. gr. graywacke formation, gneiss formation, &c.

Def. This term often embraces several distinct rocks, when there is reason to suppose them the result of the same geological period.

Fig. 22. will give an idea of the English lias formation lying between the oolite formation above, and the red sandstone formation beneath.

Fig. 22.

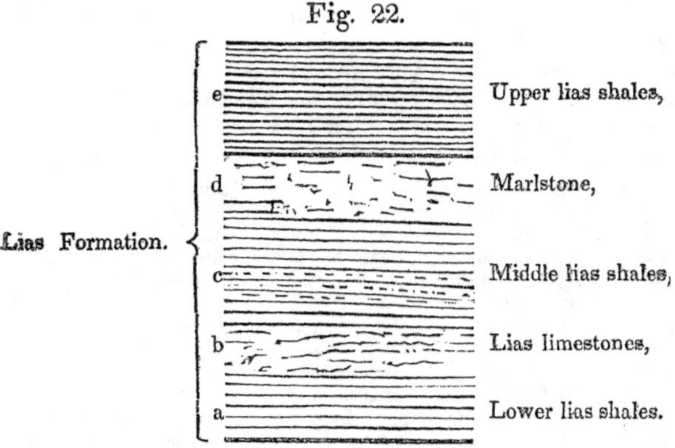

Def. The English word *group,* and the French *terrain,* are nearly synonymous with formation. The term *series* is also very convenient in description.

Def. When the planes of stratification are parallel to one another in different formations, the stratification is said to be *conformable:* when not parallel, it is *unconformable.*

Descr. The stratification in different formations is usually unconformable, as is shown in the position of the secondary and primary formations, in Fig. 12.

Inf. It is hence inferred that the stratified rocks were elevated at different epochs: in other words, those formations which are the most highly inclined, must have been partially elevated before the others were deposited upon them.

Descr. These numerous elevations of the strata have produced in them a great variety of cracks, fissures, and slides.

Def. When the continuity of the strata is interrupted by a fissure, so that the same stratum is higher on one side than on the other, or has been slidden laterally, that fissure is called a *fault,* or a *trouble,—a slip,—a dyke,* &c: as *a, b, c, d,* Fig. 23.

Fig. 23.

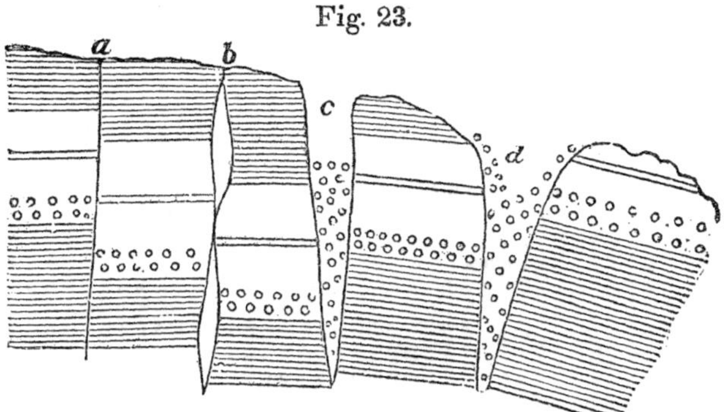

Rem. A fault is generally filled with fragments of rocks, clay, &c. as *d.* It often occasions great trouble in the working of mines, because, when it is reached, it is impossible to decide whether the continuation of the mineral sought is above or below the level, or to the right or left.

Def. If the fissure is open and of considerable width, and is succeeded at each extremity by a wider valley, it is called a *gorge*, as *c.*

Def. If it be still wider, with the sides sloping or rounded at the bottom, a *valley* is produced; as *d.*

Prin. In a similar way most of the valleys of primitive countries were formed.

Def. The line forming the top of a mountain ridge, or running through a valley, along which the strata dip in opposite directions, is called an *anticlinal line*, or *anticlinal axis:* as at *a*, Fig. 24, and *b*, Fig. 25.

Fig. 24.

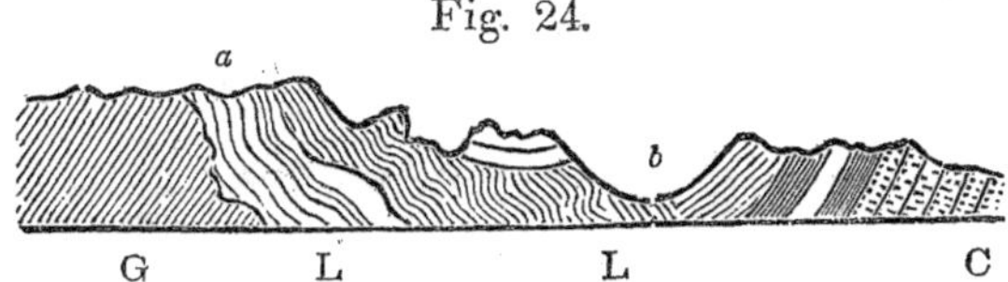

Def. When the strata dip towards this line on each side of it, is called a *synclinal line*, or *axis*, as at *b*, Fig. 24.

Fig. 25.

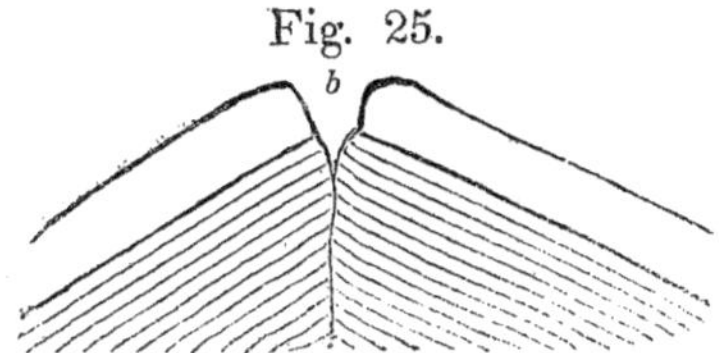

Def. When the strata dip from any point in all directions outwards, (as around the crater of a volcano,) the dip is said to be *quaquaversal.*

Classification of Rocks.

Rem. Numerous attempts have been made to classify the rocks. But none of the arrangements hitherto proposed, possess so decided a superiority over the others as to be adopted in every particular. In the present state of the subject, all that can be done is to give the outlines of the most important systems of classification, leaving the reader to take his choice among them. I shall first describe some of the larger groups of rocks, which have long been admitted by most geologists to exist in nature. The names by which they are designated are indeed objectionable, because they were originally founded upon uncertain hypothesis: and many excellent writers have entirely excluded them from their works. But as they still continue to be employed by most geologists, and as they can now hardly mislead even a tyro in the science, and are moreover founded in nature, I shall retain the most of them in the following familiar description of the earth's crust.

Descr. If we suppose ourselves placed in a meadow, which has resulted from the successive deposits of annual floods, and begin a perpendicular excavation into the earth, we shall pass through the different classes of rocks in the following order.

Def. For a few feet only,—rarely as many as 100, we shall pass through layers of loam, sand, and fine gravel, arranged in nearly horizontal beds. This deposite, from an existing river, is denominated *alluvium.*

Def. All deposits from causes now in action, which have taken place since the present order of things commenced on the globe, are usually regarded as alluvial.

Def. The second formation which we shall penetrate, is composed of coarse sand and gravel, with fine sand and even sometimes clay, containing, however, large rounded masses of rock called bowlders; the whole mixed together, yet often distinctly, and horizontally stratified. This formation, evidently the result of glacio-aqueous agency is called *drift.* It is distinguished from alluvium, first by its inferior position; secondly, by the marks of a more powerful agency; and thirdly, by extending over regions where no existing streams or other causes now in action could have produced it.

Def. The third series of strata which we penetrate in descending into the earth, is composed of layers of clay, sand, gravel, and marl, with occasional quartzose and calcareous beds more or less consolidated; all of which were deposited in waters comparatively quiet and in separate basins. They also contain many peculiar organic remains, and sometimes dip at a small angle, though usually they are horizontal. These strata are called *tertiary.*

Obs. It is only a few years since all the formations above described were regarded as alluvium.

Descr. The formations which we penetrate after passing through the tertiary, are composed for the most part of solid rocks. They are, however, mostly made up of sand, clay, and pebbles, bound together by some sort of cement. With these are interstratified many varieties of limestone; and throughout the whole series is found a great variety of the remains of animals and plants, very different from those in the tertiary strata. These groups of rock sometimes lie horizontal; but are usually more or less elevated, so as to make them dip at various angles. They are called *secondary rocks.*

Obs. It will be seen that all the fossiliferous rocks below the tertiary are here included in the secondary class. Many geologists, even to the present time, have separated some of the lower groups into a class named *transition*, because they appear as if produced when the earth was in a transition state from desolation to a habitable condition, and have a texture partly mechanical and partly chemical. I do not attempt to define such a class, because I cannot fix upon any characters by which it can be distinguished from the secondary strata.

Def. The stratified rocks below the secondary, are distinguished by the absence of organic remains, by having a structure more or less crystalline, and by being more highly inclined. They are called *primary rocks.* This term has also been applied to the unstratified crystalline rocks. Mr. Lyell has proposed, as a substitute for primary, the term *hypogene;* meaning "*nether-formed rocks*, or rocks which have not assumed their present form and structure at the surface." *Elem. Geol. Vol.* 1, *p.* 20.

Rem. D'Halloy, Dr. Macculloch, Prof. Phillips, Mr. Lyell, and some others, include some of the fossiliferous rocks,—as clay, slate and graywacke,—in the primary class, as may be seen on the tabular view at the end of this section.

Descr. Immediately beneath the primary stratified rocks, we find the unstratified ones.

Inf. As this is found to be the case wherever the stratified rocks have been penetrated, it is inferred that the internal parts of the globe, beneath a comparatively thin crust, are made up of unstratified rocks: at least to a very great depth.

Descr. Among the primary rocks there is no settled order of superposition. Perhaps gneiss most commonly lies immediately above granite; but the other members of the series are frequently found also in the same position.

Descr. Among the fossiliferous rocks there exists an invariable order of superposition.

Exception. In a few cases, internal forces have not merely lifted upon their edges, but actually overturned strata of considerable thickness. The section in Fig. 24, was taken in the Alps, and exhibits a case of this kind. G, is gneiss, L, L, limestone, C, conglomerate, locally called *nagleflue.* Now the limestone is really an older rock than the conglomerate; and yet it lies above the conglomerate, because the whole series has been tossed over, so as to bring the newer rocks beneath the older.

Descr. In some instances the strata have been folded together on a vast scale, and in such a manner as to bring some of the newer rocks beneath the older. Fig. 26, is a section of this character. Originally the strata were probably folded, as is shown by the curved lines passing from 1 to 1, 2 to 2, and so on. But their upper parts have been denuded, so that the present surface is *a, a.* The oldest strata are now found to be 6, 6; and they correspond outward on each side of these; as 5, 5; 4, 4; &c. Such an example as this has been called a *folded axis.*

Fig. 26.

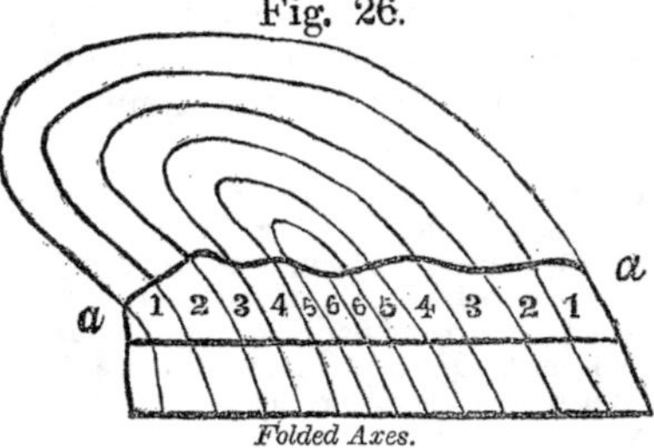

Folded Axes.

Exam. I have no small reason to believe that a similar folding and overturning of the strata have taken place on a vast scale in the United States; although the phenomena have not yet been studied as carefully as would be desirable; and some of our geologists explain the inverted dip on other principles. Along the western part of the Green and Hoosac Mountains, in New England, occur interstratified beds of gneiss, mica slate, talcose slate, clay slate, limestone, and older Silurian rocks, which are either perpendicular, or have a high easterly dip: and yet the oldest members of the series are found along the eastern side of this belt, and the strata become newer and newer as we go westerly: that is, the oldest rocks lie apparently above the newer ones. These appearances present themselves nearly the whole distance from Connecticut river to Hudson river; a breadth of nearly fifty miles. Now suppose in early times the strata between those rivers to have been somewhat elevated, with Hoosac Mountain, or the Green Mountains as their principal axis, as is shown in fig. 27; where L stands for limestone, M, for mica slate, T, for talcose slate, S, for clay slate, and Gra. for graywacke. While yet in a yielding state, imagine powerful forces to act in opposite directions at the two extremities of these strata: that is at Connecticut and Hudson rivers; where we have reason to believe extensive faults exist: while at the same time we may suppose an upward strain upon the strata from gas or melted matter beneath. The effect would be to bend and fold together the strata, as they are shown upon fig. 28. If afterwards they were denuded by water, to the depth of the irregular line A A, their character, dip, and outline would correspond essentially to what we find between Connecticut and Hudson rivers. It appears further, from the Geological reports of Professors Mather on New York, of Prof. Henry D. Rogers on New Jersey and Pennsylvania, of Prof. William B. Rogers on Virginia, and of Prof. Troost on Tennessee, that these same rocks, with similar inversion of their dip, occur in all those states, forming a considerable part of the Appalachian Mountains; and that in fact they extend almost uninterruptedly from Canada to Alabama; a distance of nearly 1200 miles. And if the above theory of the folding and inversion of this belt of rocks be correct in the latitude of Massachusetts, it is without doubt true over this vast extent of country. Nor can we doubt that agencies suffi-

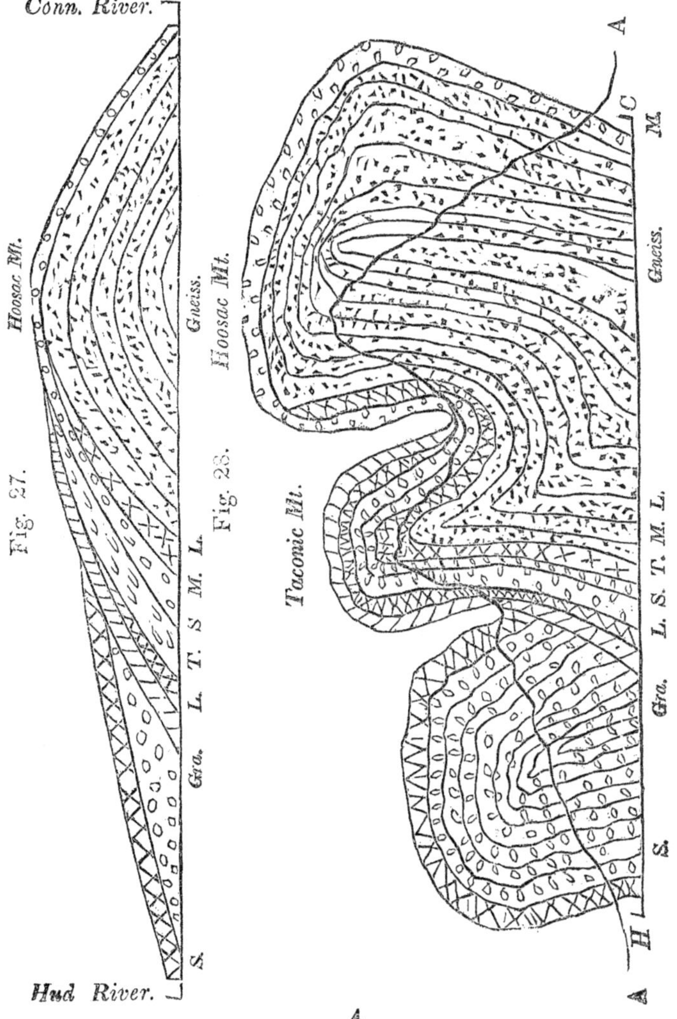

Fig. 27.

Fig. 28.

ciently powerful have operated often in nature to produce these effects; although it is difficult for the mind to become familiar with such stupendous forces. (*See this theory farther elucidated in the Final Report on the Geology of Massachusetts, Vol. 2. p. 577, and in the first Anniversary Address before the Association of American Geologists at Philadelphia*, 1841. *Also, and especially, in a paper by the Professors Rogers, on the Physical Structure of the Apalachian Chain in Trans. Amer. Ass. Geologists*, 1843.)

Descr. Sometimes the strata, after descending in an inverted position from 1000 to 1500 feet, curve in such a direction as to bring them into their proper position; as is shown in Fig. 29, taken in the Alps.

Fig. 29.

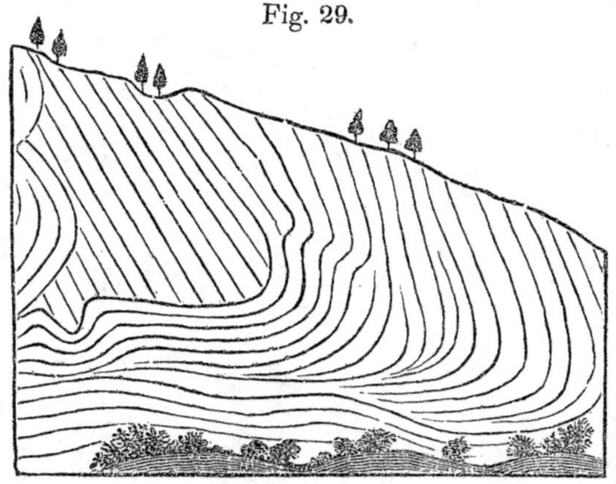

Descr. One or more rocks are frequently wanting in the secondary series, which brings those of very different ages into contact: but the order of arrangement is never thereby disturbed.

Exam. Thus in Fig. 30, on the left side of the central mass of granite (A,) we have the primary (B,) the secondary (C,) and tertiary (D,) in regular order: but on the other side, the secondary is wanting; and the tertiary (D,) lie directly upon the primary (F,) as well as upon the granite, while a deposite of drift (E,) comes in contact with all the formations.

Other Systems of Classification.

Rem. Most of the systems of classification of the stratified rocks, described in the following paragraphs, are exhibited synoptically on the Tabular View appended to this Section.

Descr. Dr. Macculloch divides the strata into four principal classes. His *Alluvial Class* embraces alluvial and drift: his *Tertiary Class* is the same as that already described: his *Secondary Class* extends no farther downward

Fig. 30.

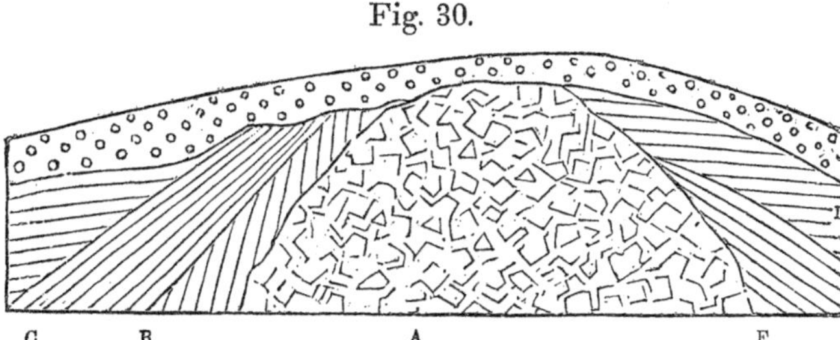

than to the bottom of the old red sandstone; and the remainder of the fossiliferous rocks, with the stratified non-fossiliferous ones constitutes his *Primary Class.* He also distributes the unstratified rocks through the two latter classes, and adds a fifth, or the *Volcanic Class.* Reckoning the stratified and unstratified rocks together, he divides the whole into ten groups, which he also denominates classes, by the designation *Protolith.*, *Deutolith*, &c. This latter arrangement he denominates the *Natural System*, and the former the *Artificial System. Macculloch's System of Geology, Vol.* 2. *p.* 78.

Descr. In Rev. W. D. Conybeare's arrangement, alluvium, drift, and the tertiary strata, are called the *Superior Order:* the rocks from the chalk to the coal measures, form his *Supermedial Order:* the coal measures, carboniferous limestone, and old red sandstone, form his *Medial Order:* the remaining fossiliferous rocks constitute his *Submedial Order:* and the stratified primary groups his *Inferior Order.* The unstratified rocks are distributed among the stratified, according to their supposed age. This system has the merit of being both simple and free from all hypothetical allusions. *Conybeare and Wm. Phillips's Geology of England and Wales, Vol.* 1. 1822.

Descr. De la Beche divides all rocks into two great classes, the *Stratified* and *Unstratified.* The latter he treats as a single family: the former, he subdivides into ten groups. The first is called the *Modern Group*, and corresponds to alluvium: the second is the *Erratic Block Group*, corresponding to drift: the third, the *Supercretaceous Group*, embracing the tertiary strata: the fourth, the *Cretaceous Group* including only the chalk and some associated strata: the fifth, the *Oolitic Group*, comprehending the oolite and the lias: the sixth, the *Red Sandstone Group*, or the new red sandstone: the seventh, the *Carboniferous Group*, containing the coal measures, carboniferous limestone, and the old red sandstone: the eighth, the *Graywacke Group*, or the graywacke formation: the ninth, the *Lowest Fossiliferous Group*, or fossiliferous slates lying below the graywacke: and the tenth, the *Primary Stratified Non-Fossiliferous Rocks.* This arrangement is also natural and satisfactory. *De la Beche's Man. Geol. p.* 38. *third edition.*

Descr. Mr. Lyell embraces the rocks in *Periods* and *Groups.* The first is the *Post Pliocene Period*, including alluvium and drift. The second is the *Tertiary Period*, subdivided into the newer and older *Pliocene*, the *Miocene*, and the *Eocene.* The third he denominates *Secondary*, which extends to the bottom of the old red sandstone: next succeeds his *Primary Fossiliferous Period*, which includes all the remaining fossiliferous rocks. His *Metamorphic Rocks* embrace all the stratified non-fossiliferous groups. The unstratified rocks are distributed through these several classes; and he has likewise made a division of those unstratified rocks, that exist below the stratified ones, into *Primary Plutonic, Secondary Plutonic, Tertiary Plu-*

tonic, and Recent Plutonic, reckoning in a descending order. *Elements of Geology, Vol.* 2. *p.* 170.

Descr. Omalius d'Halloy, a French geologist, in 1831, proposed the following system of arrangement. He first divides all rocks into *Neptunian*, or the stratified, and the *Plutonian*, or unstratified: then under his first class, or *Modern Formation*, he places alluvium; under *Tertiary Formations*, he puts drift and the tertiary strata; under *Ammonean Formations*, he includes the subjacent rocks as far as the coal measures; and under *Hemilysian Formations*, all the remaining stratified rocks. The unstratified ones he divides into two classes; the first, embracing granite and porphyry, is called *Agalysian;* the second, embracing basalt, trachyte, and lava, is called *Pyroidal.* His three first classes he also denominates *Secondary;* and the remaining class of stratified rocks, with the first division of the unstratified, *Primordial. Elements de Geologie, par J. J. D'Omalius D'Halloy, Paris*, 1831.

Descr. Prof. Alexander Brongniart, another distinguished French geologist, in 1829, proposed to embrace all the rocks under the *Jovian Period*, or the existing era; and the *Saturnian Period*, or the era preceding the last revolution of the globe. His first period embraces only alluvium which he divides into the *Alluvial, Lysian, and Pyrogenous Formations.* His second period embraces, first, the *Stratified or Neptunian Formations*, and secondly, the *Massive or Typhonian Formations.* The first of these is divided into the *Clysmian Formations*, or drift; the *Izemian Formations*, or the tertiary strata, and the secondary as far down as the mountain limestone; *Hemilysian Formations*, embracing all the remaining fossiliferous strata; and the *Agalysian Formations*, which include all the primary stratified rocks. His Typhonian class he divides into the *Plutonian*, and *Vulcanian Formations. Tableau des Terrains, &c. par Prof. Al. Brongniart, Paris*, 1829.

Descr. Rozet, another French author, in 1835, divided all rocks into two great *Series*, the first embracing the stratified, and the second the unstratified rocks. His first series he refers to six geognostic epochs, the first embracing alluvium, the second drift, the third the tertiary strata, the fourth the subjacent rocks as deep as the coal measures, the fifth the remaining fossiliferous rocks, and the sixth the non-fossiliferous stratified rocks. *Traite Elementaire de Geologie, par M. Rozet, Paris*, 1835.

Descr. Dr. Mantell proposes a chronological arrangement of the rocks. His two great classes are the *Fossiliferous Strata*, and the *Metamorphic Rocks:* the latter embracing the unstratified rocks as well as the stratified primary. Under *Modern and Ancient Alluvium*, he places alluvium and drift. Next come the *Tertiary Strata:* then, as the first group of the secondary formations, the *Chalk or Cretaceous System.* The second group is the *Wealden;* the third, the *Oolite;* the fourth, the *Lias;* the fifth, the *Saliferous Strata;* the sixth, the *Carboniferous System, or Coal;* the seventh, the *Silurian System*, or upper members of the graywacke series; and the eighth, the *Cambrian or Graywacke System.* His metamorphic rocks comprehend three groups. 1. *Mica Schist.* 2. *Gneiss.* 3. *Granite.—Mantell's Wonders of Geology, Vol.* 1, *p.* 178, *London*, 1838.

Descr. Professor John Phillips divides all rocks into the stratified and unstratified: and then, like De la Beche and Dr. Buckland, he does not attempt to distribute the latter among the former, but treats of each class separately. The stratified class he thus subdivides. Alluvium and drift are placed under *Superficial Accumulations*, and denominated, *Alluvial Depositions* and *Diluvial Depositions.* The tertiary strata he divides into the *Crag, Freshwater Marls, and London Clay.* The secondary strata, which extend to the bottom of the old red sandstone, he divides into the

Cretaceous System, the Oolitic System, the Saliferous or Red Sandstone System, and the Carboniferous System. Next succeed his *Primary Strata*, which embrace the *Silurian System, the Cambrian or Graywacke System, the Skiddaw or Clay Slate System, the Mica Schist System, and the Gneiss System.* The subdivision of these systems may be seen in the accompanying table. *Treat. on Geol. from the Encyclopædia Britannica, Vol.* 1. *Edinburg*, 1838. *Also a Treat. on Geol. in* 2 *Vols. London*, 1837 *and* 1839, *in Cabinet Library.*

Descr. The most important changes in the classification of the stratified rocks that have of late been proposed, are those by Dr. Murchison, and Professor Sedgwick, in the group that has long been known under the general name of graywacke. The former gentleman, after years of study, has produced a splendid quarto, with a geological map, upon the upper members of these strata, which he denominates the *Silurian Group*, because it is well developed in the ancient British kingdom of the Silures. The lower members of the graywacke with some slates, Prof. Sedgwick denominates the *Cambrian System*, because fully developed in North Wales. *Phillips's Treatise on Geology, Vol.* 1. *p.* 56. It seems now, however, to be pretty well ascertained, that the Silurian and Cambrian systems cannot be separated by their organic remains, and must therefore be considered as one group. (*Murchison's Anniversary Address before the Lond. Geol. Soci. for* 1843.) Accordingly they are so represented in the classification by Professor Ansted, in the Tabular View annexed. His Tertiary Period extends from the top of the rock series to the chalk. The chalk he calls the Newer Secondary ; the Wealden and Oolite the Middle Secondary ; the Upper New Red Sandstone or Triassic System, the Older Secondary : the Lower New Red, or Permian System, and the Carboniferous System, form his Newer Palæozoic Period; the Devonian System, or Old Red Sandstone, his Middle Palæozoic ; and the Silurian and Cambrian Systems, (divided into Upper and Lower,) his Older Palæozoic ; while the non-fossiliferous rocks he denominates the Metamorphic. This classification corresponds essentially with the views of Professor Sedgwick and Mr. Murchison. *See Ansted's Geology*, 2 *Vols., London*, 1844. The smaller groups of rocks, adopted in this work, correspond with those of Professor Ansted.

Descr. The System of the New York State Geologists in the Table, embraces only the rocks found in that State, where the Cretaceous, Wealden Oolitic, and Liassic groups, are wanting. Their Quaternary system embraces Alluvium and Drift; next comes the Tertiary ; next the New Red Sandstone, embracing the Triassic and Permian Systems; next the Carboniferous; next the Old Red System; next the New York System, (two groups of which, according to Mr. Hall, correspond to the Devonian System of England—*see Hall's State Report, p.* 20,) embracing the European Silurian System, and subdivided into four principal, and twenty-eight minor groups; next the Taconic System, which, according to Professor Emmons, is fossiliferous, and distinct from the New York System, but according to others, it is the latter metamorphosed by heat; finally, the Primary or Hypogene System.

Descr. The System by Prof. W. B. Rogers of Virginia, and H. D. Rogers of Pennsylvania, embraces the rocks of the United States. They begin with alluvium and drift; and dividing the Tertiary into four groups (Post-Pleiocene, Pleiocene, Miocene, and Eocene,) they denominate the whole the Cainozoic Period. The Cretaceous rocks, (divided into Upper, Middle, and Lower,) and the Oolitic Coal Series of eastern Virginia, they call the Mesozoic Period; the Upper New Red, the Triassic Series; and all the period between the Trias and Clay Slate, they call the Apalachian Palæozoic Day, which they divide into nine portions, which refer to the time of day, and whose correspondence with the New York System will be seen upon the Tabular View.

Descr. The last system in the Tabular View, presents a classification founded upon Palæontology, or Organic Remains, which I venture to suggest as sometimes convenient. The first column marks out the vertical limits of the different systems of organic life that have appeared on the globe, so far as they have been well ascertained. The lowest, embracing all the rocks below the Silurian, is called Azoic, that is, destitute of organic beings; the second extends to the top of the Silurian, and is called Protozoic, that is, the first organic series; the third embraces the Devonian, Carboniferous, and Permian systems, and is called the Deutozoic; the third, or Tritozoic, embraces only the Trias; the fourth, or Tetrazoic, only the Lias and Oolite; the fifth, or Pentezoic, the Chalk; the sixth, or Hectozoic, the Tertiary Strata; the sixth embraces Drift, and is called Heterozoic, because the organic remains are derived almost exclusively from other formations; the seventh is the Heptazoic, and embraces alluvium only. The second column, expresses the animal or substance most characteristic of the different groups. Thus the Heptazoic is Homoniferous, or contains the remains of man; the Heterozoic is Fragmentiferous, or abounds in fragments; the Hectozoic is Mammaliferous, or contains the remains of mammalia; the Pentezoic is Foraminiferous, or abounds in Foraminifera; the Tetrazoic is Dinosauriferous, (containing land Saurians,) Pterosauriferous, (containing flying Saurians,) and Enaliosauriferous, (containing marine Saurians;) the Tritozoic is Ichniferous, because abounding in the tracks of animals; the Deutozoic is Ichthyferous, because containing peculiar fish, and carboniferous, because abounding in coal; the Protozoic is Cephalopodiferous, because abounding in Cephalopod shells; Polypiferous, because abounding in Corals; Trilobiferous, because abounding in Trilobites; and Brachiopodiferous, because abounding in Brachiopod shells; the Azoic series is Crystaliferous, because abounding in Crystals.

Rem. A cursory view of the table is apt to convey the impression that almost everything relating to the classification of rocks is unsettled, and that there is scarcely any agreement among the different systems. Some explanations and inferences, therefore, seem desirable, to present the subject in its true light.

Prin. In judging of a classification of natural objects, it is important that we distinguish natural from artificial characters. Thus, in botany, plants may be divided into classes and orders depending upon the number and situation of the stamens and pistils of their flowers; or upon the anatomical structure of the plant. By the first arrangement we shall bring together plants the most unlike in their general properties; and therefore, there is no necessary connexion between those properties and the number and situation of the stamens and pistils, and hence such characters are artificial or arbitrary. But those plants which are alike in anatomical structure, correspond in most of their properties; and such characters, therefore, are natural.

Inf. 1. In applying this principle to rocks, we find first, that their division into stratified and unstratified is natural: that is, it brings together those kinds whose origin and other important characters are similar. Now we shall find that this division enters into nearly all the more recent systems of classification that have been described.

Inf. 2. In the division of the rocks into fossiliferous and non-fossiliferous, all geologists agree: and in fact there is scarce a possibility of disagreement on this point. So that here we have another important natural character as the basis of classification.

Inf. 3. In nearly all the systems of classification, the larger formations coincide; which is a presumptive proof that they are natural; since so many different observers agree in forming their boundaries. These formations ought perhaps to be regarded as the *species* in geology.

Inf. 4. Classification founded upon the relative age of different rocks, is entirely natural, because all observers agree that they were produced at different times. But as superposition and organic remains are the only safe criteria of relative age, there is ground for a diversity of opinion in assigning places to the different formations; since these criteria can be ascertained sometimes only imperfectly.

Inf. 5. Characters dependent upon theoretical considerations are artificial, since few of the theories are so certainly settled as not to be liable to considerable modification. Hence such terms as primary, transition, secondary, tertiary, drift, &c. are objectionable, if they are not understood to refer simply to superposition.

Rem. Neology is often a greater evil in science than the continued use of objectionable terms; continued, I mean, until terms are proposed which are so decidedly good as to force themselves into use. It is partly on this ground that the terms primary, transition, secondary, and tertiary, still continue in use. But it is partly, also, because, apart from theoretical views, there does exist in nature some foundation for a division of the rocks into groups of this sort. Still those more numerous groups or formations into which the best geologists now divide rocks, are much more natural; and it is to be hoped that they will soon come into general use, so as to exclude the terms primary, secondary, &c.

Inf. 6. Characters founded upon lithological distinctions are artificial, for the same reason that those derived from the number of stamens and pistils are bad in botany.

Inf. 7. Discrepancy in classification often springs from carrying the subdivisions of a formation too far, for the same reason that characters in botany and zoology could not be depended on, that were derived from the varieties of a species.

Inf. 8. Finally, it appears that in all the essential principles of the classification of rocks, geologists are nearly agreed. They all admit one class to be stratified and another unstratified :—one portion of the stratified rocks to be fossiliferous and another portion not fossiliferous. And they generally agree, also, as to the extent of the different distinct formations ; although some would make their number greater than others,—just as it is in respect to species in mineralogy, botany, and zoology. Now these three principles are all that are essential for classification ; and some of the best geologists, as may be seen by the table, limit themselves to these. But if others choose to subdivide the formations still farther, and to refer the groups to primary, secondary, &c. classes, even though they differ widely here, it must not be hence inferred that they are at variance in respect to the essential principles of classification.

SECTION II.

THE CHEMISTRY AND MINERALOGY OF GEOLOGY.

Descr. Of the fifty-four simple substances hitherto discovered, sixteen constitute, by their various combinations, nearly the whole of the matter yet known to enter into the composition of the globe. They are as follows, arranged in three classes, according to their amount ; the first in each class being most abundant.

1. *Metalloids, or the bases of the earths and alkalies.*

1. Silicium. 2. Aluminium. 3. Potassium. 4. Sodium 5. Magnesium. 6. Calcium.

2. *Metals Proper.*

1. Iron. 2. Manganese.

3. *Non Metallic Substances.*

1. Oxygen. 2. Hydrogen. 3. Nitrogen. 4. Carbon. 5. Sulphur. 6. Chlorine. 7. Fluorine. 8. Phosphorus. *De la Beche's Researches in Theoretical Geology, p.* 22. *Amherst*, 1837.

Descr. The metallic substances mentioned above, united with oxygen, constitute the great mass of the rocks, consolidated and unconsolidated, accessible to man. Oxygen also forms twenty per cent. of the atmosphere, and one third part by measure of water. Hydrogen forms the other two thirds of this latter substance; and it is evolved also from volcanos, and is known to exist in coal. Nitrogen forms four fifths of the atmosphere, and enters into the composition of animals, living and fossil. It is found also in coal. Carbon, however, forms the principal part of coal; and it exists likewise in the form of carbonic acid in the atmosphere, though constituting, only one thousandth part. *Liebig's Organic Chemistry, p.* 74. *First Amer. Edition*, 1841. and it forms an important part of all the carbonates, and is produced wherever vegetable and animal matters are undergoing decomposition. Sulphur is found chiefly in the sulphurets and sulphates that are so widely disseminated. Chlorine is found chiefly in the ocean, and in the rock salt dug out of the earth. Fluorine occurs in most of the rocks, though in small proportion Still less is the amount of phosphorus, though widely diffused in the rocks and soils, and abundant in organic remains.

Descr. Nearly all the simple substances above mentioned have entered into their present combinations as binary compounds; that is, they were united two and two before forming the present compounds in which they are found. The following constitute nearly all the binary compounds of the accessible parts of the globe.

1. Silica. 2. Alumina. 3. Lime. 4. Magnesia. 5. Potassa. 6. Soda. 7. Oxide of Iron. 8. Oxide of Manganese. 9. Water. 10. Carbonic Acid.

Obs. It is meant only that these binary compounds, and the sixteen simple substances that have been enumerated, constitute the largest part of the known mass of the globe: for many other binary compounds, and probably all the known simple substances are found in small quantity in the rocks; but not enough to be of importance in a geological point of view.

Descr. It has been calculated that oxygen constitutes 50 per cent. of the ponderable matter of the globe, and that its crust contains 45 per cent. of silica, and at least 10 per cent. of alumina. Potassa constitutes nearly 7 per cent. of the unstratified rocks, and enters largely into the composition of some of the stratified class. Soda forms nearly 6 per cent. of some basalts and other less extensive unstratified rocks; and it enters largely into the composition of the ocean. Lime and magnesia are diffused almost universally among the rocks in the form of silicates and carbonates,—the carbonate of lime having been estimated to form one seventh of the crust of the globe; at least three per cent. of all known rocks are some binary combination of iron, such as an oxide, a sulphuret, a carburet, &c.; manganese is widely diffused, but forms much less than one per cent. of the mass of rocks.

Descr. The following table presents an approximate estimate of the mean amount of metallic bases and of oxygen in some of the important rocks. *Phillips's Treatise on Geology, Vol.* 1, *p.* 24.

100 parts of	Granite—52	Metallic Basis	48 Oxygen.
"	Basalt—57	"	43 "
"	Gneiss—53	"	47 "
"	Clay Slate—54 ?	"	46 "
"	Sandstone—49 to 53	"	47 to 51
"	Limestone—52	"	48 "

Descr. The following table shows the approximate amount of silica and alumina in the most important rocks.

Granite,	69,40 Silica.	12,34 Alumina.
Greenstone,	54,86	15,56
Basalt,	52,00	14,12
Compact Felspar,	55,50	21,00
Gneiss,	70,96	15,20
Mica Slate,	67,50	14,26
Hornblende Rock,	54,86	15,56
Talcose Slate,	78,15	13,20

See De la Beche's Researches in Theoretical Geology, p. 29 *and* 30. *Also Traite Elementaire de Mineralogie, par S. F. Beudant, Tome* 1, *p.* 112. *Paris,* 1830.

Descr. Seven or eight simple minerals constitute the great mass of all known rocks. These are 1, Quartz; 2, Felspar; 3, Mica; 4, Hornblende and Augite; 5, Carbonate of Lime; 6, Talc, embracing chloric and soapstone; 7, Serpentine. Oxide of iron is also very common; but it does not usually show itself till the decomposition of the rock commences.

Obs. The student of geology should become very thoroughly conversant with these minerals in all their modifications: for in the rocks their characters are often very obscure.

Descr. Other minerals forming rocks of small extent, or entering so largely into their composition as to modify their character, are the following: sulphate of lime, diallage, cnloride of sodium (common salt), coal, bitumen, garnet, schorl, staurotide, epidote, olivine, pyrites.

Descr. A few of these minerals exist in so large masses as to be denominated rocks; ex. gr. quartz, carbonate of lime, &c. but in general, from two to four of them are united to form a rock; ex. gr. quartz, felspar and mica, to form granite. In some instances the simple minerals are so much ground down, previously to their consolidation, as to make the rock appear homogeneous; ex. gr. shale and clay slate.

Descr. Water constitutes a part of nearly all rocks; but in most cases it appears to be mechanically combined; for with one or two exceptions, it does not exist in the simple minerals that enter into the composition of rocks.

Obs. In the simple minerals that have been enumerated, analysis has detected water only in the following.

Sulphate of Lime (Gypsum),	19,88 per cent.
Serpentine,	12,75
Diallage,	8,20
Talc,	4,20
Pyroxene (a few varieties),	3,74
Mica,	2,65
Quartz,	1,62
Hornblende,	0,55

Geological Situation of Useful Rocks and Minerals.

Prin. The rocks and minerals useful in an economical point of view, are in a few instances found in almost every part of the rock series: but in a majority of cases, they are confined to one or more places in that series.

Examples.

Granite, Syenite, and Porphyry: found intruded among all the stratified rocks as high in the series as the tertiary strata: but they are almost entirely confined to the primary rocks. *De la Beche's Manual, p.* 94.

Greenstone and Basalt are found among and overlying all the primary and secondary rocks: but they are mostly connected with the secondary strata. *Macculloch's System of Geology, Vol.* 2, *p.* 102.

Lava, some varieties of which, as *Peperino*, are employed in the arts, being the product of modern volcanos, is found occasionally overlying every rock in the series.

Clay: the common varieties used for bricks, earthen ware, pipes, &c, occur almost exclusively in the tertiary strata: but we have reason to think some of them belong to the period of drift. Porcelain clay results from the decomposition of granite, and is found in connection with that rock.

Marl, or a mixture of carbonate of lime and clay, is confined to the alluvial and tertiary strata: and differs from many varieties of limestone, only in not being consolidated.

Limestone, from which every variety of marble, one variety of alabaster and every sort of quicklime are obtained, is found in almost every rock, stratified and unstratified, below drift. In the oldest stratified rocks and in the unstratified, it is highly crystalline; and in the newest strata (ex. gr. chalk) it is often not at all crystalline. The most esteemed marbles are obtained from the newer primary and older secondary strata.

Serpentine occurs chiefly in connection with the older stratified rocks. This is generally the case in N. England. It is found, however, with some secondary rocks, and not unfrequently with trap rock.

Sulphate of Lime, or Gypsum, which produces one variety of alabaster, and is employed for taking casts, forming hard mortar, and spreading upon land in the state of powder, occurs chiefly in the new red sandstone series. It is found also in the lias, oolite, green sand, and tertiary strata. In this country it is found associated with the oldest of the secondary (transition) rocks.

Rock Salt (Chloride of Sodium) is frequently found associated with gypsum in the new red sandstone. It occurs also in the supercretaceous or tertiary strata; as at the celebrated deposite at Wieliczka in Poland; and in Sicily, and Cardona (Spain), in cretaceous strata: in the Tyrol, in the Oolites; and in Durham, England, salt springs occur in the coal formation. In the United States they issue from the Silurian rocks. *Buckland's Bridgewater Treatise, Vol.* 1, *p.* 72. *De la Beche's Manual, p.* 246.

Descr. If vegetable matter be exposed to a certain degree of moisture and temperature, it is decomposed into the substance called *peat*, which is dug from swamps, and belongs to the alluvial formation.

Lignite or Brown Coal, the most perfect variety of which is jet, is found chiefly in the tertiary strata; sometimes in the higher secondary; and appears to be peat which has long been buried in the earth, and has undergone certain chemical changes, whereby bitumen has been produced. It generally exhibits the vegetable structure.

Bituminous Coal appears to be the same substance which has been longer buried in the earth, and has undergone still farther changes; though their precise nature is not well known. Its principal deposite is in that part of the secondary series called the coal formation, or coal measures. But it occurs in small quantity in the new red sandstone series, in England, Poland, and Massachusetts: and in Scotland it is worked in the lias limestone. A thick bed of it has also been found in the plastic clay of the tertiary strata in Hesse. *Conybeare's Report* (1832) *on Geology to the British Association, p.* 390. *Also, Philosophical Magazine, Vol.* 2. *New Series, p.* 101 *and* 108.

Obs. The French geologists describe a variety of coal intermediate in its characters and position between cannel coal and lignite: but it is doubtful whether on strict mineralogical principles even lignite can be separated from bituminous coal. *See Tableaux des Terrains par Al. Brongniart, and Traite Elementaire de Mineralogie par F. S. Beudant.*

The principal deposite of anthracite in Europe, is in the graywacke formation; and it is supposed that this substance is common coal which has undergone still farther changes and lost its bitumen. In this country, however, there is reason to suppose that the vast deposite of anthracite in Pennsylvania, the largest in the world, is associated with the common coal measures. *See Prof. H D. Rogers' Second Annual Report on the Geological Exploration of the State of Pennsylvania,* 1838.

The anthracite of Rhode Island, and the southeastern part of Massachusetts is in a rock whose exact place in the series has not been satisfactorily determined. But probably it may belong to the coal formation. In Worcester, the anthracite occurs in a sort of bastard mica slate. On the continent of Europe, it occurs also in mica slate, in primary limestone and in gneiss. *Macculloch's System of Geology, Vol. 2, p. 296.*

Prof. Alexander Brongniart describes a true anthracite as occurring in the plastic clay of Mount Meissner in Hesse. This, however, appears to have been formed from bituminous coal by the action of igneous rocks; and such cases have led some geologists to suppose that anthracite was always thus produced. It occurs in small quantities in almost all the stratified rocks from the oldest to the plastic clay.

Graphite, Plumbago, or Black Lead, appears to be anthracite which has undergone still farther mineralization: at least, in some instances, when coal has been found contiguous to igneous rocks, it is converted into plumbago; and hence such may have been the origin of the whole of it In the Alps plumbago is found in a clay slate that lies above the lias *Annales des Sciences Naturelles, Tome XV.* 1828. *p.* 377. It is found also in the coal formation. *Traite Elementaire de Mineralogie par F. S. Beudant, Tome 2, p.* 263.

Descr. All the varieties of coal that have been described, occur in the form of seams, or beds, interstratified with sandstones and shales: and most usually there are several seams of coal with rocks between them; the whole being arranged in the form of a basin. Fig. 31, is a sketch of the great coal basin of South Wales, in Great Britain; which contains twenty-three beds of coal; whose united thickness is ninety-three feet. When we consider how much this arrangement facilitates the exploration and working of coal, we can hardly doubt but it is the result of Divine Benevolence.

Fig. 31.

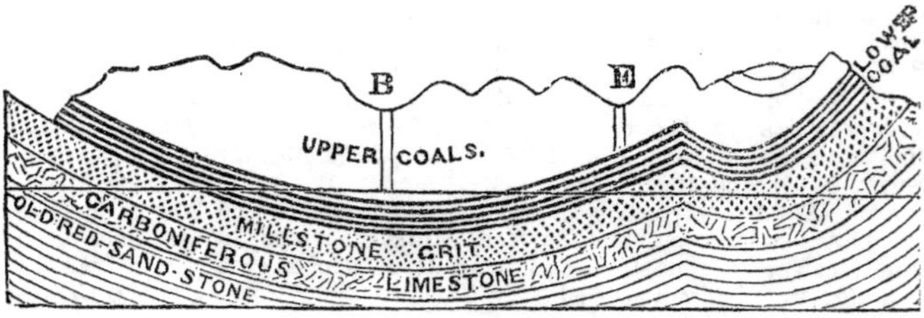

Descr. The *Diamond,* which is pure crystallized carbon, has been found associated with new red sandstone at Golconda, India, and with talcose slate in Brazil. Both these rocks have been subject to high heat, and hence perhaps the crystallization of the carbon. *Edinburgh Journal of Science, Vol. X. p.* 184. *Conybeare's Report on Geology, p.* 395 *and* 398. In general, the diamond is found in drift; having been removed from its original situation: and we may always presume that every mineral existing in the older rocks will be found also in drift; because their detritus must contain them.

Inf. It has been inferred from the preceding facts, that all the varieties of carbon above described, had their origin in vegetable matter; and that heat and water have produced all the varieties which we now find. *Macculloch's System of Geology, Vol.* 2. *p.* 297.

Almost all the precious stones, such as the sapphire, emerald, spinel, chrysoberyl, chrysoprase, topaz, iolite, garnet, tourmaline, &c., are found exclusively in the oldest and most crystalline rocks. Quartz in the various forms of rock crystal, chalcedony, carnelian, cacholong, sardonyx, jasper, &c., is found sometimes in the secondary strata, and especially in the trap rocks, associated with the secondary formations.

Descr. Some of the metals, as platinum, gold, silver, mercury, copper, bismuth, &c. exist in the rocks in a pure, that is, a metallic state; but usually they occur in the state of oxides, sulphurets, and carbonates, and are called ores. It is rare that any other ore is found in sufficient quantity to be an object of exploration on a large scale.

Descr. These ores occur in four modes: 1. In regular interstratified layers, or beds. 2. In veins or fissures, crossing the strata and filled with ore united to some gangue or matrix. 3. In irregular masses. 4. Disseminated in small fragments through the rocks.

Descr. Iron is the only metal that is found in all the formations in a workable quantity. Among all its ores, only four are wrought for obtaining the metal: viz. the magnetic oxide, the specular or peroxide, the hydrated peroxide, and the protocarbonate.

Manganese occurs in the state of a peroxide and a hydrate;—and is confined to the primary rocks; except an unimportant ore called the earthy oxide, which exists in earthy deposites.

The most important ores of copper are the pyritous copper and the carbonates. These are found in the primary rocks, and as high in the secondary series as the new red sandstone; in one instance in tertiary strata. *Wonders of Geology, Vol.* 2, *p.* 651.

The only ore of lead of much importance is the sulphuret. This generally occurs in the primary rocks both stratified and unstratified; but it exists also in the newer rocks as high in the series as the lias.

The deutoxide of tin is the principal ore of that metal. This is most commonly found in the oldest formations of gneiss, granite, and porphyry: also in the porphyries connected with red sandstone. It is found likewise in quantity sufficient to be wrought in drift.

Of zinc the most abundant ore is the sulphuret, which is commonly associated with the sulphuret of lead, or galena. Other valuable ores are the carbonate, silicate, and the oxide, which occur in secondary rocks.

The most common ore of antimony, the sulphuret, has hitherto been found chiefly in granite, gneiss, and mica slate.

5

The principal ore of mercury, the sulphuret, occurs chiefly in new red sandstone:—sometimes in a sort of mica slate.

Silver in its three forms of a sulphuret, a sulphuret of silver and antimony, and a chloride, has been found mostly in primary and transition slates:—sometimes in a member of the new red sandstone series; and in one instance in tertiary strata. *Wonders of Geology, Vol. 2, p.* 651.

Gold and platinum always occur in a metallic state; and they have usually been explored in drift. They are often associated, however, with the older rocks; and in this country especially, a gold deposite has been traced from Canada to the southern part of Georgia, and the metal is embraced in the talcose slate formation, in veins, usually of quartz. It is found also rarely in graywacke, and even in tertiary strata.

Cobalt, bismuth, arsenic, &c. are usually found associated with silver, or copper; and of course occur in the older rocks. The other metals, which, on account of their small economical value, and minute quantity, it is unnecessary to particularize, are also found in the older strata; frequently only disseminated, or in small insulated masses.

Obs. An excellent and much more extended view of the geological situation of useful minerals, may be found in Beudant's *Traite Elementaire de Mineralogie, Tome Premier, Livre Quartieme: Paris,* 1830.

Inf. It appears from the facts that have been detailed respecting the situation of the useful minerals, that great assistance in searching for them may be derived from a knowledge of rocks and their order of superposition.

Illus. No geologist, for instance, would expect to find valuable beds of coal in the oldest crystalline rocks, nor in the tertiary strata; but in the secondary fossiliferous rocks alone: and even here, he would have but feeble expectations in any other rock except the coal formation. What a vast amount of unnecessary expense and labor would have been avoided, had men, who have searched for coal, been always acquainted with this principle, and able to distinguish the different rocks! Perpendicular strata of mica and talcose slate would never have been bored into at great expense, in search of coal: nor would schorl have been mistaken for coal, as it has been!

By no mineral substance have men been more deceived, than by iron pyrites: which is very appropriately denominated *fools' gold.* When in a pure state, its resemblance to gold in color is often so great, that it is no wonder those unacquainted with minerals, should suppose it to be that metal. Yet the merest tyro in mineralogy can readily distinguish the two substances; since native gold is always malleable, but pyrites never. This latter mineral is also very liable to decomposition, and such changes are thereby wrought upon the rocks containing it, as to lead the inexperienced observer to imagine that he has got the clue to a rich depository of mineral treasures; and probably nine out of ten of those numerous excavations that have been made in the rocks of this country, in search of the precious metals, had their origin in pyrites, and their termination in disappointment, if not poverty. This ore also, when decomposing, sometimes produces considerable heat, and causes masses of the rock to separate with an explosion. Hence the origin of the numerous legends that prevail respecting lights seen, and sounds heard, in the mountain where the supposed treasure lies, and which so strongly confirm the ignorant in their expectation of finding mineral treasures. Now all this delusion would be dissipated in a moment, were the eye of a geologist to rest on such spots, or were the elementary principles of geology more widely diffused in the community.

Another common delusion respects gypsum; which is as often sought among the primary, as in the secondary and tertiary rocks: although it is doubtful whether primary gypsum has ever been found. A few years since, however, a farmer in this country supposed that he had discovered gypsum on his farm, and persuaded his neighbours that such was the case. They bought large quantities of it, and it was ground for agriculture, when accidently it was discovered that it was only limestone: a fact that might have been determined in a moment at first, by a single drop of acid.

Caution. It ought not to be inferred from all that has been said, that because a mineral substance has been found in only one rock, it exists in no other. But in many cases we may be almost certain that such and such rocks cannot contain such and such minerals. Of these cases, however, the practised geologist can alone judge with much correctness, and hence the importance of an extensive acquaintance with geology in the community. An amount of money much greater than is generally known, has been expended in vain for the want of this knowledge.

Obs. The chemical changes which rocks have undergone since their deposition, as well as the operation of decomposing agents to which they are now exposed, properly belong to the chemistry of geology. But these points will be deferred to subsequent sections; because they will there be better understood.

SECTION III.

THE LITHOLOGICAL CHARACTERS OF THE STRATIFIED ROCKS.

Def. The *lithological* character of a rock embraces its mineral composition and structure as well as its external aspect, in distinction from its zoological and botanical characters, which refer to its organic remains.

Rem. I shall describe the stratified rocks under the names and in the order used in the first column of the Tabular Synopsis, of the different classifications, given at the close of Section I. I have also arranged them into the larger Groups of Alluvium, Drift, Tertiary, Secondary, Palæozoic, and Primary, because these divisions are used by the best English and American writers on the subject, and because there is undoubtedly more or less foundation for them in nature.

1. ALLUVIUM.

Descr. The following stratified deposits are the result of alluvial agency.

1. Soil.
2. Sand.
3. Peat.
4. Marl.
5. Calcareous Tufa, or Travertin.
6. Coral Reef.
7. Siliceous Sinter.
8. Siliceous Marl, or deposits of the skeletons of Infusoria.
9. Bitumen.
10. Sulphate of Lime.
11. Hydrate of Iron.
12. Bog Manganese.
13. Chloride of Sodium (Sea Salt.
14. Sandstones, Conglomerates, and Breccias.

Prin. *Soil* is disintegrated and decomposed rock, with such a mixture of vegetable and animal matter that plants will grow in it.

Proof. 1. We see almost every where the rocks crumbling down into soil. 2. Chemical analysis shows that the soils are composed generally of silica, alumina, lime, magnesia and iron, in about the same proportion as they are found in the rocks. Silica is much the most abundant ingredient. 3. The presence of organic matter is easily proved by burning it off.

Descr. Vast accumulations of *sand*, the result of alluvial agency, occur not merely in the bed of the ocean and in lakes, but also upon the dry land, where they are called *dunes* or *downs*. These are composed almost entirely of silica; and being destitute of organic matter, cannot sustain vegetation.

Descr. The manner in which *peat* is formed has already been explained in general terms. (Section II.) When perfectly formed, it is destitute of a fibrous structure, and is, when wet, a fine black mud: and when dry, a powder. It consists chiefly of the decomposed organic matter called *geine* or *humic acid*, with crenic and apocrenic acids, phosophates, &c. part of which are soluble, and a part insoluble, in water. These deposits of peat are sometimes 30 or 40 feet thick; but they are not formed in tropical climates on account of the too rapid decomposition of the organic matter.

Descr. *Alluvial marl* is usually a fine powder, consisting of carbonate of lime, clay, and soluble and insoluble geine; and is found usually beneath peat in limestone countries; sometimes at the bottom of ponds. It is produced partly by the decay of the shells of molluscous animals, and partly by the deposition of the carbonate of lime from solution in water. It contains numerous small fresh water shells, and has received the name of *shell marl*.

Method of detecting calcareous marl. The great value of this substance in agriculture, and the confusion that prevails in its description, render it desirable to point out a test by which it can be distinguished. That test is an acid of some sort, the common mineral acids, oil of vitriol, aqua fortis, and muriatic acid being the best; but strong vinegar will answer. If the substance effervesce, when the acid is applied, we may be sure that it is genuine marl: otherwise not.

Other kinds of marls. Several other substances that contain no carbonate of lime have often been denominated marl by agriculturists and not without reason; for they have produced effects analogous to those of calcareous marl. But it seems very desirable that terms should not be applied too loosely, and I propose the following designations for these substances.

Calcareous marl: that which contains carbonate of lime in any quantity.

Siliceous marl: that in which silica predominates, and no calcareous matter is present.

Alluminous marl: that in which clay predominates, and no calcareous matter is present.

Green sand marl: that which contains green sand. This is the substance that has been of late employed with signal success as a fertilizer of land in New Jersey, Virginia, Delaware, &c. If it contain any carbonate of lime, the compound term *Calcareo-green sand marl*, might be employed.

Method of searching for alluvial marl. The presence of marl beneath a peat bog can be determined with a good degree of certainty, by plunging a pole,—the rougher the better, through the peat, until it reaches the solid bottom of the morass; and, on withdrawing it, some of the marl if any exist, will adhere to the surface; though a coating of the black mud may cover it.

Descr. Calcareous tufa or tavertin, is a deposite of carbonate of lime, made by springs containing that substance in solution. It forms a solid limestone, sometimes even crystalline, and of considerable extent; so as to be used for architectural purposes. Thermal waters produce it most abundantly, as in Central France, Hungary, Tuscany, and Campagna di Roma: but it is also deposited by springs of the ordinary temperature, as at Saratoga and in the Appenines. *Dr. Daubeny's Report on Mineral and Thermal Waters. p.* 56. *London*, 1837. *Also De la Beche's Manual of Geol. p.* 158. *Also Lyell's Prin. Geol. Vol.* 2. *p.* 198. Tavertin is also precipitated by rivers, as in Tuscany; and at the mouths of rivers on the coast of Asia Minor. *Lyell's Prin. Geology, Vol.* 1. *p.* 397. Very similar are the concretionary calcareous deposits formed in caverns: those depending from the roof are called *stalactites*, and those on the floor, *stalagmites*.

Descr. Coral reefs are extensive deposits of carbonate of lime, formed by myriads of polyparia, or radiated animals, in shallow water, in the south seas. They form the habitations of these animals: and of course are organic in their structure.

Descr. Siliceous sinter, or tufa, is a deposit of silica, made by water of thermal springs, which sometimes hold that earth in solution. Successive layers of sinter and clay frequently occur, and these are sometimes broken up and recemented so as to form breccia. *Prof. J. W. Webster in the Edinburgh Philosophical Journal, Vol. VI.*

Descr. Siliceous marl, or the *fossil shields of infusoria.* **Beneath the beds of peat and mud in the primary regions of this** country, a deposit often occurs from a few inches to several feet thick, which almost exactly resembles the calcareous marl that is found in the same situation. When pure, it is white and nearly as light as the carbonate of magnesia: but it is usually more or less mixed with clay. It is found by analysis to be nearly pure silica; and it turns out to be almost entirely composed of the siliceous shields, or skeletons, of those microscopic animals called *infusoria*, or *animalcular*, which have lived and died in countless numbers in the ponds at the bottom of which this substance has been deposited.

Rem. The discovery of this curious fact (concerning which more will be said in a subsequent section,) in relation to this country, was made by Prof. Bailey of West Point. *American Journal of Science, Vol.* 35. *p.* 118. Analogous substances occur in Europe: and the most of that just described, appears to be identical with the *Bergmehl* of Prof. Ehrenberg.

Descr. Some springs produce large quantities of *bitumen* in the form of naptha and asphaltum. Their localities and extent will be described in a subsequent section.

Descr. Although *sulphate of lime* very generally exists in the waters of springs, yet it is rarely deposited. One or two examples only are mentioned, where a deposit of this salt has been made: as at the baths of San Philippo in France. *De la Beche's Manual, p.* 158.

Descr. *Hydrate of iron or bog ore*, is a common and abundant deposit from waters that are capable of holding it in solution; and it appears also, that this ore is often made up of the shields of infusoria, which are often ferruginous. *Wonders of Geology Vol.* 2. *p.* 660.

Descr. *Bog manganese*, also, by a somewhat similar process, is frequently deposited in the form of the earthy oxide, or *wad*, in low grounds: and it can hardly be doubted but it is an alluvial product. *Report on the Geology of Massachusetts*, 2*d Edition, p.* 130.

Descr. *Chloride of sodium* or *rock salt*, is very rarely deposited from its solution in water so as to be visible, though some have supposed that this deposition does take place extensively at the bottom of such seas as the Mediterranean. It is said, however, to accumulate in some of the cavities of the rocks along the shores of the Mediterranean, in such quantities as to be collected by the inhabitants.

Rem. Rev. Dr. J. Perkins, American Missionary at Ooroomiah, in Persia, states that the common opinion among the people on the shores of Lake Ooroomiah, is, that frequently it deposits salt upon its bottom. At any rate, there is a pond at Gavalan, on the shore of the lake, and separated from it by a narrow sand bank, whose bottom is covered by beautiful white salt, three or four inches thick near the shore, and four or five feet thick n the middle. This pond covers from 75 to 100 acres.

Rem. 2. The waters of lake Elton in Asiatic Russia, and of other lakes adjoining the Caspian Sea, have deposited thick beds of rock salt at their bottom. *Daubeny's Report on Mineral and Thermal Waters for* 1836, *p.* 7. The same is true of lake Indersk on the steppes of Siberia. *Ure's Geology, p.* 373.

Descr. Alluvial sandstone, conglomerate, and breccia, are formed by the cementation of sand, rounded pebbles, or angular fragments, by iron, or carbonate of lime, which is infiltrated through the mass in a state of solution. They are not very common, nor on a very extended scale.

Def. When sand is cemented, the solid mass is called *sandstone:* rounded pebbles produce a *conglomerate* or *plum pudding stone;* and angular fragments, a *breccia.*

Def. The varieties of alluvium, that have been described may be regarded as a formation in the geological sense: and the period during which such a group is in the progress of deposition, that is, until some important change takes place in the material or mode of production, is called a *geological period:* and the point of time when the change occurs, is called an *epoch.*

2. DRIFT (FORMERLY DILUVIUM.)

Rem. There is more diversity of opinion respecting the origin of this formation, than on almost any other subject of geology. Hence it has received a great variety of names; such as *diluvium, drift, bowlder formation, erratic block group, &c.* The term diluvium literally implies that it has been the result of a deluge, and is therefore, objectionable. The term *drift* appears at present to be preferable, and is therefore used.

Descr. The great mass of drift is composed of sand and gravel of different degrees of comminution, mixed together con fusedly. This gravel is often not derived from the rocks beneath it, but from those at a distance of several miles, and in this country usually from ledges which lie in a north-westerly direction. The surface of this gravel is often scooped out into deep basin-shaped depressions, and raised into corresponding elevations, the difference of level being sometimes 20 or 30 and even 100 or 200 feet.

Descr. Scattered through this gravel, are rounded masses of rock larger than pebbles, which are called *bowlders:* and as they are frequently found a great distance from the place of their origin, they are also denominated *erratic blocks* and *lost rocks.* Oftentimes these bowlders lie insulated upon the surface, and even upon the crests of mountains.

Def. When they happen to be thus insulated upon other

rocks, and so poised that a small force will make them oscillate, they are called *rocking stones.*

Fig. 32, exhibits a rocking stone in the west part of Barre, Mass.

Fig. 32.

Rocking Stone: Barre.

Fig. 33, shows a rocking stone in Fall River, Mass., poised upon granite and weighing 160 tons.

Fig. 33.

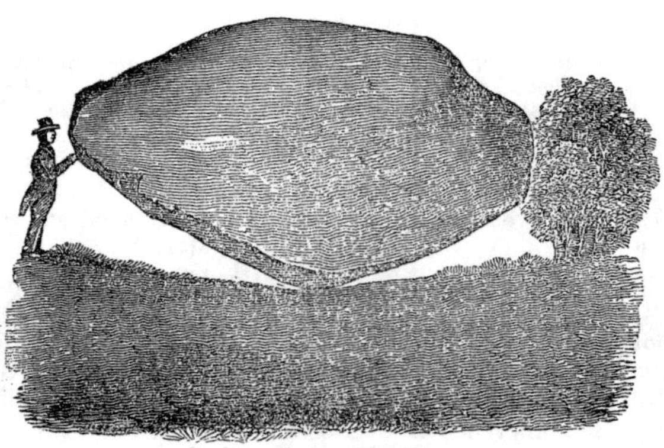

Rocking Stone: Fall River.

Descr. On many plains through which no existing stream now passes, and lying over the gravel above described, we find thick beds of sand and clay, deposited in a much more quiet manner than common drift, and yet apparently near the close of the same period. The usual order of the series is, first and lowest, the coarse materials, then clay, then sand. *Traite Elementaire Geologie, par M. Rezot, p.* 256. *Tome* 1.

Descr. Sometimes the drifted sand and gravel that have been described, are consolidated into sandstone and conglomerate, by the infiltration of iron or carbonate of lime ; as in Pownal, Vermont. *Report on the Geology of Massachusetts, Vol.* 2, *p.* 354. *Also Tableaux des Terrains, p.* 66.

Descr. Many of the most valuable of the precious stones and metals are found in drift; such as the diamond, the sapphire, the topaz, the ruby, and the zircon ; as well as platinum, gold, and tin. Platinum, gold, and the diamond are explored almost exclusively in this formation. *Tableaux des Terrains, p.* 115.

3. TERTIARY STRATA.

(*Supercretaceous Group of De la Beche. Cainozoic Period,* Rogers.)

Historical Rem. Until the publication by Cuvier and Brongniart of their memoir on the tertiary strata around the city of Paris in 1810, these formations were confounded with alluvium. Since that period, other similar deposits have been studied with diligence and success ; and it is found that tertiary strata occupy more than half the surface of Europe; and in this country they embrace much of the level region in the eastern part of the middle and southern states.

Descr. The tertiary rocks have been divided into four distinct groups of marine strata, distinguished by important peculiarities in their organic remains, and separated from one another, by strata which contain fresh water and terrestrial remains. *Buckland's Bridgewater Treatise, Vol.* 1. *p.* 76.

Rem. Marine strata are easily distinguished from those of fresh water origin, by the occurrence in the former of animals peculiar to the ocean, and in the latter of those peculiar to fresh water.

Descr. Mr. Lyell has divided these strata into four groups, to which he gives the names *Eocene, Miocene, and Older and Newer Pliocene.* In the first, the number of shells identical with living species is very small, only 3.5 per cent. In the second group, reckoning upwards, it is 17 per cent. : in the older pliocene, 35 to 50 per cent., and in the newer pliocene, 90 to 95 per cent. And by this character are the groups distinguished. *Lyell's Elements of Geology, Vol.* 1. *p.* 280. Other geologists object to these characters as too indefinite. *De la Beche's Theoretical Geology, Chap. XVII. Phillips's Edinburgh Treatise on Geology, p.* 180.

Descr. The tertiary rocks are in general distinctly stratified, and the strata are usually horizontal. In some cases, however, (as in the Isle of Wight and at Gay Head,) they are inclined at a large angle.

Prin. All the stratified rocks appear to have been originally deposited from water.

Proof. The manner in which the ingredients of these rocks are arranged, viz. in parallel strata and laminæ, is precisely like that of the subaqueous deposits which are now forming in many localities, so that these latter need only to be hardened into stone, (when they are not already consolidated,) and in some cases rendered more crystalline, in order to be converted into the former. And by no other agent that we know of, by which rocks are formed, is a stratified and schistose arrangement produced. Again, the materials composing these stratified rocks, viz. clay, sand, and carbonate of lime, are very similar to those deposits which water is now producing. And further, the organic remains which many of these rocks contain, can be accounted for only on the supposition that the rocks enveloping them were deposited from water.

Descr. Rocks are deposited by water in two modes: first, as mere sediment, by its mechanical agency, in connection with gravity: secondly, as chemical precipitates from solution.

Def. The first kind of rocks is called *mechanical* or *sedimentary* rocks; the second kind, *chemical deposits*.

Descr. As a general fact, the lower we descend into the rock series, we meet with less and less of a mechanical and more and more of a chemical agency in their production. The primary stratified rocks have generally been regarded as destitute of every mark of a mechanical origin except their parallel arrangement; but in fact, the greater part of them are made up of the fragments of crystals more or less worn and cemented together.

Rem. I possess specimens of mica slate, talcose slate, and quartz rock, from various parts of New England, which are made up of fragments as distinctly rounded by attrition, as those of any fossiliferous conglomerate; and these pebbles are cemented by similar materials in a finer state. Most of these specimens are associated with highly inclined strata of the oldest primary rocks in New England. They are good examples of what are called *metamorphic rocks. Phillips's Geology, p.* 75.

Descr. In the fossiliferous rocks we sometimes find an alternation of mechanical and chemical deposits: but for the most part, these rocks exhibit evidence of both modes of deposit, acting simultaneously.

Rem. It is difficult to conceive how any rock can be consolidated without more or less of chemical agency, except perhaps in that imperfect consolidation which takes place in argillaceous mixtures by mere desiccation. Even in the coarsest conglomerate there must be more or less of chemical union between the cement and the pebbles.

Descr. In the tertiary rocks a mechanical agency decidedly predominates: nevertheless, several beds are the result of chemical precipitation; as gypsum, limestone, and rock salt.

Descr. The varieties of rocks composing the tertiary strata are concretionary, tufaceous, argillaceous, and siliceous; or limestone, marl, plastic clay, siliceous and calcareous sands, green sand, gypsum, lignite, rock salt, and buhrstone.

4. SECONDARY ROCKS.

Def. Under *Secondary Rocks* are included all those from the Tertiary downwards to the base of the Trias, or Upper New Red Sandstone. The Cretaceous System is considered as the Newer Secondary: the Wealden, Oolite, and Lias, as the Middle Secondary; and the Triassic System as the Older Secondary.

1. *Cretaceous System.*

Descr. In Europe this formation is usually characterized by the presence of chalk in the upper part, and sands and sandstones in the lower. In this country, chalk is wanting: yet it seems to be well established that the Ferruginous sand formation is the equivalent of the chalk formation of Europe. *Dr. Morton in Journal of Academy of Natural Sciences, Vol. VI. Also American Journal of Science, Vol. XVII. p. 274, and XVIII. p. 243, and XXIV. p. 128.*

Descr. The cretaceous system is thus arranged by Dr. Fitton, leaving out the Wealden, which, in this work, is regarded as a distinct formation.

Chalk.	Upper, Lower, Marly,
Green Sand.	Upper Green Sand, Gault, Lower Green Sand.

Observations on some of the Strata between the Chalk and Oxford Oolite in the South East of England, by W. H. Fitton, p. 105. *London,* 1836.

Descr. *Chalk* is a pulverulent carbonate of lime, and its varieties have resulted from the impurities that were deposited with it. The upper beds are remarkable for the great quantity of flints dispersed through them, generally in parallel position.

Descr. *Green Sand* is a mixture of arenaceous matter, with a peculiar green substance greatly resembling chlorite, or green earth.

Composition. The coloring matter of green sand has been analyzed with much care by several distinguished chemists with the following results.

	French Green Sand. *By M. Berthier.*	*English Do.* *By Prof. Turner.*	*Massachusetts Do.* *By Dr. S. L. Dana.*	*N. Jersey, Do.* *By Prof. H. D. Rogers.**
Silica,	50.0	48.5	56.700	49.27
Protoxide of Iron,	21.0	22.0	20.100	24.67
Alumina,	7.0	17.0	13.320	7.71
Water,	11.0	7.0	7.000	5.91
Potassa,	10.9	traces		9.99
Lime,			1.624	5.08
Magnesia,		3.8	1.176	
Manganese,		traces and loss=	0.080	

* The mean of eight analyses.

See Dr. Fitton on the Strata below the Chalk, p. 109. *Also, Prof. H. D Rogers's Report on the Geological Survey of New Jersey, p.* 74, *et seq. Also Final Report on the Geology of Massachusetts.*

Use of Green Sand. This substance has been applied within a few years in this country with great success as a manure, especially in N. Jersey. If its fertilizing power depends on the potassa alone, the English and Massachusetts deposits would be of no value; but if, as some suppose, the oxide of iron and the other ingredients assist in this respect, it may prove of great importance.

Descr. *Gault or Galt,* is a provincial name for a blue clay or marl, forming an interstratified bed in the green sand of England.

2. *Wealden Formation.*

This formation embraces, 1. *The Weald Clay;* 2. *Hastings Sand;* 3. *Purbeck Strata.* They were first described in the South East of England, chiefly in the *wealds* or woods of Sussex and Kent, and are composed of beds of limestone, conglomerate, sandstone, and clay, which abound in the remains of fresh water and terrestrial animals, and appear to have been deposited in an estuary that once occupied that part of England. Similar beds occur in Scotland, and in a few places on the European Continent.

Rem. Some of the most remarkable facts in fossil geology have been derived from this formation, which will be found described in *Dr. Mantell's Illustrations of the Geology of Sussex, &c.* And in his *Geology of the South East of England;* also in *Dr. Fitton's Observations on the Strata below the Chalk;* and in *Dr. Mantell's Wonders of Geology,* 2 *Vols.* 1838.

3. *Oolitic System.*

Descr. In many of the rocks of this series, small calcareous globules are imbedded, which resemble the roe of a fish, and hence such a rock is called *roestone* or *oolite.* But this structure extends through only a small part of this formation, and it occurs also in other rocks.

Descr. The oolite series consist of interstratified layers of clay, sandstone, marl, and limestone. The upper portion, or that which is oolite proper, is divided into three systems or groups, called the upper, middle, and lower, separated by clay or marl deposits.

The Upper Oolites in England consist of, 1. *The Portland Stone;* 2. *Portland Sand;* 3. *Kimmeridge Clay.* The Middle Oolites embrace, 1. *The Upper Calcareous Grit;* 2. *Coral Rag;* 3. *Lower Calcareous Grit;* 4. *Oxford Clay;* 5. *Kelloway's Rock.* The Lower Oolites are divided into, 1. *Cornbrash;* 2. *Forest Marble;* 3. *Great Oolite and Bradford Clay;* 4. *Carbonaceous Gritstones and Shales, Stonesfield Slate and Fuller's Earth;* 5. *Inferior Oolite;* 6. *Calcareo-Siliceous Sand. Ansted's Geology, Vol.* 1. *p.* 357. The Oolite is widely distinguished in other parts of the world except in North America.

Rem. The Oolite Group is remarkable for the vast amount of its organic remains, and for the extraordinary character of many of them.

4. *Lias.*

Descr. Lias is a rock usually of a bluish color like common clay; and it is indeed highly argillaceous, but at the same time generally calcareous. Bands of true argillaceous limestone do, indeed, occur in it, as well as of calcareous sand. It nas been usual to describe it as a member of the oolitic series. But it is widely diffused; is very marked in its characters, and contains peculiar and very interesting organic remains.

5. *Triassic System, or Upper New Red Sandstone.*

Rem. Until quite recently, all the stratified rocks between the lias and the coal measures, have been regarded and described as a single group, under the name of New Red Sandstone, Saliferous System, &c. But it is founa that the fossils of the two lower divisions of this system correspond witu those in the Newer Palæozoic Series, much nearer than with those of the lias, or oolite; while those in the three upper divisions, approach nearest to the fauna and flora of the secondary rocks.

Descr. In Continental Europe the Triassic System is divided into three distinct groups, and hence its name: 1. *Variegated Marl, (Marnes irisees of the French, and Keuper of the Germans:)* or indurated clays of different colors, with a predominance of red. With this, gray sandstone and yellowish magnesian limestone are interstratified. Beds of gypsum and rock salt are common. This is the upper member of the series. 2. *Muschelkalk,* a gray compact limestone, occasionally dolomitic. This does not occur in Great Britain, nor, as we know, in this country; and hence a good deal of difficulty in distinguishing the upper from the lower divisions. 3. *Red or variegated Sandstone, (Gres Bigarre* of French writers, and *Bunter Sandstein* of the Germans.) Its colors are white, red, blue, and green; which often give a strikingly variegated appearance to the rock. Its composition is chiefly siliceous and argillaceous, with occasional beds of gypsum and rock salt, which, however, are often wanting.

Rem. This rock, especially where the Muschelkalk is wanting, is very deficient in organic remains. But it is in this formation chiefly that those remarkable footmarks are found which have of late excited no small interest.

6. *Palæozoic Rocks.*

Def. This class of Rocks extends from the base of the Triassic System to the bottom of the fossiliferous rocks, embracing deposits of vast thickness. The lower portion, or the Devonian and Silurian Rocks, were formerly called the *Transition Class*, and sometimes *Graywacke.* The subdivisions of the Palæozoic Strata are, 1. The Newer Palæozoic, or the Lower New Red and the Carboniferous System. 2. *The Middle Palæozoic*, or Old Red Sandstone; and 3. the *Older Palæozoic* or the Silurian.

1. *The Lower New Red Sandstone, or Permian System.*

Descr. In Germany and England the Lower New Red System presents two very distinct groups of rocks: the highest is a Magnesian Limestone, and the lowest Sandstone. In Germany the limestone is called Zechstein, and its subdivisions are *Letten*, *Stinkstein*, and *Rauwacke*, with some shaly beds beneath of argillaceous, bituminous, and arenaceous schist, called *Kuperscheifer* or copper slate, because sometimes worked for copper. The sandstone beneath the schists in Germany, (and frequently also in England and America,) is called *Rothe-todte-liegende*, or *red dead lier: red* as to color, *dead* as to metallic ores, and *underlying* the copper slate.

Descr. The Permian system consists of numerous strata of great extent in Russia, (700 miles long and 400 broad,) in the ancient kingdom of Permia, made up of common and magnesian limestones, with gypsum and rock salt conglomerates, red and green gritstones, shales, and copper ore, lying in a trough of the carboniferous strata, and below the Triassic system. Hence it has been very probably referred to the Lower New Red System.

Rem. It is not yet settled whether the Lower New Red Sandstone occurs in North America, though some of its most remarkable fish,—the Palæonisci, are common here.

2. *Carboniferous System.*

Descr. This system derives its name from the great amount of carbon, or coal found in it: it being in fact the only deposit, with a few exceptions, that contains good coal in workable quantity. It is made up of two distinct varieties or groups of rocks. 1. *Coal Measures.* These consist of irregularly interstratified beds of sandstone, shale, and coal. Frequently these are deposited in basin-shaped cavities; but not always. These rocks abound in faults produced by igneous agency; whereby the continuity of the beds of coal is interrupted, and the difficulty of exploring for coal increased in some respects; but in other respects facilitated; so that upon the whole, these faults are decidedly beneficial. 2. *Carboniferous Limestone.* A gray compact limestone, traversed by veins of calcareous spar, and frequently abounding in organic remains. Encrinites are sometimes so abundant that the rock is called *encrinal limestone.* It is also called *mountain limestone,* and *metalliferous limestone,* since in England, as well as in North America, it abounds in lead ore.

Rem. 1. The coal measures exist in almost every country of much extent, and form one of the most important sources of national wealth and happiness. In England not less than 6.000.000 tons of coal are yearly raised from the mines of Northumberland and Durham; at which rate they will be exhausted in about 250 years. In South Wales, however, is a coal field of 1200 square miles, with 23 beds, whose total thickness is 95 feet; and this will supply coal for 2000 years more. *Bakewell's Geology, p.* 125. In Great Britain about 15.000 steam engines are in operation by the use of coal, with a power equal to that of about 2.000.000 of men. The machinery moved by this power has been supposed equivalent to that of between 300.000.000 and 400.000.000 men by direct labor. Well may Dr. Buckland say, "we are almost astounded at the influence of coal and iron and steam upon the fate and fortunes of the human race." *Bridgewater Treatise, Vol.* 1. *p.* 535.

Rem. 2. No part of the world contains coal fields so extensive as the United States, and the provinces of Nova Scotia and New Brunswick. In these provinces is a triangular coal field, covering more than 10.000 square miles. The Appalachian Coal Field, extending from New York to Alabama is 720 miles long, and contains nearly 80.000 square miles. The Indiana Coal Field, which is 360 miles long, covers about 55.000 square miles. Another occurs in Michigan, 150 miles long, and covers 12.000 square miles. Still farther west, in Missouri, and towards the Rocky Mountains, are other fields of vast extent, not to mention the smaller deposits in eastern Virginia, Rhode Island and Massachusetts. In some parts of these fields, the aggregate thickness of the beds of coal is not less than 100 feet; and even single beds occur from 20 to 50 feet thick. In whatever else, therefore, our country fails, her coal can never be exhausted, and steam will make it available all over the land.

3. *Old Red Sandstone, or Devonian System.*

Descr. The Middle Palæozoic Rocks consist of a deposit of vast thickness and extent, which, in Scotland and England, has long been known as the Old Red Sandstone; but lately it has been denominated the Devonian System, by Professor Sedgwick and Mr. Munchison, because largely developed in Devonshire, and proved by them with great labor and singular sagacity to be contemporaneous with the Old Red Sandstone of other regions. In Scotland, this formation is not less than 10,000 feet thick. In England it is divided into three groups: 1. *Old Red Conglomerate,* which is the uppermost division: 2. *Cornstone and Marl,* or argillaceous marly beds, alternating with sandstone, and sometimes with impure limestone: 3. *Tilestone,* or fusile beds, used sometimes for tiles. In other countries other varieties of rock are found in this formation, such as red and green flags, black calcareous slaty masses, and magnesian limestone. But the organic remains identify the formation in different regions. These consist of numerous peculiar corals, radiata and molluscs; but more especially of peculiar fishes.

Descr. This formation is widely developed on the continent of Europe, as in Belgium and Westphalia, France and Spain. In Russia it covers more surface than the whole of Great Britain, not less than 150,000 square miles. In the United States it occupies extensive tracts, as may be seen by a reference to the Geological Map connected with my outline of the Geology of the Globe. *Ansted's Geology, Vol.* 1. *Chaps. VII. VIII. IX. X. Murchison's Geology of Russia, Vol.* 1. *Chap. IV.*

4. *Silurian System, Murchison. Graywacke of Others.*

Descr. The vast deposits which are now described under this name, embrace all the fossiliferous rocks below the Devonian, or Old Red Sandstone. A few years since Professor Sedgwick gave an account of fossiliferous rocks still lower, under the name of Cumbrian and Cambrian Systems; from the name of the region where they are developed in Great Britain. But it seems to be now agreed among the best English geologists, that the organic remains in the Cumbrian, Cambrian, and Silurian groups, are too much alike to allow them to be separated from one another and they are accordingly all united in the Silurian System. *Murchison's Anniversary Address before the London Geolog. Society,* 1843. *Also Ansted's Geology, Vol.* 1. *p.* 93, *et seq.* Professor Emmons, however, contends that the fossils in the Cambrian and Silurian Systems are distinct, and that what he calls the Taconic System in this country, is the equivalent of the Cambrian in Europe. *The Taconic System, p.* 5. *Albany,* 1844.

Descr. The Silurian System is divisible into two great groups, which are probably about as distinct from each other as the other groups that have been described.

1. *Upper Silurian Rocks.*

Descr. This series embraces the Ludlow and Wenlock rocks of Murchison's Silurian System, and the Upper Cumbrian and Cambrian rocks of Sedgwick. The Ludlow and Wenlock Formations consist of shales and limestones conformably interstratified, with the following names, beginning with the lowest. 1. Wenlock Shale. 2. Wenlock Limestone. 3. Lower Ludlow Shale. 4. Aymestry Limestone. 5. Upper Ludlow Shale. The Cumbrian and Cambrian rocks are mostly argillaceous and calcareous slates, frequently fossiliferous and alternating with porphyries and feldspar rock of igneous origin. These slates occasionally pass into flagstones and thick sandy beds of the rock called by the Germans *Grauwacke*, which is a gray micaceous sandstone. *Ansted's Geology, Vol.* 1. *Chaps. II. III. and IV. Murchison's Silurian System, Vol.* 1 *p.* 195, *London,* 1839.

2. *Lower Silurian Rocks, Murchison. Protozoic Group of Sedgwick.*

Descr. The two most important varieties of the Lower Silurian Rocks, as described by Mr. Murchison in Great Britain, are the Caradoc Sandstone and the Llandeilo Flags. These rest upon the strata of the Cambrian Series, and pass into them insensibly. The entire series, including also a portion of the Cumbrian System, is highly argillaceous, or arenaceous; the clay being changed into slates, or shales, with cleavage planes, and the sand into fissile sandstones. Interstratified, however, among these argillaceous and arenaceous strata, are occasional bands or lumps of calcareous matter, containing organic remains, which occur also in the slates. *Murchison's Silurian System, Vol.* 1. *p.* 216. *Ansted's Geology, Vol.* 1. *p.* 103.

Descr. The Silurian and Cambrian Rocks are described as of great thickness, certainly many thousand feet. But their exact thickness has not yet been ascertained. In the Tabular View, Section I., I have given the thickness at 20,000 feet.

Descr. The Silurian Rocks, embracing also the Cumbrian and Cambrian, occupy large areas in Belgium, Germany, Scandinavia, and Russia, as well as in North and South America. Their great extent in the United States may be seen in the Geological Map of N. America above spoken of. Their thickness here is certainly as great as in Europe; and they appear to be more fully developed in this country than in Europe; since we have several varieties here not found there. Mr. Conrad thinks there are good grounds for a threefold division of these rocks in this country; the groups being as distinct from one another as they are from the Devonian. *Jour. of Acad. Nat. Sci. Phil.* 1842, *Vol. VIII. part* 2. *p.* 233. As will be seen by a reference to the Tabular View at the end of Section I., the New York Geologists divide the Silurian System into 28 distinct groups: though the three uppermost, according to Mr. Hall, (*Geological Report, p.* 20*; and Lyell's Travels in North America, Vol.* 2. *p.* 216,) correspond to the Devonian System of Professor Phillips. The Upper Silurian embraces the groups from the Hamilton to the Clinton; and the Lower Silurian from thence to the Potsdam Sandstone inclusive. Mr. Hall considers the Cambrian Rocks to be embraced in the New York System, (*Report, p.* 20;) but as we have seen, Prof. Emmons regards his Taconic System, which underlies the New York System, as identical with the Cambrian System. Others, as Prof. W. B. and H. D. Rogers, and Prof. W. W. Mather, regard the Taconic System as merely certain members of the New York System, that have been subjected to powerful igneous metamorphic action.

6. *Primary Rocks in part. Hypogene and Metamorphic Rocks in part of Lyell.*

Rem. 1. These terms are all objectionable; because they are founded upon hypotheses, or theories that may not prove true, or only in part true. The rocks to be described (including granite, porphyry, &c.) were called *Primary*, because they were supposed to have been produced before the deposition of the fossiliferous strata already described: whereas it now appears that some of these rocks, especially the unstratified, have in some instances been formed at a later period. The term *Hypogene* (*nether-formed*, or formed beneath the earth's crust,) is intended to express this fact. But it represents all the non-fossiliferous rocks as thus formed; which would not be admitted by many geologists. The term *Metamorphic* implies that these rocks have beeen altered since their original production; which would be admitted to a greater or less extent by all geologists: but the same is true of the rocks already described, especially of the older ones. These terms, therefore, must be regarded as only provisional, to be changed when better ones shall be proposed. *Lyell's Elements of Geology, Vol.* 1. *p.* 18. *Also his Travels in North America, Vol.* 1. *p.* 106.

Rem. 2. To find the oldest fossiliferous rock has been, and still is a problem of great interest with geologists. In Europe, the prevailing opinion seems now to be that the Lower Silurian is the true Protozoic Group: though Mr. Lyell contends that even the primary rocks may have been once fossiliferous, and that the organic remains have been obliterated by metamorphic action. In this country the Potsdam sandstone of the New York System, is usually regarded as the oldest rock containing animals and plants. We have seen that Professor Emmons believes in a fossiliferous series still older; *viz. the Taconic. Prof. H. D. Rogers' Address before Geol. Assoc. at Washington in* 1844. *Emmons' Taconic System; Hall's, Vanuxem's, and Mather's Reports on the N. York Survey.*

Rem. 3. As the non-fossiliferous or primary rocks have no settled order of superposition, different writers will describe them in different orders. I shall give them in the order in which they most usually occur, especially in this country.

1. *Clay slate or argillaceous slate.* This rock is composed of fine argillaceous matter which has a fissile structure, and in the oldest varieties its surface is more or less shining from chloritic or plumbaginous matter. Its principal deposit has already been described, as a part of the Cumbrian and Cambrian Systems. But it occurs frequently interstratified with mica slate and quartz rock; and must, therefore, be regarded as a non-fossiliferous primary rock. Yet on the other hand, it also occurs interstratified with fossiliferous graywacke. There

seems, therefore, a necessity for regarding clay slate as belonging both to the fossiliferous and non-fossiliferous strata. The farther we recede from the line separating these two classes of rocks, towards the oldest, the more highly glazed does the clay slate become, until it passes at length insensibly into mica slate, talcose slate, or hornblende slate. But receding from that line in the other direction, its surface becomes more dull, and its texture looser, until it forms what is usually termed *shale;* and if we follow it still higher up in the series, it becomes gradually changed into unconsolidated clay.

Rem. A variety of clay slate used for whetstones and hones is called *whetstone slate.* Some of the best stones, however, are compact felspar. The common notion that they are petrified wood, is utterly groundless. *Graphic slate or drawing slate,* is a variety of clay slate that contains several per cent. of carbon. The best novaculite, or honestone, in this country, comes from Wisconsin and Arkansas.

2. *Quartz rock.* This rock is essentially composed of quartz, either granular or arenaceous. The varieties result from the intermixture of mica, felspar, talc, hornblende, or clay slate. In these compound varieties the stratification is remarkably regular; but in pure granular quartz, it is often difficult to discover the planes of stratification. It is interstratified with every one of the primary rocks, and also with graywacke: in which last case it often assumes a decidedly mechanical structure: and even when a member of the primary series, this structure is sometimes visible. *Macculloch's Principles of Geology, Vol. 2. p.* 174. *Also Geological Classification, p.* 317.

Rem. The arenaceous varieties of this rock form good *firestones;* that is, stones capable of sustaining powerful heat. Some varieties of mica slate are still better. Gneiss of an arenaceous composition is also employed; as are several varieties of sandstone of different ages. The firestone of the English green sand, is a fine siliceous sand cemented by limestone. *Fitton on the Strata below the Chalk, p.* 137.

3. *Hornblende slate.* Hornblende predominates in this rock; but its varieties contain felspar, quartz, and mica. When it is pure hornblende, its stratification is often indistinct, and it passes, by taking felspar into its composition, into a rock resembling greenstone. It occurs in every part of the primary series; but its more common associations are argillaceous slate, mica slate and gneiss; into which it passes by insensible gradations.

Variety. Dr. Macculloch describes *actynolite schist,* as distinct from hornblende slate: but as mineralogists now regard the two minerals as only one species, it is unnecessary to separate the rocks.

4. *Talcose slate.* The talc in this rock, which is the essential ingredient, and is sometimes in a pure state, is usually

mixed with quartz and mica, and sometimes with limestone, felspar, and hornblende. It is associated sometimes with argillaceous slate, and even graywacke: but usually, at least in the United States, with mica slate, and rarely with gneiss.

Varieties. Chlorite slate is only a variety of talcose slate, in which the talc is almost pulverulent and compact, of a green color, and in much large quantity than the quartz. *Steatite* is often nothing but schistose talc, whicl is adherent enough to be wrought, and at other times it is somewhat granular, and slightly indurated. This is the valuable stone so extensively used for furnaces, fire-places, aqueducts, &c., under the name of *soapstone* or *free stone.*

Obs. Most of the beds of steatite in New England, lie at the junction o talcose and hornblende and mica slate.

5. *Primary limestone.* Limestone that alternates with primary strata is called *primary.* Dr. Macculloch considers sucl alternation the only decided proof that a limestone is primary *Principles of Geology, Vol.* 2. *p.* 209. Others, as De la Beche make its primary character to depend more upon its crystalline state; and hence assert that it occurs interstratified witl fossiliferous rocks. *Manual of Geology, p.* 435. It is generally white and crystalline, resembling loaf sugar so much as to b called *saccharine.* But in some situations it is dark colored, by being penetrated with other rocks, and also nearly compact.

Rem. When this rock occurs in the unstratified class, and also in some o the older stratified ones, it is often nearly or quite destitute of stratificatior (Ex. gr. the limestone beds in syenite in Newbury and Stoneham, and i gneiss at Bolton, Massachusetts: also in hornblende slate in Smithfiel R. I.: and in granite, in St. Lawrence and Essex County, N. Y.) Henc it has been proposed to put primary limestone into the unstratified class *Prof. Emmons's Report on the Geology of the Second District of New Yor* 1842, *p.* 37. In many cases, however, it is most distinctly stratified: as fc instance, in the bed lying between strata of gneiss on Cole's Brook, in th west part of Middlefield, in Massachusetts. The interesting examples give by Prof. Emmons in St. Lawrence County, in his report above referred t do indeed prove that this rock may exist sometimes in the form of veins i granite. But looking at all the facts on the subject, they seem more satisfa torily explained by supposing primary limestone a metamorphic rock, whic may therefore be found both stratified and unstratified, than by regarding it a always unstratified and of igneous origin. This view of the subject, howeve scarcely differs in reality from that of Professor Emmons, although he seen to regard the difference as very great (*Emmons' Final Geological Report on th Second District of New York, p.* 40.) For we admit the rock to have bee melted, and hence it might be called igneous as much as granite, which wa certainly sometimes produced by the melting or metamorphism of gneiss, o other rock. The same is true of trap rock, syenite, and eminently of serpentine, which in New England frequently retains some of the original planes o stratification and lamination. The only difference between us, then, seen to be that Prof. E. does not admit that primary limestone was, or is, eve stratified.

6. *Mica Slate.* This is a slaty mixture of mica and quartz i

which the former predominates. Garnet and staurotide are often so abundant in it, over extensive tracts, as properly to be regarded as constituents: hence the varieties, *garnetiferous* and *staurotidiferous* mica slate. This is one of the most common and best characterized of the primary rocks.

8. *Gneiss.* The essential ingredients in this rock are quartz, felspar and mica. Hornblende is occasionally present. These ingredients are arranged more or less in laminæ, and the rock is stratified. Where it passes into granite, however, (which is composed of the same ingredients,) the stratification, as well as the laminar arrangement, become exceedingly obscure; and it is impossible to draw a definite line between the two rocks. Gneiss, as well as mica slate, is remarkable in some places for tortuosities and irregularities exhibited by the strata and laminæ: while in other places these same rocks are equally distinguished for the regularity and evenness of the stratification, by which they are rendered excellent materials for economical purposes.

Varieties. Gneiss sometimes contains crystals of felspar, which give it a spotted appearance; and this is called *porphyritic gneiss.* When talc takes the place of mica, the rock is called *protogine.*

Rem. Gneiss is a rock of great extent in the United States: especially in New England.

Eurite or compact felspar. Dr. Macculloch describes a stratified rock associated with gneiss in Scotland, composed chiefly of compact felspar. De la Beche regards this as eurite, although most writers consider eurite as a member of the unstratified class.

Prin. If all the stratified rocks have been deposited from water, as we have seen, the layers must have been originally nearly horizontal.

Proof. Deposits now taking place rarely have an inclination greater than 10° over any considerable extent of surface: though in some favorable circumstances, as when sand accumulates outward on a steep shore, the strata may be inclined as much as 40°. But a little care will enable any one to distinguish such cases from the effects of subsequent elevation; and it still remains true, as a general fact, that deposits now forming have only a slight inclination.

Inf. Hence if we get the perpendicular thickness of a series of strata we ascertain the character of the crust of the globe to that depth.

Expl. If we measure the breadth of a series of upturned strata, on a line at right angles to their strike, and ascertain their dip, we have given the hypothenuse and angles of a right angled triangle to find the perpendicular, which is the thickness of the strata. If the strata are perpendicular, a horizontal line across their edges give their thickness.

Facts. By measurements and calculations of this sort, it has been ascertained that the total thickness of the fossiliferous strata in Europe, is not less than 6 or 7 miles. In Pennsylvania fossiliferous rocks beneath the top of the coal measures, are 40,000 feet, or more than 7.5 miles, in thickness. *Prof. Rogers's Report on the Geology of Pennsylvania for* 1838, *p.* 82.

Descr. In the peninsula of Tauris, Pallas describes a continued series of primary strata, inclined 45°, over a distance of 86 miles; which would give a perpendicular thickness of more than 68 miles. *Lyell's Prin. Geol. Vol.* 2. 443. In New England, as for instance, on the railroad between Westfield and Pittsfield we have strata of primary rocks, for the most part nearly perpendicular, not less than 20 miles in thickness.

Rem. 1. It ought to be recollected, that the primary strata have been subjected to far more numerous disturbances than the secondary and tertiary; and, therefore, all such measurements as the above, are liable to give results not a little erroneous: since the strata may be so shifted as to be measured twice. In the example last quoted there may exist, as shown on page 37, one or more folded axis. Such sections, however, as those mentioned above, indicate, after all allowances are made, a great perpendicular thickness.

Rem. 2. Dr. Buckland estimates the total thickness of all the stratified rocks in Europe to be at least ten miles. *Bridgewater Treatise, Vol* 1. *p.* 37.

Inf. We see from these statements how groundless is the opinion, that geologists are able to ascertain the structure of the earth only to the depth that excavations have been made which is less than a mile; especially, when we recollect, that the unstratified rocks are uniformly found beneath the stratified and since their igneous origin is now generally admitted, it can hardly be doubted that they come from very great depths; so that probably the essential composition of the globe is known almost to its centre.

SECTION IV.

LITHOLOGICAL CHARACTERS AND RELATIVE AGE OF THE UNSTRATIFIED ROCKS.

Prin. The differences among the unstratified rocks result from two causes. 1. A difference in chemical composition. 2 The diversity of circumstances under which they were produced

Descr. All the varieties of those rocks pass into one another by insensible gradations, even in the same mountain mass; giving rise to endless varieties, which cannot be described minutely in a treatise like the present.

Descr. The two predominant and characteristic minerals in the unstratified rocks, are felspar, and augite, or hornblende.

Rem. The recent researches of Rose and Mitscherlich, render it probable that augite and hornblende are only varieties of the same mineral species, which acquire their different crystalline forms and other characteristic differences, in consequence of a difference in the rate of cooling from a state of fusion :—the former crystallizing rapidly, and the latter slowly. Rose fused hornblende, and found that on cooling it took the form of augite. *Lyell's Elements of Geology, Vol. 2. p. 192. Phillips's Treatise on Geology, Vol. 2. p. 54.*

Descr. The following arrangement of the unstratified rocks, founded upon the relative quantity of felspar and augite or hornblende, which they contain, has been suggested by Prof. Phillips : *Treatise on Geology, Vol. 2. p. 57*, and is liable only to the objection, that we have not a knowledge of the composition of the older rocks sufficiently perfect, to make it certain that they are all put into the right place in the classification.

Division 1. *Felspathic.*

Felspar alone, or but slightly mixed with augite, hornblende, hypersthene, diallage, &c.

Ancient.	*Modern.*
Granite and most Porphyries.	Trachyte.

Division 2.

Felspar in nearly equal proportions with augite, hornblende, hypersthene, &c.

Ancient.	*Modern.*
Syenite and Greenstone.	Graystones of Scrope.

Division 3.

Augite, hornblende, hypersthene, or dillage, predominates over felspar (or olivine.)

Ancient.	*Modern.*
Basaltic series of most authors.	Basaltic series of Scrope.

Descr. On the same principles, that is, mineralogical constitution, Mr. Scrope has divided the products of extinct and active volcanos into three kinds : 1, *Trachyte*, which is felspathic : 2, *Graystone*, or a mixture of felspar and iron : 3, *Basalt*, which is augitic. Girardin (*Considerations General sur les Volcans, p.* 13,) divides these products into the *trachytic formation* (terrain,) *the basaltic formation, and the lava formation.*

Def. The melted matter that is ejected from a volcano, or remains within it, is called *lava.* Hence it is not improper to apply the term to any rock that is proved to have been in a melted state. But it is usual to confine it to the more modern unstratified rocks, such as have been ejected from a crater.

Rem. The igneous origin of all the unstratified rocks is now so generally admitted, that we may take it for granted : and make it the basis of classification. The proof, however, will be presented in a subsequent section.

Descr. Lava cooled rapidly, and not under pressure, forms glass, or scoria: but cooled slowly, and under pressure, it becomes crystalline. Now the older unstratified rocks, such as granite, syenite, porphyry, and greenstone, are more or less crystalline: whereas basalt, trachyte, and other products of existing volcanos, are compact or cellular. Nor have we any but presumptive proof, that the former class are now produced by igneous action. Hence it is inferred, that they were cooled under a vast pressure of the ocean and its subjacent beds: and hence they are called *plutonic rocks:* whereas the latter are denominated *volcanic rocks. Phillips's Treatise on Geology, Vol.* 2, *p.* 52. *Lyell's Elements of Geology, Vol.* 1, *p.* 11.

Prin. There is strong reason to believe that in some instances as in Saxony, and in Sutherlandshire, and Arran in Scotland, granite has been protruded through the strata after it became solid. *Lyell's Elem. Geol. Vol.* 2, *p.* 370. Solid basalt was protruded in a similar manner, according to Von Buch, in the year 1820, in the island of Banda, in great quantities. *Description des Isles Canaries, &c., par L. de Buch, p* 412. *Paris,* 1836. *Also Am. Bib. Repos. Jan.* 1840, *p.* 34.

Obs. The most important of the unstratified rocks will now be describe in an order as nearly chronological (beginning with the oldest) as the presen state of our knowledge will admit.

1. *Granite.*

Descr. The essential ingredients of this rock are quartz felspar, and mica. Its prevailing colors are white and flesh colored. In some cases the materials are very coarse, the crystalline fragments being a foot or more in diameter. In othe cases, they are so fine as to be scarcely visible to the nake eye: and between these extremes, there exists an almost infinit variety. The fine grained varieties are best for economica uses: but the coarser varieties abound most in interesting simple minerals.

Varieties. Graphic granite is composed of quartz and felspar, in whic the former has an arrangement which makes the surface of the rock exhib the appearance of letters, as in Fig. 33. When granite contains distinc crystals of felspar, it is called *porphyritic.* When the ingredients are blende into a finely granular mass, with imbedded crystals of quartz and mica, it ha been called by the French writers, *eurite. Pegmatite* is a granular mixtur of quartz and felspar.

2. *Syenite.*

Def. Syenite is composed essentially of felspar, quartz, an hornblende, the first predominating. When mica is also presen the compound is frequently denominated *syenitic granit Traite Elementaire de Geologie, par M. Rozet, Tome* 1, *p.* 48

Obs. 1. A great deal of confusion and diversity of opinion has existed in respect to the nature and position of syenite. Macculloch makes it to consist of felspar (compact or common) hornblende and quartz; and he limits it to the overlying or trap family, and considers the analogous compounds associated with granite, as merely varieties of the latter. In N. England, such a distinction would be very difficult, since the same continuous formation of syenite, is sometimes connected, on the one hand, with granite, and on the other with porphyry and greenstone. *Macculloch's Classification of Rocks, p.* 512.

Fig. 33.

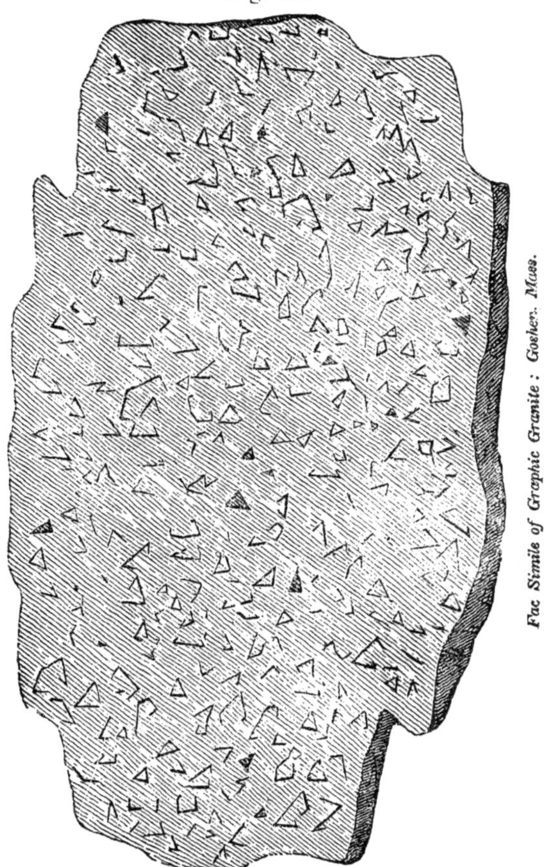

Fac Simile of Graphic Granite: Goshen. Mass.

Obs. 2. When it was ascertained that the famous rock from Syene in Upper Egypt (so much employed in ancient monuments,) from which the name of syenite was derived, was nothing but granite with black mica, and also that Mount Sinai in Arabia was composed of genuine syenite, a French geologist proposed to substitute *Sinai* for syenite:—but the suggestion, which was certainly a good one, has not been adopted.

Obs. 3. Most of the syenite so famous in N. England for architectural purposes, as that from Quincy and Cape Ann, is composed of felspar, quartz and hornblende, the latter frequently disappearing.

3. *Porphyry.*

Def. Rocks with a homogeneous, compact, or earthy base, through which are disseminated crystalline masses of some other mineral of contemporaneous origin with the base, are denominated *porphyry.* True classical porphyry, such as was most commonly employed by the ancients, has a base, of compact felspar, with imbedded crystals of felspar. When the base is greenstone, pitchstone, trachyte, or basalt, the porphyry is said to be greenstone porphyry, pitchstone porphyry, trachytic porphyry, and basaltic porphyry. The base is sometimes clinkstone, or claystone, and the imbedded crystals may be felspar, augite, olivine, &c.

Inf. Hence the term porphyry designates only a certain form of rock, but does not refer to any particular kind of rock. When porphyry is spoken of in general terms however felspar, porphyry is usually meant.

Obs. The name porphyry signifies *purple*, πορφυρα, such having been the most usual color of the ancient porphyries: but this rock exhibits almost every variety of color. It is the hardest of all the rocks; and when polished, is probably the most enduring.

Descr. *Claystone* is an earthy compact stone of a purplish color, appearing like indurated clay. *Compact felspar*, sometimes called *petrosilex*, is a hard compact stone of various colors; fusible before the common blow pipe, and often translucent on the edges, like hornstone. Its predominant ingredient appears to be felspar. (*clinkstone* or *phonolite ;*) or fissile *petrosilex*, a greenish or greyish rock, dividing into slabs or columns, ringing under the hammer, and apparently a variety of compact felspar. *Hornstone* is a compact mineral, often translucent like a horn : of various colors : in hardness and fracture approaching flint : infusible before the blow pipe : and hence composed chiefly of silica. *Cornean* is between hornstone and compact felspar, compact and homogeneous; supposed to consist of felspar, quartz and hornblende. All these substances form the basis of porphyry; and hence we have clinkstone porphyry, hornstone porphyry, claystone porphyry, &c. When black augite forms the base of porphyry, it is called *melaphyre.*

4. *Greenstone.*

Descr. Several unstratified rocks, whose principal ingredients are felspar and hornblende or augite, are called *trap Rocks*.

from the Swedish word *trappa*, a stair: because they are often arranged in the form of stairs or steps. Although the term trap is loosely applied, most writers limit it to the varieties of rock called greenstone, syenitic greenstone, basalt, compact felspar, clinkstone, pitchstone, wacke, amygdaloid, augite rock, hypersthene rock, trap-porphyry, pitchstone porphyry, and tufa. Macculloch includes claystone and syenite. *System of Geology, Vol.* 2. *p.* 80.

Descr. Greenstone is ordinarily composed of hornblende and felspar, both compact and common, the former in the greatest quantity.

Descr. The term *dolerite* has been used by the geologists of continental Europe, as equivalent to greenstone. But according to Rose, dolerite consists of black augite and Labrador felspar: to which Leonhard adds iron. *Diorite* is another name for a variety of greenstone, which Rose says is composed of albite and hornblende in grains.—But albite and hornblende are sometimes called *andesite*.—Dr. Macculloch calls those varieties of greenstone which have a green color, *augite rock;* because augite is the predominant ingredient; but the augite rock of Leonhard is almost wholly augite. When hypersthene takes the place of hornblende, he calls the compound *hypersthene rock*. *System of Geology, Vol.* 2. *p.* 108, 110. When greenstone is composed almost entirely of hornblende, the rock is denominated *hornblende rock*. When the grains of felspar and hornblende are quite coarse, it is called *syenitic greenstone*, which often takes quartz into its composition, and passes into granite.—All the above rocks are frequently porphyritic: and hence we have augitic, or pyroxenic porphyry, dioritic porphyry, &c.

5. *Trachyte.*

Descr. Trachyte is of a whitish or grayish color, usually porphyritic by felspar crystals, and essentially composed of glassy felspar, with some hornblende, mica, titaniferous iron, and sometimes augite. *Beudant's Traite de Mineralogie, Tome* 1. *p.* 566. *Lyell's Elements of Geology, Vol.* 2. *p.* 199. Its name is derived from the Greek, τραχυς, *rough*, from its harshness to the touch. It was an abundant product of volcanic action during the tertiary period, and usually appears to be older than basalt, although trachitic lavas have continued to be ejected down to the present day. Trachyte occurs in Auvergne and Hungary, and in vast quantities in South America: but not in the United States. It constitutes the loftiest summits of the Cordilleras. *Humboldt's Geognostical Essay on the Superposition of Rocks, p.* 423.

Descr. Trachyte in an earthy condition, as it occurs in the Pays de Dome, in Auvergne, is called *domite*. Trachyte is usually porphyritic, and hence we have *trachytic porphyry.*

6. *Basalt.*

Descr. This rock appears to be composed of augite, felspar and titaniferous iron; and sometimes olivine in distinct grains. Its color is black, bluish, or grayish; and its texture compact and uniform;—more so than greenstone. Augite is the predominant ingredient. Probably in some cases, hornblende takes the place of augite; but from the nature of these two minerals, this can be regarded as of little importance. Basalt passes insensibly into all other varieties of trap rocks. *De la Beche's Manual of Geology, p.* 453. *Lyell's Elements, Vol.* 2, *p.* 197.

Rem. It is often asked whether basalt occurs in the United States. The lithological characters of some of our trap rocks can hardly be distinguished from those of basalt: yet it is not probable that any of our trap rocks are as recent as the basalt of Europe; and hence our geologists usually refer them to greenstone.

7. *Amygdaloid.*

Descr. This term, like porphyry, is not confined to any one sort of rock; but indicates a certain form, which extends through all the trap family. Amygdaloid abounds in rounded cavities, like the scoriæ and pumice of modern lavas, and these are often filled with calcareous spar, quartz, chalcedony, zeolites, and other minerals, which have taken the shape of the cavity: so that the rock appears as if filled with almonds, and hence the name from the Latin, *amygdala*, an almond. These cavities, however, have sometimes been lengthened by the flowing of the matter while melted, so that cylinders are found several inches long. When they are not filled, the rock is said to be vesicular.

Descr. A soft variety of trap rock resembling indurated clay, is called *wacke*, which may or may not be vesicular. From its resemblance to the toad probably, it is called in Derbyshire, *toadstone*.

Rem. The slaty graywacke in the vicinity of Boston, as at Brighton and Hingham, is converted into decided amygdaloid, without losing wholly its laminated structure. The same is the case with the red sandstone lying beneath the greenstone near Connecticut river. In the latter case, however, the cavities are rarely filled.

Prismatic, or Columnar Structure.

Descr. One of the most remarkable characteristics of the trap rocks, is their columnar structure, This consists in the occasional division of their substance into regular prisms, with sides varying in number from three to eight, usually five or six, whose length is sometimes not less than 200 feet. They are sometimes jointed; that is, divided crosswise into blocks from one to several feet in length: whose extremities are more or less convex or concave, the one fitting into the other. Usually these columns stand nearly perpendicular, and when worn away on the side, they present naked walls which appear like the work of art. They stand so closely compacted together, that though perfectly separable, there is no perceptible space between them. The thickness of the columns varies from one to five feet.

Rem. 1. The concave extremity is usually uppermost. But at Titan's Pier, at the foot of Mount Holyoke, in Hadley, some of the columns are convex at the top. The following sketch, Fig. 34, shows their appearance at the Giant's Causeway in Ireland.

Fig. 34.

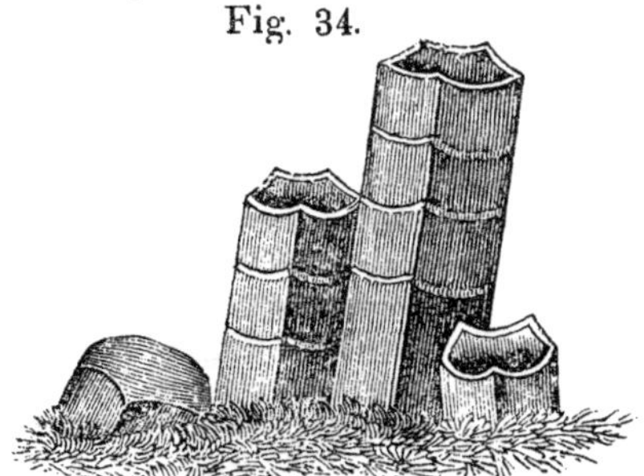

Rem. 2. The columnar and trappose forms of basalt and greenstone have produced some of the most remarkable scenery on the globe. Fingal's Cave in the island of Staffa, (one of the Western Islands of Scotland,) and the Giant's Causeway in the north of Ireland, are almost too well known to need description. Staffa is composed entirely of basalt with a thin soil, and its shores are for the most part a steep cliff, 70 feet high, formed of columns. The cave is a chasm (in these columns,) 42 feet wide, and 227 feet long, formed by the action of the waves. The following sketch, Fig. 35, will convey an idea of the situation of the cave and of the general structure of the island.

Fig. 35.

Fingal's Cave: Staffa.

Fig. 36, gives some idea of the appearance of an overhanging group of greenstone columns at a place on Mount Holyoke in Massachusetts, which I have denominated *Titan's Piazza.* The lower end of the columns, several rows of which project over the observer's head, are exfoliated in such a manner as to present a convex surface downwards.

Fig. 36.

Titan's Piazza: Mt. Holyoke.

Descr. The Giant's Causeway consists of an irregular group of pentagonal columns, from one to five feet thick, and from 20 to 200 feet high, jointed as usual. Where the sea has had access to them, their upper portions are worn away, while the lower part remains extending an unknown distance beneath the waves, and seeming the ruin of some ancient work of art, too mighty for man, and therefore referred to the giants. Here also is a cave of considerable extent. *Geological Transactions, Vol.* 4 *New Series.*

Rem 3. When a trap vein, or dyke, is columnar, the columns often lie horizontal, or rather perpendicular to the sides of the vein: and thus is produced a wall of stones, regularly fitted to one another and laid up, apparently by man: while often a decomposition of the surfaces of the blocks, produces a powder resembling disintegrated mortar. A wall of this sort was formerly discovered in Rowan County, North Carolina, which projected above the rock which it traversed, in consequence of the decay of the rock, and it was for a long time confidently believed to be a work of human skill, proving the former existence there of a powerful and civilized people. Dykes of this description are very common in the State of Maine.

Rem. 4. Greenstone columns standing upright, or leaning only a few degrees, are quite common in North America; and form some of our most interesting scenery. The most extensive formation of this sort appears to be in the country west of the Rocky Mountains, where the Columbia river passes through mountains of trap, (not improbably of basalt,) from 400 to 1000 feet high; and where several successive rows of columns are superimposed one upon another, separated by a few feet of amygdaloid, conglomerate, or breccia. *Parker's Journal of an Exploring Tour beyond the Rocky Mountains, p.* 208. *Ithaca*, 1838.

The Palisadoes on the banks of Hudson river, are another example of greenstone columns. They exist also on Penobscot river; and very perfect examples occur on Mount Holyoke and Tom, on Connecticut river, an example of which has been given in Fig. 36.

Prin. The columnar structure of the trap rocks, has resulted from a sort of crystallization while they were cooling under pressure from a melted state.

Proof 1. Precisely similar columns are found in recent lavas. *Wonders of Geology, Vol.* 1. *p.* 248, 250, *and Vol.* 2. *p.* 640. 2. Mr. Gregory Watt melted 700 pounds of basalt and caused it to cool slowly; when globular masses were formed, which enlarged and pressed against one another until regular columns were the result. *Bakewell's Geology, p.* 146.

8. *Serpentine.*

Descr. Serpentine is a mottled rock, the predominant color green, containing about 40 per cent. of magnesia. It sometimes retains traces of original stratification or lamination; so that Dr. Macculloch regards it as metamorphic, and places it both in the stratified and the unstratified class of rocks. It is elegant as an ornamental rock, though not much used in this country, where it exists in immense quantities.

Descr. Diallage rock, which is the *euphotide* of the French, the *gabbre* of the Italians, and some *ophiolites* of Brongniart, is essentially composed of felspar and diallage: but it sometimes contains serpentine, mica, and quartz. Diallage and serpentine are very nearly allied. *Ophite* is a green porphyritic rock with a base of hornblende and felspar, (the former greatly predominating,) and containing crystals of hornblende. It passes into serpentine by a mixture with talc. *Ophicalce* (French), is composed of limestone and serpentine, with talc and chlorite. Under this rock is arranged the beautiful *verd antique marble*, such as occurs in Newbury and Middlefield in Massachusetts, and at New Haven and Milford in Connecticut. *Brongniart's*

Tableau des Terrains, &c. p. 325. *Cipolin* is a saccharine limestone which contains mica, or talc, as a constituent. This forms several interesting varieties of marble. *Rozet's Traite Elementaire de Geologie, p.* 181. *tome* 1.

9. *Lava.*

Descr. Lava, as remarked in another place, embraces all the melted matter, ejected from volcanos: and the two minerals felspar and augite, constitute almost the entire mass of these products. When the former predominates, light colored lavas are the result: when the latter, the dark varieties. The former are called *felspathic* or *trachytic*, and the latter, *augitic* or *basaltic* lavas.

Rem. Other simple minerals occur in lava. Thus in the products of Vesuvius alone, not less than 100 species have been detected; but they form so inconsiderable a part of the whole mass, as not to deserve consideration in a general view like the present.

Descr. *Trachytic lava* corresponds in most of its characters to the trachyte of the older igneous rocks. When cooled under pressure, solid rock results; but when cooled in the air, it is porous, fibrous, and light enough to swim on water, as is the case with pumice, large masses of which are found sometimes in the midst of the ocean. Sometimes it is porphyritic, like the older trachytes.

Descr. In like manner the basaltic or augitic lavas exceedingly resemble the more ancient basalt; and are in fact the same thing, produced under circumstances a little different. When cooled under pressure, compact basalt is the result; but cooled in the open air, they are scoriaceous or vesicular, and are usually called scoriæ.

Descr. *Graystone lava* is a lead grey or greenish rock, intermediate in composition between basaltic and trachytic lavas: but the felspar predominates, being more than 75 per cent. When albite takes the place of common felspar, the lava is denominated *andesitic*.

Descr. *Vitreous lava* has a fracture like glass. *Obsidian* seems to be merely melted glass. *Pitchstone* is less glassy, with an aspect more like pitch. It is usually composed of felspar and augite, and often passes into basalt. Its composition however varies.

Descr. The small angular fragments and dust of pumice, (which is vesicular trachytic lava,) and of scoriæ, (which is vesicular basaltic lava,) which are produced by an eruption, falling into the sea, or on dry land, and mixing with sand, gravel, shells, &c. and hardened by the infiltration of carbonate of lime or other cement, constitute the substance denominated ***tuff***. When this rock occurs with trap, it is called ***trap tuff;***

and when with modern lava, *volcanic tuff.* If it contain large and angular fragments, it is called *volcanic breccia.* When the fragments are much rolled, the rock is a *tufaceous conglomerate.* The basaltic tuffs are denominated by the Italian geologists, *peperino.* A kind of mud is poured out of some volcanic craters, which forms what is called *trass.*

Descr. Sometimes, especially at the great volcano of *Kairauea,* on the Sandwich Islands, when lava is thrown into the air, the wind spins it out into threads, resembling flax, and drives it against the sides of the crater. This is called *volcanic glass;* and by the natives of Sandwich Islands, *Pele's hair;* Pele having formerly been regarded as the presiding divinity of the volcano of Kairauea.

Descr. Other substances ejected from volcanos are fragments of granite and other rocks, scarcely altered; cinders and ashes of various degrees of fineness, which are often converted into mud by the water that accompanies them; also sulphur in a pure state; various salts and acids; and several gases; among which are the hydrochloric, sulphurous, and sulphuric acids; alum, gypsum, sulphate of iron and magnesia, chloride of sodium and potassium, of iron, copper, and cobalt; chlorine, nitrogen, sulphuretted hydrogen, &c. &c.

Descr. The unstratified rocks, as a general fact, are more fusible than the stratified; and of the unstratified, the fusibility increases in passing from granite along the scale to modern lava. This is owing to the fact that the quantity of lime, and sometimes of alkali, is greater in the more recent rocks; for these substances act as a flux.

Relative Age of the Rocks.

Prin. In the stratified rocks the relative age of the different groups is determined by their superposition; the lowest being the oldest: but in the unstratified rocks, there is reason to believe a reverse order exists: that is, the oldest member of the series lies immediately beneath the stratified rocks; the next oldest beneath this; and so on, till we reach the lava of existing volcanos; which probably comes from a greater depth in the earth than any other unstratified rocks.

Illus. Fig. 37, will more clearly illustrate this proposition.

Prin. The ages of the unstratified compared with those of the stratified rocks, are determined by ascertaining how far the former have intruded upwards among the latter.

Illus. If for instance, we never find the veins of a particular igneous rock shooting upward higher than the primary rocks, we may infer that it is older than the secondary strata, but newer than the primary; because the latter must have existed prior to the intrusion of the unstratified rock.

And so, if an igneous rock is intruded only into the primary and secondary strata, we may infer that it is older than the tertiary strata, and newer than the secondary; and so on with the groups still higher. Hence the igneous rocks, a, a, Fig. 37, formed during the deposition of the primary strata, whose veins extend no higher than those strata, may be called, (to adopt the phraseology of Mr. Lyell,) the *primary plutonic:* those during the deposition of the secondary strata, b, b, whose veins do not enter the tertiary series, the *secondary plutonic:* those during the deposition of the tertiary strata, c, c, the *tertiary plutonic:* and lava from active volcanos, d, d, the *recent plutonic.*

Fig. 37.

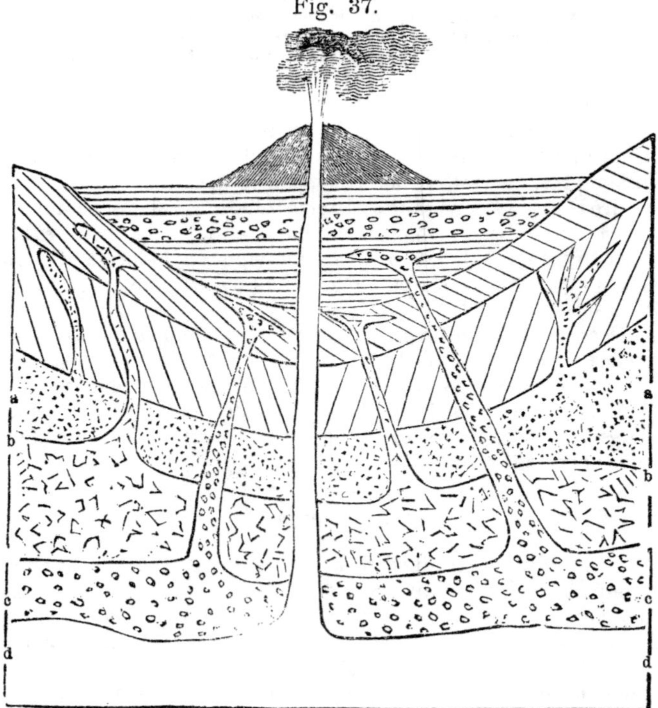

Section of the Relative Age of the Unstratified Rocks.

Descr. In reality, however, we do not find varieties of unstratified rocks whose veins are thus distinctly confined to each of the great classes of rocks, though there is evidence that volcanic agency was active during all the periods of their deposition. But the same igneous rock appears to have been ejected at different epochs. Granite, however, seems to have greatly predominated during the first or primary period; and is found only occasionally during the secondary period; though in a few instances (at Weinbohla) syenitic granite has been protruded through the chalk, but never among the tertiary strata. Porphyry appears to have been mostly confined to the period of the latest primary, and the older secondary (transition) rocks. Trap rock predominated in the secondary and tertiary periods, while volcanic rocks, in the common acceptation of the term, began to be protruded during the tertiary period, and continue to the present time.

Rem. It is obvious, that with the exception of lava, the above rule for determining the relative age of the stratified and unstratified rocks, does not show us when the latter began to be erupted, but only when their eruptions ceased.

Inf. From the phenomena that have been detailed respecting the unstratified rocks, it has been inferred that the condition of the earth, both internal and external, must have been different at different epochs; so as at one period to be peculiarly favorable for the production of granite and syenite, at another of porphyry, at another of trachyte, at another of basalt, and finally at another of the lava of extinct and active volcanoes; and hence, that the older igneous rocks, (ex. gr. granite, syenite, &c.) are no longer produced, except perhaps in the deep recesses of the earth.

Proof. 1. The greater abundance of granite and syenite associated with the primary than with the newer strata, and of trap and volcanic rocks with the higher formations. 2. The almost entire identity between the chemical constitution of granite and the primary stratified deposits, indicates some general and common cause for the origin of both: while the difference of ultimate constitution between granite and the newer stratified rocks, particularly in the greater quantity of lime in the latter, indicates a difference of origin. 3. The gradual and insensible passage, on an extensive scale, of granite into gneiss, hornblende slate and mica slate, indicates some general cause for their production, and that the diversity existing between them has resulted from slightly modifying circumstances: while no such transition of any consequence between granite and the newer stratified rocks has ever been discovered. 4. Granite and the trap rocks differ so much in chemical constitution, as to show that they must have originated from different masses of matter. Thus granite contains about 20 per cent. more of silica than greenstone; about 3 per cent. less of alumina; 8 per cent. less of magnesia; 7 per cent. less of lime; and two per cent. less of oxide of iron. 5. The correspondence between the chemical composition of the fossiliferous stratified and the trappean and volcanic rocks: that is, we find in both classes a diminution of silica and an increase of alumina, magnesia, and lime. 6. In consequence of containing much more of silicate of lime, the trap rocks are more fusible than the granitic: so if we admit that the internal temperature of the earth has diminished, we might expect that the former would remain in a melted state after the latter had all been consolidated. *De la Beche's Theoretical Geology*, p. 305.

Opposite Hypothesis. Mr. Lyell maintains that unstratified rocks of every description may now be forming at various depths in the earth, even in the same relative quantity as at any former period; and that the different degrees of pressure and other circumstances under which the melted matter is cooled, are sufficient to explain the differences of composition and lithological characters which they exhibit. It is not pretended, however, that there is any positive evidence in favor of this supposition: that is, none of these recent Plutonic rocks have ever been seen. *Lyell's Elements Geology, Vol. 2, p.* 347.

Geological Maps and Sections.

Descr. Common or physical maps form the basis of geological ones: and when the former are inaccurate, the latter must be so too. The chief difference between them is, that on a geological map the different rocks found in the region delineated are shown either by dots, crosses, circles, &c. or more usually by colors. The only exception is, that when the nature of the subjacent rock can be determined, diluvium is usually omitted.

Descr. Some geological maps designate only the classes of rocks; but these are very imperfect, and the best maps show the extent of each rock.

Descr. The dip of the strata, (which of course determines the strike,) is sometimes shown upon a geological map. This is usually done by an arrow, which points in the direction of the inclination. If the strata are perpendicular, it may be represented by the lines crossing at right angles, one of which is shorter than the other. If the two lines are equal, so as to form a cross, they indicate horizontal strata. An anticlinal axis is shown by a straight line crossed by an arrow with two heads. Where the strata undulate a good deal, the body of the arrow may be crooked. *De la Beche's Manual of Geology, p.* 602.

Rem. 1. The best geological maps hitherto published in Europe, are Greenough's map of England and Wales; Elie de Beaumont's and Dufrenoy's France; Hoffman's Northwestern Germany; and Oeynhausen, La Roche, and Von Dechen's Rhine. In this country, Maclure's Geological Map of the United States, although it exhibits only the great classes of rocks, yet considering the early period at which it was executed, must be regarded as very valuable and a work of immense labor. The geological surveys now going on in most of the States, have already produced maps of Massachusetts, Rhode Island, New York, New Hampshire, New Jersey, and Tennessee; and others are in a state of great forwardness.

Rem. 2. A map has lately been constructed by an eminent French geologist, M. Boue, which gives an outline of the geology of the entire globe. It was constructed by employing all the materials known up to this date, and then by reasoning from the known to the unknown. And although it must necessarily, be very imperfect, yet it doubtless gives a fair outline of the geology of the globe. The author of this work copied and simplified this map, with a view of appending it to this volume. He also constructed a more detailed geological map of North America, derived mainly, as to the country west of Missouri, and north of the great lakes, from the map of Boue; but as to the United States proper, from the maps of Hall, Lyell, and others: and he had intended to append this, also, to the present work. But he found it would make the work too cumbersome, and too expensive. They will both, therefore, be published with a brief description of the geology of the globe, in a small separate work, under the title of *An Outline of the Geology of the Globe; and especially of the United States; with two Geological Maps.* It would be very desirable that all who study this work, should possess these

Maps; as a few moments' inspection of them will give a better idea of the world's geology, than the best description can do.

Descr. A geological section represents a vertical cut in the earth's crust, so as to exhibit to the eye the rocks in their natural and relative situation. The most valuable sections of this sort are those copied from cliffs, on the sea-coast, or the banks of rivers. But usually it is necessary to construct them from what we can learn of the rocks and their dip at the surface; presuming that they continue the same to the depth of the section. Such sections, therefore, are somewhat ideal: but if carefully constructed, we may be sure that we are not far from the truth.

Descr. It is usually necessary to employ two scales in constructing sections: one for heights, and the other for horizontal distances: otherwise the sections must be of great extent, or the heights would be scarcely perceptible. On the other hand, two scales produce distortion: so that great caution is necessary, not only in the construction of sections but in drawing inferences from them.

SECTION V.

PALÆONTOLOGY, OR THE SCIENCE OF ORGANIC REMAINS.

Def. In all the stratified rocks above the primary, more or less of the relics or traces of animals and plants occur, sometimes called petrifactions, but more commonly, *organic remains.*

Def. That branch of geology which gives the history of these remains, was formerly denominated *oryctology*; but is now called *palæontology.*

Rem. Palæontology is now usually limited by authors to fossil animals. But I prefer a meaning extended enough to embrace all organic remains, whether animal or vegetable; which is certainly consistent with the etymology of the term (παλαιος, ουτα, λογος.) *Roberts' Dict. of Terms in Geology: Ansted's Geology, Vol.* 1, *p.* 54. *Geology of Russia, by Murchison, &c., Vol.* 2.

1. *General Characters of Organic Remains.*

Descr. In a few instances, animals have been preserved entire in the more recent rock.

Exam. About the beginning of the present century, the entire carcass of an elephant was found encased in frozen mud and sand in Siberia. It was covered with hair and fur, as some elephants now are in the Himalayah mountains. The drift along the shores of the Northern Ocean, abounds with bones of the same kind of animals: but the flesh is rarely preserved. *Cuvier's Essay on the Theory of the Earth, p.* 253, *New York*, 1818. *De la Beche's Manual of Geology, p.* 200. In 1771, the entire carcass of a rhinoceros was dug out of the frozen gravel of the same country. *Bakewell's Geology, p.* 331.

Descr. Frequently the harder parts of the animal are preserved in the soil or solid rock, scarcely altered.

Rem. Many well authenticated instances are on record, in which toads, snakes, and lizards, have been found alive in the solid parts of living trees, and in solid rocks, as well as in gravel, deep beneath the surface. But in these instances the animals undoubtedly crept into such places while young, and after being grown could not get out. Being very tenacious of life, and probably obtaining some nourishment occasionally by seizing upon insects that might crawl into their nidus, they might sometimes continue alive even many years. But such examples cannot come under the denomination of organic remains. See an interesting paper on this subject by Dr. Buckland, in the *American Journal of Science, Vol.* 23, *p.* 272.

Descr. Sometimes the harder parts of the animal are partially impregnated with mineral matter; yet the animal matter is still obvious to inspection.

Descr. More frequently, especially in the older secondary rocks, the animal or vegetable matter appears to be almost entirely replaced by mineral matter, so as to form a genuine *petrifaction.*

Rem. Probably in every case, however, a chemical process would show the presence of considerable organic matter. *Parkinson's Organic Remains of a Former World, Vol.* 2. *p.* 284.

Descr. Sometimes after the rock had become hardened, the animal or plant decayed and escaped through the pores of the stone, so as to leave nothing but a perfect *mould.*

Descr. After this mould had been formed, foreign matter has sometimes been infiltrated through the pores of the rock, so as to form a *cast* of the animal or plant when the rock is broken open. Or the cast might have been formed before the decay of the animal or plant.

Descr. Frequently the animal or plant, especially the latter, is so flattened down that a mere film of mineral matter alone remains to mark out its form.

Descr. All that remains of an animal sometimes is its track impressed upon the rock.

Descr. The mineralizer is most frequently carbonate of lime; frequently silica, or clay, or oxide or sulphuret of iron, and sometimes the ores of copper, lead, &c.

2. *Nature and Process of Petrifaction.*

Def. Petrifaction consists in the substitution, more or less complete, by chemical means, of mineral for animal or vegetable matter. *De la Beche's Theoretical Geology, Chapter* 13.

Descr. The process of petrifaction goes on at the present day to some extent, whenever an animal or vegetable substance is buried for a long time in a deposit containing a soluble mineral substance that may become a mineralizer.

Exam. 1. Clay containing sulphate of iron, will, in a few years, or even months, produce a very perceptible change towards petrifaction in a bone buried in it. *Bakewell's Geology*, p. 19. Some springs also hold iron in solution; and vegetable matters are in the process of time thoroughly changed into oxide of iron. This is seen often where bog iron ore is yearly depositing.

Exam. 2. M. Goppert placed fern leaves carefully in clay, and exposed the clay for some time to a red heat, when the leaves were made to resemble petrified plants found in the rocks. *Wonders of Geology*, *Vol.* 2. *p.* 561.

Hypothetical Exam. 3. M. Patrin and Brongniart suggest that the petrifying process may sometimes be effected "suddenly by the combination of gaseous fluids with the principles of organic structures." *Wonders of Geology*, *Vol.* 2. *p.* 559. Some facts render this probable. For stems of a soft and succulent nature are preserved in flint; and the young leaves of a palm tree in a state just about to shoot forth, have been found completely silicified. *Lyell's Elements of Geology*, *Vol.* 1. *p.* 82.

3. *Means of determining the Nature of Organic Remains.*

Prin. The first requisite for determining the character of organic remains, is an accurate and extensive knowledge of zoology and botany. This will enable the observer to ascertain whether the species found in the rocks are identical with those now living on the globe.

Prin. The second important requisite is a knowledge of comparative anatomy: a science which compares the anatomy of different animals and the parts of the same animals.

Rem. 1. This recent science reveals to us the astonishing fact, that so mathematically exact is the proportion between the different parts of an animal, "that from the character of a single limb, and even of a single tooth, or bone, the form and proportion of the other bones, and the condition of the entire animal may be inferred."—"Hence, not only the frame-work of the fossil skeleton of an extinct animal, but also the character of the muscles, by which each bone was moved, the external form and figure of the body, the food, and habits, and haunts, and mode of life of creatures that ceased to exist before the creation of the human race, can with a high degree of probability be ascertained." *Buckland's Bridgewater Treatise*, *Vol.* 1. *p.* 109. See also *Cuvier's Ossemens Fossiles*, *Tome* 1. *p.* 47, *Troisième Edition.*

Rem. 2. It is clear from the preceding statement, that no individual can hope to possess in himself all the requisites for successfully determining organic remains. For the field is too large for any one to hope to become familiar with all its parts. Hence, at this day, it is customary for the geologist to resort for aid to the botanist, the zoologist, and the comparative anatomist.

4. *Classification of Organic Remains.*

Prin. Organic remains may be divided, according to their origin, into three classes: 1. Marine. 2. Freshwater. 3. Terrestrial.

Rem. 1. The last class appear in most instances where they occur, to have been swept down by streams from their original situation into estuaries; where they were mixed with marine relics. Sometimes, perhaps, they were quietly submerged by the subsidence of the land.

Rem. 2. The following table will show the origin of the remains in the different groups of fossiliferous rocks.

Cambrian and Silurian Systems (Graywacke.)	Marine. Rarely Terrestrial

Old Red Sandstone.	Marine
Carboniferous Limestone.	Do.
Coal Measures.	Terrestrial Estuary Deposits,
and submerged land. Rarely perhaps fresh water deposits.	
New Red Sandstone Group.	Marine
Oolitic Group.	Mostly Marine.
but in a few instances,	Terrestrial.
Wealden Rocks.	Estuary Deposit.
Cretaceous Group.	Marine.
Tertiary Strata.	Marine and Fresh Water.
Diluvium.	Terrestrial.

Inf. It appears from the preceding statements that by far the greatest part of organic remains are of marine origin. Nearly all the terrestrial relics indeed, and many of fresh water origin, have been deposited beneath the waters of the ocean.

5. *Amount of Organic Remains in the Earth's Crust.*

Descr. The thickness in feet of the fossiliferous strata in Great Britain, as given in the tabular view of the stratified rocks, with the exception of the Silurian and Cambrian systems, which I give on the authority of Prof. Phillips, is as follows:

Tertiary,	2000 feet.
Chalk,	1100 do.
Wealden,	900 do.
Oolite,	2000 do.
Lias,	700 do.
Upper New Red,	900 do.
Lower New Red,	800 do.
Carboniferous System,	5700 do.
Old Red Sandstone,	10.000 do.
Silurian Rocks,	7470 do.
Cambrian Rocks,	9000 do.

Total, 40.570 feet ; or $7\frac{2}{3}$ miles.

Rem. 1. We have already seen that Prof. Rogers makes the fossiliferous rocks in this country below the top of the coal measures, 40.000 feet thick.

Descr. Organic remains occur more or less in all the fossiliferous strata whose thickness has been given. As a matter of fact, they have been dug out several thousands of feet below the present surface.

Descr. In the Alps, rocks abound in organic remains from 6000 to 8000 feet above the level of the sea : in the Pyrenees, nearly as high ; and in the Andes and the Himalayas, at the height of 16.000 feet.

Descr. Frequently beds or layers of rock, many feet in thickness, appear to be made up almost entirely of the remains of animals or plants : indeed, whole mountains, hundreds and even thousands of feet high, are essentially composed of organic matter.

Descr. Prodigious accumulations of the relics of microscopic animals are frequently found in the rocks.

Exam. 1. From less than 1.5 ounce of stone, in Tuscany, Soldani obtained 10.454 chambered shells, like the Nautilus:—400 or 500 of these weighed only a single grain; and of one species it took 1000 to make that weight. These were marine shells. *Buckland's Bridgewater Treatise, Vol.* 1. *p.* 117.

Exam. 2. In fresh water accumulations a microscopic crustaceous animal called the cypris, often occurs in immense quantities; as in the Hastings sand and Purbeck limestone in England, where strata 1000 feet thick are filled with them: and in Auvergne, where a deposit 700 feet thick, over an area 20 miles wide and 80 in length, is divided into layers as thin as paper by the exuviæ of the cypris. *Same Work, p.* 118.

Exam. 3. But perhaps the most remarkable example is that derived from the recent discoveries of the Prussian naturalist Ehrenberg, respecting the fossil remains of animalcula. In one place in Germany is a bed 14 feet thick, made up of the shields of animalcula so small, that it requires 41.000.000.000 of them to form a cubic inch; and in another place, a similar bed is 28 feet thick. In Massachusetts, are numerous beds composed of the siliceous shields of infusoria (of a somewhat larger size than those mentioned above,) many feet in thickness; and similar beds occur all over New England and New York. Recently deposits of these *carapaces* or shields, have been discovered by Prof. Wm. B. Rogers in the tertiary strata of Virginia, extending over large areas, and from 12 to 25 feet thick! *Report on the Geology of Virginia*, for 1840, *p.* 28.

Descr. It is a moderate estimate to say, that two thirds of the surface of our existing continents are composed of fossiliferous rocks; and these as already stated, often several thousand feet thick.

Rem. 1. This estimate might, without exaggeration, be confined to strata that contain marine exuviæ:—that is, such as were deposited beneath the ocean.

Rem. 2. After all, the preceding statements convey but a very imperfect idea of the amount of organic relics in the rocks. To obtain a just conception of their vast amount, a person must visit at least a few localities.

6. *Distribution of Organic Remains.*

Descr. Existing animals and plants are arranged into distinct groups, each group occupying a certain district of land or water; and few of the species ever wander into other districts. These districts are called zoological and botanical provinces; and very few of the species of animals and plants which they contain, can long survive a removal out of the province where they were originally placed; because their natures cannot long endure the difference of climate and food, and other changes to which they must be subject.

Rem. Although naturalists are agreed in maintaining the existence of such provinces, yet they have not yet settled their exact number; because yet ignorant of the plants and animals in many parts of the earth.

Dr. Prichard proposes seven for the animals. *Physical History of Mankind, Vol. 1. pp. 68—97, third edition.* Mr. Swainson estimates the ornithological provinces at five. *Encyclopædia of Geography, Vol. 1. p. 267.* Bory St. Vincent makes five for the mammalia alone. *Dictionnaire Classique D'Hist. Naturelle, Tome Septieme, p. 300.* Some have reckoned as many as eleven: but what makes it very difficult to determine the point, is, that the boundaries of these provinces for different classes of animals does not always coincide. Decandolle, reckons the botanical provinces at twenty. Prof. Schouw makes twenty-two. *Encyclopædia of Geography, Vol. 1. p. 250, Am. Edition.* But Prof. Henslow estimates them approximately at forty-five. *Dr. Smith's Scripture Geology, p. 73.* On this difficult subject, see the *French Dictionary of Natural History above referred to, Article Geographie: Also Fleming's Philosophy of Zoology, Vol. 2. Also Lyell's Principles of Geology, Vol. 2. Also Prichard's Physical History of Man, Vol. 1. Encyclopædia of Geography, Vol. 1. &c. &c.*

Descr. Sometimes mountains and sometimes oceans separate these districts on the land. In the ocean they are sometimes divided by currents or shoals. But both on land and in the water, difference of climate forms the most effectual barrier to the migration of species: since it is but a few species that have the power of enduring any great change in this respect.

Descr. In some instances, organic remains are broken and ground by attrition into small fragments, like those which are now accumulating upon some beaches by the action of the waves. But often the most delicate of the harder parts of the animal or plant are preserved; and they are found to be grouped together in the strata very much as living species now are on the earth.

Illus. In a fossiliferous formation of any considerable thickness, we usually find somewhat such an arrangement as the following. The whole is divided into many distinct beds of different thickness. At the bottom, perhaps, we shall find a layer of argillaceous or siliceous rock, with few or no remains: then will succeed a layer, perhaps calcareous, full of them in a perfect state: next a layer of sand or clay, or limestone containing none: next a layer made up of the fragments of rocks, animals, and plants, more or less comminuted: next a layer of fine clay: then a layer abounding in remains. And thus shall we find a succession of changes to the top of the series.

Inf. From these facts it is inferred, that for the most part, the imbedded animals and plants lived and died on or near the spot where they are found; while it was only now and then, that there was current enough to drift them any considerable distance, or break them into fragments. As they died, they sunk to the bottom of the waters and became enveloped in mud, and then the processes of consolidation and petrifaction went slowly on, until completed.

Rem. 1. So very quietly did the deposition of the fossiliferous rocks proceed in some instances, that the skeletons and indusiæ of microscopic animals, as we have seen, which the very slightest disturbance must have crushed, are preserved uninjured; and frequently all the shells found in a layer of rock, lie in the same position which similar shells now assume upon the bottom of ponds, lakes, and the ocean: that is, with a particular part of the shell uppermost.

Rem. 2. Were the bottom of our existing oceans and lakes, where mud, sand, and gravel, have been accumulating for ages, and enveloping the animals and plants that have died there, or been drifted thither, were this to be now elevated above the waters, we should find exactly such an arrangement of organic remains, as we find in a particular formation of the solid rocks. While there would be a resemblance between the relics in different seas and lakes, there would be great specific diversity; just as we find in different groups of rocks in different countries: and hence the conclusion seems fair, that these rocks with their contents had an origin similar to the deposites now forming at the bottom of existing bodies of water.

Rem. 3. In the existing waters we find that different animals select for their *habitat* different kinds of bottom: thus, oysters prefer a muddy bank; cockles a sandy shore; and lobsters prefer rocks. So it is among the fossil remains: an additional evidence of the manner in which they have been brought into a petrified state. *Phillips's Geology, p.* 53.

Prin. There is reason to believe that the temperature of the globe in early times was much higher than at present, and of course more uniform over its surface: and hence the range of particular species of animals and plants might then have been more extensive than at present; and the number of botanical and zoological provinces less numerous; and this inference is sustained by the facts of fossil geology.

Descr. From the researches of Prof. E. Forbes in the Egean Sea, it appears, first, that increase of depth has the same kind of effect upon the marine animals, as increase of height has upon those on dry land, that is, the animals become more and more like those of a colder climate. Secondly, that most marine animals and vegetables inhabit particular localities, which at length become unfit for their abode, and they emigrate or die out. Thirdly, that species ranging widest in depth range farthest horizontally. Fourthly, below 300 fathoms, deposits of fine mud are going on without organic remains, because animals do not live there. These conclusions correspond to the manner in which organic remains occur in the rocks. *Ansted's Geology, Vol.* 1. *p.* 497. *Report of the British Association for* 1843, *p.* 2.

Prin. In comparing organic remains from different formations, it should be recollected that they may belong to the same class or order, or genus, and yet be widely different from one another: and that it is only when they are of the same *species,* that they are identical.

Descr. If we compare together the remains of the cretaceous formation, the red sandstone formation, the carboniferous system, and the Silurian formation in different parts of England, we shall find that those most remote from one another in locality, differ most widely: but almost without an exception, those in each formation are specifically distinct from all those in the other formations. *Phillips's Geology, p.* 51.

Descr. If we compare the fossils of the tertiary and secondary classes of rocks, we shall find that they have scarcely any species common, so far as has been yet ascertained, either of animal or plant. *Lyell's Prin. Geol. Vol.* 1. *p.* 205.

Descr. If we examine a formation through its whole extent, we shall rarely find that any species of organic remains is universally diffused, unless the extent of the formation be quite limited. If we compare the same formation in different countries, the specific resemblance between the organic contents will diminish nearly in the inverse ratio of the distance between them. *Phillips's Treatise on Geology, from Ency. Britt. p.* 52.

Exam. In Egypt the cretaceous rocks contain different fossils from the chalk of England: and the same is true of the chalky rocks on the southern faces of the Alps. More than a hundred species of organic relics have been described in the rocks of the United States, which are supposed to correspond with the chalk formation in Europe: yet only two or three species are identical. *Morton's Synopsis of the Organic Remains of the Cretaceous Group of the United States, p.* 83. *Philadelphia,* 1834. *Phillips's Geology, p.* 156.

Rem. In a few instances, particular species have a very wide diffusion in contemporaneous rocks. The *Belemnites mucronatus* is found in nearly every chalk deposit in Europe. The trilobite, *Calymene Blumenbachii,* and the coral, *Catenipora,* are found at most localities of Silurian limestone in Europe and North America.

Prin. Families and genera that were contemporaries, appear to have had a very wide geographical diffusion, as they have among existing animals and plants: but for the most part, species occupied but a narrow geographical area. *Phillips's Treatise on Geology, p.* 53.

Exam. Specifically unlike as are the organic contents of the cretaceous formations in Europe and North America, yet the same genera (ex. gr. Exogyra, Gryphæa, Baculites, Belemnites, Scaphites, and Ammonites) abound, and even between the species there is a close analogy.

Prin. Judging from the distribution of living animals and plants, contemporaneous formations in widely separated portions of the globe may contain organic remains very much alike, or very much unlike.

Illus. Comparing the marine animals on the coast of the United States with those on the shores of Europe, we find at least 24 species of shells common to both, and no reason can be assigned why as close a resemblance might not have existed at earlier periods. *Morton's Synopsis, p.* 83. On the other hand, how unlike are the animals and plants of New Holland and its coasts, to those of Europe or the United States.

Prin. Rocks agreeing in their fossil contents, may not have been contemporaneous in their deposition.

Proof. The causes that have produced changes of organic life may have operated sooner upon some parts of the globe than upon others, so that particular animals and plants may have continued to be deposited in some spots longer than in others.

Rem. Probably, however, such a diversity in different parts of the globe could not have continued very long, so that rocks with the same organic remains may be regarded as not differing greatly in age: and besides, as already stated, there is reason to suppose that in earlier times there was greater uniformity of climate and condition on the globe than at present.

Inf. From all that has been advanced, it appears that an identity of organic remains is not alone sufficient to prove a complete chronological identity of rocks widely separated from each other: but it will show an approximate identity as to the period of their deposition; and in regard to rocks in a limited district, it will show complete identity.

Proof. Identity of organic remains proves only the existence of similar conditions as to climate, food, &c.; but in remote regions of the globe these conditions may have existed at different periods, though not probably separated by long intervals; and therefore the identity is approximate: that is, deposits containing the same organic remains were produced at eras not widely remote from each other. But in respect to limited regions of a continent, much difference of climate could not have existed at the same time; and, therefore, an identity of organic remains proves the synchronism of the deposits containing them.

Prin. If the mineral character of two rocks agree, as well as their organic contents, their synchronism will be shown to be more probable. But on the other hand, a want of agreement in the mineral characters, ought not to be regarded as proof that they were not contemporaneous.

Proof. The mineral composition of rocks, forming in regions very remote, must have been subject to as great diversity as their organic contents. But if their mineral composition is the same, it increases the probability of their synchronal deposition.

Prin. Still stronger evidence of synchronism is obtained when rocks agree in their superposition, as well as in the characters above named. This character, indeed, when it can be applied, is very conclusive: but in remote regions it is applied with great difficulty.

Rem. The identification of strata in widely separated regions is one of the most difficult problems in geology; and one where there is great room for the play of fancy. *De la Beche's Theoretical Geology, Chapters* 11, 12, 13. *Lyell's Elements of Geology, Chapter IX. Vol.* 1.

Estimates of the Number of Species of Organic Remains already discovered.

Descr. In 1844 Dr. Sandberger (*Address before the Lyceum of Wiesbaden in Nassau*) stated the whole number of species of organic remains exhumed up to that date, and described to be about 16.000; and that for some years this number had increased about 900 annually. At this rate, the present known number would be about 18.000.

Descr. Mr. Morris (*Catalogue of British Fossils, London*, 1843,) has given an accurate list of the organic remains found in Great Britain up to the year 1843; amounting to 5140 species. The following list shows them distributed into the different classes. Assuming the whole number now known, as above stated, to be 18.000, and that in all parts of the world they are distributed in the several classes as in Great Britain, I have added a column of the approximate number on the globe.

		In Great Britain.	*On the Globe.*
Plants,		534	1872
Infusoria,		20	70
Amorphozoa,		77	270
Zoophyta,		326	1141
Echinodermata,		274	960
Foraminifera,		82	287
Annelida,		80	280
Cirrhipeda,		21	74
Insecta,		19	66
Crustacea,		165	578
Mollusca.	Conchifera Dimyaria,	729	2553
	———— Monomyaria,	323	1131
	Rudistes,	4	14
	Brachiopoda,	460	1629
	Cephalopoda,	473	1656
	Gasteropoda,	822	2877
	Pteropoda,	7	24
	Heteropoda,	29	102
Pisces,		540	1891
Reptilia,		94	329
Aves,		8	28
Mammalia,		53	186

Descr. In the following table the British fossils are distributed into the different formations: and I have added the corresponding numbers on the whole globe, on the same principle as in the preceding list. In a few instances—marked by an asterisk—I have given the numbers actually described, instead of the hypothetical numbers. Thus, 155 species of Infusoria have been described by Ehrenberg in the tertiary strata of the United States; although none are given in Morris's Catalogue of British Fossils. *Pritchard on Infusoria; London*, 1841. *Also, Ehrenberg's Verbeitung und Einflus Mikroscopischen Lebens in Sud und Nord Amerika; Berlin*, 1843. *Also, Ehrenberg über drei Lager von Gebirgsmassen aus Infusorien als Meeres-Absatz in Nord Amerika, &c.; Berlin*, 1844. In the Permian system I have given the species of plants known in all Europe, or stated in *Murchison's Russia, Vol. 2. p. 13. London*, 1845. In the Devonian and Silurian systems I have given the plants (seaweeds) as described in the New York Reports. Although such plants are abundant in the Devonian and Upper Silurian systems, very few of the species seem as yet to have been described.

Formations.		Plants.	Infusoria.	Amorphozoa and Zoophyta.	Echinodermata.	Foraminifera.	Annelida.	Cirrhipeda.	Insecta.	Crustacea.	Mollusca.	Reptilia.	Pisces. (Fishes.)	Aves. (Birds.)	Mammalia.
	Living Species on the Globe.	134,000	1100						120,000		12,000	1000	8000	6000	1000
Drift.	In Great Britain.													5	27
	On the Globe.													17	94
Tertiary.	In Great Britain.	119		58	15	47	20	11	4	6	771	16	62	1	22
	On the Globe.	420	155*	206	54	166	71	40	14	21	2792	57	220	3	78
Chalk and Green Sand.	In Great Britain.	12	19	118	66	35	21	6		11	441	6	56		
	On the Globe.	43	67	418	234	124	74	21		40	1563	21	197		
Wealden.	In Great Britain.	9							5	5	26	12	14	1	
	On the Globe.	31							17	17	92	43	50	3	
Oolite.	In Great Britain.	109		47	36		11		4	4	463	19	70		3
	On the Globe.	396		167	128		40		14	14	1638	67	248		12
Lias.	In Great Britain.	8			10		1		3	3	157	24	100		
	On the Globe.	28			35		3		12	12	557	85	355		
Trias.	In Great Britain.	4									16	10	20	tracks	
	On the Globe.										57	35	71	25	
Permian.	In Great Britain.	1		4							15	3	15		
	On the Globe.	27*		14							54	12	54		
Carboniferous.	In Great Britain.	277		67	91		7		3	28	451		110		
	On the Globe.	982		238	323		25		12	99	1600		390		
Devonian.	In Great Britain.			35	20					9	239		53		
	On the Globe.	3*		123	71					31	847		188		
Silurian.	In Great Britain.			85	30		11	3		81	265		8		
	On the Globe.	4*		302	106		40	12		286	939		28		

Rem. From a variety of causes, which cannot be here described, I have been unable to distribute with entire accuracy the British fossils in the several formations: and the numbers hence deduced for the whole globe, must be regarded as only vague approximations to the truth. These numbers, moreover, must not be considered as attempting to show the actual number of species in the rocks, but only as an approximation to the number already discovered. And yet we know that the supposition which makes Great Britain a type of the whole globe in this respect, must be very inaccurate; yet I give these numbers for the want of more exact data.

Inf. 1. From the preceding table we learn that all the important classes of animals and plants are represented in the different formations.

Inf. 2. Hence we learn that the hypothesis of Lamarck is without foundation, which supposes there has been a transmutation of species from less to more perfect, since the beginning of organic life on the globe: that man, for instance, began his race as a *monad*, (a particle of matter endowed with vitality,) and was converted into several animals successively; the ourang-outang being his last condition, before he became man. *Lyell's Prin. Geol. Vol.* 3, *Book* 3, where this subject is treated ably and fully. "The Sauroid fishes," says Dr. Buckland, (*Bridgewater Treatise, Vol.* 1, *p.* 294,) "occupy a higher place in the scale of organization than the ordinary forms of bony fishes; yet we find examples of Sauroids of the greatest magnitude and in abundant numbers, in the carboniferous and secondary formations, whilst they almost disappear and are replaced by less perfect forms in the tertiary strata, and present only two genera among existing fishes.—In this, as in many other cases, a kind of *retrograde* development, from complex to simple forms may be said to have taken place."

Inf. 3. We learn, however, that in the earlier periods of the world, the less complex and less perfect tribes of animals and plants greatly predominated, and that the more perfect species became more and more numerous up to the creation of the present races.

Inf. 4. Vegetable life must have commenced on the globe nearly as early as animal life.

Proof. Vegetables are necessary for the food of animals. The lowest rocks, however, contain but few plants, and these mostly marine. A few land plants, however, have been found in the Devonian, and even the Upper Silurian, in New York. *Vanuxem's Final Report on the Third District, p.* 184.

Inf. 5. Dry land, capable of sustaining vegetation, must

have existed soon after the deposition of the fossiliferous rocks commenced.

Proof 1. Terrestrial vegetable remains occur in rocks of the Silurian group. 2. The detrital character of the rocks of that group makes the existence of dry land during its deposition almost certain; since rocks entirely beneath the waters are but slightly worn away by oceanic currents.

Descr. The family of coniferous plants is found in the earliest rocks containing plants, and at each successive change in the physical condition of the globe, the number of its genera and species increased, until it forms among existing plants about one three hundredth part of the whole flora, or nearly 200 species. Palms also occur, though sparingly, in all the formations.

Descr. The 300 species of fossil plants found in and beneath the carboniferous strata, are two-thirds tree ferns and gigantic Equisetaceæ. Ten Coniferæ, and plants intermediate between these and Lycopodiaceæ, viz. Lepidodendra, Sigillaria, and Stigmariæ, together with 10 Monocotyledonous plants, form the remainder. *Mantell's Wonders of Geology, Vol.* 2, *p.* 568.

Descr. Of the 100 species found between the carboniferous strata and the tertiary groups, one third are ferns; and most of the remainder are Cycadeæ, Coniferæ, and Liliaceæ. More of the first named family have already been found fossil, than exist at present on the globe. They form more than one-third of the entire fossil flora of the secondary formations; but less than the 2000th part of the existing flora.

Descr. The plants of the tertiary strata approximate closely to the existing flora.

Descr. Below the new red sandstone vascular cryptogamiæ, or flowerless plants, greatly predominate, while dicotyledonous plants are rare. In the secondary strata above the coal, there is an approach to equality between these two classes: in the tertiary strata the latter predominates; and in the existing flora, two-thirds are of this class. *Buckland's Bridgewater Treatise, Vol.* 1, *p.* 520.

7. *Periods in which different plants and animals began to appear on the Globe, and in which some of them became extinct.*

Prin. In general, plants and animals began to exist first on the globe during the period when the lowest rock in which their remains are found, was deposited.

Proof. 1. Those animals and plants are excepted that are too frail to be preserved in the rocks: but in respect to all others, no reason can be assigned why their remains should not be found along with those of other organic beings existing at the same period. Particular species, from being less numerous, or being less likely to get enveloped in deposits formed by water (as birds for instance,) may be rarely found in the rocks: and therefore, we should not be hasty to infer that a species did not exist, because we have not discovered its remains. But if a formation has been pretty extensively examined, the presumption is strong that few new species will be found in it.

2. Comparative anatomy here comes in to our aid. For it is found that certain types of organic existence characterize particular geological periods: and having ascertained the type for any particular period, we may infer with great certainty that an animal or plant of a very different type will not be found among the organic remains of that formation. Thus, we find in general the fossils of the carboniferous group to have been adapted to a climate of a tropical character: and to expect to find in that group, animals or plants adapted to a temperate climate, would be unreasonable; because the two tribes could not have existed in the same climate.

Descr. The following is the order in which some of the most important animals and plants have first appeared on the globe: in other words, the epoch of their creation. It may indeed, be hereafter found, when the rocks have been more extensively examined, that some appeared earlier.

Silurian and Cambrian, or Graywacke Period.	Echinodermata, Annelida, Zoophyta, Crustacea Cirrhipeda. *Marine Shells.* *Crustacea,* (Trilobites.) *Fishes*—Placoidians and Ganoidians, (Sauroids and Sharks,) also those with heterocercal tails. *Flowerless Plants.* Marine. *Flowering Plants.* Terrestrial.
Devonian Period.	Fishes, (Cephalaspis, Chirolepis, &c.) abundan and peculiar.
Carboniferous Period.	Peculiar Fishes: *Arachnidans,* such as Scor pions; *Insects,* as Curculionidæ: *Fresh Water Shells; Infusoria: Dicotyledonous Plants* Coniferæ, Cycadeæ: *Monocotyledonous Plants,* Palmae, Scitaminæ. *Batrachians,* (tracks in Pennsylvania.)
Red Sandstone Period. Trias and Permian.	Tracks of Birds, Tortoises, and Chirotheria or gigantic Batrachians. (*Labyrinthodon.*) *Reptiles:* Monitor, Phytosaurus, Ichthyosaurus, Plesiosaurus, Thecodontosaurus, Palæosaurus. *Crustacea:* Palinurus. *Fishes:* Palæoniscus, &c. *Dicotyledonous Plants* Voltzia &c.

Oolitic Period.

Mammalia: (Marsupials) Thylacotherium, and Phascolatherium, (Didelphys of Buckland.)
Reptiles: Saurocephalus, Saurodon, Teleosaurus, Streptospondylus, Megalosaurus, Lacerta neptunia, Ælodon, Rhacheosaurus, Pleurosaurus, Geosaurus, Macrospondylus, Pterodactylus, Crocodile, Gavial, Tortoise.
Fishes: Pycnodontes and Lepidoides. (Dapedium, &c.) with homocercal tails.
Arachnidans: Spiders.
Insects: Libellulæ, Coleoptera.
Crustacea: Pagurus, Eryen, Scyllarus, Palæmon, Astacus.
Plants: Cycadeæ, (Pterophyllum, Zamia,) Coniferæ, (Thuytes, Taxites,) Lilia, (Bucklandia.)

Wealden Period.

Birds: Grallæ, (Tilgate Forest.)
Reptiles: Iguanodon, Leptorynchus, Trionyx, Emys, Chelonia.
Fishes: Lepidotus, Pycnodus, &c. Fresh water and Estuary shells.

Cretaceous Period.

Insects.
Reptiles: Mososaurus, &c.
Fishes: Ctenoidians and Cycloidians.
Crustacea: Arcania, Etyæa, Coryster.
Plants: Confervæ, Naiades.

Tertiary Period.

Mammalia: 1. *Eocene Period*, 50 *species:*—Palæotherium, Anoplotherium, Lophiodon, Anthracotherium, Cheroptamus (allied to the hog), Adapis (resembling the hedgehog,) *Carnivora:* Bat, Canis (Wolf and Fox) Coatis, Racoon, Genette, Dormouse, Squirrel. *Reptiles:* Serpents.
Birds: Buzzard, Owl, Quail, Woodcock, Sea Lark, Curlew, Pelican, Albatros, Vulture.
Reptiles: Fresh Water Tortoises.
Fishes: seven extinct species of extinct genera.
2. *Miocene Period:* Ape, Dinotherium, Tapir, Chalicotherium, Rhinoceros, Tetracaulodon, Hippotherium, Sus, Felis, Machairodus, Gulo, Agnotherium, Mastodon, Hippopotamus, Horse.
3. *Pliocene Period:* Elephant, Ox, Deer, Dolphin, Seal, Walrus, Lamantin, Megalonyx, Megatherium, Glyptodon, Hyæna, Ursus, Weasel, Hare, Rabbit, Water Rat, Mouse, Dasyurus, Halmaturus, Kangaroo, and Kangaroo Rat.
Birds: Pigeon, Raven, Lark, Duck, &c.
Fishes: (in the formation generally) more than 100 species now extinct which belong to more than 40 extinct and as many living genera.
Insects: 162 genera of Diptera, Hemiptera, Coleoptera, Aptera, Hymenoptera, Neuroptera, and Orthoptera.*

* Bronn's Lethæa Geognostica, p. 811.

Tertiary Period.	*Shells:* In the Newer Pliocene Period, 90 to 95 per cent. of living species; 35 to 50 per cent. in the older Pliocene: 17 per cent. in the Miocene; and 3.5 in the Eocene; amounting in all, extinct and recent, to 4000 species. *Plants:* Poplars, Willows, Elms, Chesnuts, Sycamores, and nearly 200 other species: seven-eighths of which are monocotyledonous or dicotyledonous.
Alluvial Period.	*Man,* and most of the other species of existing animals and plants. *Gigantic Birds,* Dinonnis, &c.

Human Remains.

Prin. The remains of men have not been found in any deposit older than alluvium, except in a few cases where they have probably been introduced into drift subsequent to its deposition. *Lyell's Prin. Geol. Vol.* 1. *p.* 249, 282; *and Vol.* 3. *p.* 204, 236, 298.

Proof. In the earlier periods of geology, the fossil bones of other animals were often mistaken for those of man. Thus the *Homo diluvii testis* of Scheuchzer, was ascertained by Cuvier to be nothing but a great salamander. At the present day, no practised geologist maintains that human remains have been found below drift; although some writers on geology still defend that opinion. See *Penn's Comparative Estimate of the Mosaical and Mineral Geologies, Vol.* 2. *p.* 124. *Fairholme's Geology of Scripture, p.* 219. *Comstock's Geology. p.* 263. But some geologists on the continent of Europe are of opinion that the bones of man are found so mixed with those of extinct quadrupeds, as in certain caverns in France, and the province of Liege, that all must have been deposited at the same time; that is, during the deposition of the most recent tertiary strata. Others suppose that these human remains must have been introduced subsequently. *Buckland's Bridgewater Treatise, Vol.* 1. *p.* 103. Upon the whole, no evidence has yet been afforded by geology, that man existed on the earth earlier than during the alluvial or historic period. *Dr. Smith's Scripture and Geology, p.* 396. *Second Edition, London,* 1840.

Objection. Some writers contend that when Asiatic countries have been examined more thoroughly, the remains of man may be found in all the fossiliferous rocks: and that they do not thus occur in Europe and America, because he had not spread into these parts of the world till a long time after his creation. But on this subject it may be observed, 1. That so far as the countries of Asia have been geologically examined, their organic relics correspond, as to distribution and general character, with those of Europe and America; and hence the presumption is, that in all that quarter of the globe, the mammiferous animals will not be found much below the tertiary strata. 2. Comparative anatomy strengthens this presumption, by showing conclusively, that most of such animals as now inhabit the globe, could not have lived when the same physical conditions existed that were necessary for the creatures found in the lower rocks.

Rem. The remarkable specimens of human skeletons found imbedded in solid limestone rock on the shores of Guadaloupe, deserve attention in this connection. At first view they may seem genuine examples of man in a fossil state. But they belong to the alluvial formation, and probably were buried there only a few hundred years ago. For the same rock contains shells of existing species, as well as arrows and hatchets of stone, and pottery. It is said that a battle took place on this spot about the year 1710, between the Caribs and Gallibis.—One of these specimens is in the British Museum in London, the other in the Garden of Plants in Paris. *Buckland's Bridgewater Treatise, Vol.* 1. *p.* 104. *Cuvier's Theory of the Earth, by Jameson and Mitchell, p.* 235, *N. York*, 1818. *Cuvier's Discourse on the Revolutions of the Surface of the Globe: Philadelphia*, 1831. *p.* 82.*

8. *Vertical Range of Animals and Plants in the Strata.*

Descr. Not only did different species, genera, and families of animals commence their existence at very different epochs in the earth's history, but some of them soon became extinct; others continued longer, and some even to the present time.

Descr. Species rarely extend from one formation into another; but genera frequently continue through several formations; and a few, even through the whole series of strata: and are still found among living animals and plants. Orders are still more extensive in their vertical range; and all the great classes, as has been shown, extend through the whole series. Very many genera, however, and some orders, are limited to a single formation. Others, after disappearing through one or more formations, at length re-appear.

Illus. The following tables will give an idea of the vertical distribution of several orders and genera. *Phillips's Treatise on Geology, Vol.* 1. *p.* 76. *et seq.*

The following table exhibits the distribution of several orders of Zoophyta; their presence being indicated by stars, and their absence by blanks.

Systems.	Spongiæ.	Lamelliferæ.	Crinoidea.	Echinida.	Stellerida.
Tertiary.	*	*	*	*	*
Cretaceous.	*	*	*	*	*
Oolitic.	*	*	*	*	*
Saliferous.			*		*
Carboniferous.		*	*	*	
Silurian.	*	*	*		*

Rem. I have placed Stelleridans in the saliferous system on the authority of Dr. Buckland, (*Bridgewater Treatise, Vol. I. p.* 416,) and in the Silurian system on the authority of Prof. Troost, in his *Fifth Report on the Geology of Tennessee, p.* 12, 1840.

The following genera of shells, very abundant at present on the globe, have a very limited range downwards.

	Cypraea.	Conus.	Voluta.	Strombus.	Murex.	Fusus.	Cerithium.	Mitra.	Pleurotoma.
Living Species.	135	181	66	45	75	67	87	112	71
In Tertiary System.	19	49	32	9	89	111	220	66	156
In Cretaceous do.			2	1	2				
In Oolitic do.					1				
In Saliferous do.									
In Carboniferous do.									
In Graywacke do.									

The following genera are very unequal in their vertical range. One of them, the terebratula, extends through the whole series of formations and still lives.

	Orthis.	Producta.	Spirifer.	Terebratula.	Trigonia.	Pholadomya.	Plagiostoma.	Inoceramus.	Gryphaea.
Living Species.				15	1	1			1
In Tertiary Strata.				18		1			3
In Cretaceous System.				57	12	1	13	19	7
In Oolitic do.			6	49	14	16	17	1	17
In Saliferous do.		7	5	14	7		8		1
In Carboniferous do.	15	76	74	44		1		1	
In Divonian.	35	14	50	49					
In Silurian.	64	22	29	50					

Out of the multitude of Cephalopods, or chambered shells, that swarmed in the ancient seas, only two species have continued to the present time; as may be seen by the following table of their vertical range.

	Bellerephon.	Orthocera.	Belemnites.	Nautilus.	Ammonites.	Hamites.	Scaphites.	Baculites.	Nummulites.
Living Species.				2					
In Tertiary Strata.				4					
In Cretaceous System.			8	9	57	28	4	5	3
In Oolitic do.			75	13	164	2	1		
In Saliferous do.				2	3				
In Carboniferous do.	36	52		40	59				
In Devonian.	21	47		2	85				
In Silurian.	15	46		4					

Palæontological Chart.

In order to bring under the eye a sketch of the vertical range of the different tribes of animals and plants, that have appeared on the globe from the earliest times, the chart which faces the title page, has been constructed. The whole surface is divided into seven strips, to represent Geological periods: viz. the lowest, the graywacke period: the next, the carboniferous period: the next, the saliferous period: the next, the oolitic period: the next, the cretaceous period: the next, the tertiary period: and the highest, the historic period, or that now passing. The animals and plants are represented by two trees, having a basis or roots of primary rocks, and rising and expanding through the different periods, and showing the commencement, developement, ramification, and in some cases the extinction, of the most important tribes. The comparative abundance or paucity of the different families, is shown by the greater or less space occupied by them upon the chart; although there can of course be no great exactness in such representations. The numerous short branches, exhibited along the sides of the different families, are meant to designate the species, which almost universally become all extinct at the conclusion of each period. Hence the branches are contracted in passing from one period into another, and then again expand, to show that the type of the genera and orders alone survive. Where a tribe, after having been developed during one period, disappears entirely during the next or several succeeding periods, but at length reappears; a mere line is drawn across the space where it is wanting.

While this chart shows that all the great classes of animals and plants existed from the earliest times, it will also show the gradual expansion and increase of the more perfect groups. The vertebral animals, for instance, commence with a few fishes; whose number increases upward: but no traces of other animals of this class appear, till we rise to the saliferous group, when we meet with the tracks of chirotheria, tortoises, and birds. But not till we reach the oolitic period, do we meet with the bones of the mammalia: and then only two species of marsupialia. No more of this class appear till we reach the tertiary strata, where they are developed in great numbers, approaching nearer and nearer to the present races on the globe as we ascend, until, in the historic period, the existing races, ten times more numerous, complete the series with MAN at their head, as the CROWN of the whole; or as the poet expresses it, "the diapason closes full in man."

In like manner if we look at that part of the chart which shows the developement of the vegetable world, we shall see that in the lowest rocks, the flowering plants are very few, and consist mostly of coniferæ and cycadeæ: links as it were, between the flowering and the flowerless plants. It is not till we ascend to the tertiary period, that the willows, elms, sycamores, and other species that form the forests of the temperate zone, appear. But lower down in the series, a few monocotyledonous plants are seen, such as lilies and palms; which, however, do not seem to have been greatly multiplied till we reach the tertiary period. Still more fully developed do we find them in the historic period; where 1000 species of PALMS,—the CROWN of the vegetable world, have been found.

To refer to another example of a somewhat different character: take the saurian animals, which began to appear during the saliferous period. In the next period above, or the oolitic, their developement is very great; so that they seem to have been the rulers of the animal creation. But above this period, they gradually decrease, until among existing animals all their representatives, except the crocodile and the alligator, are on a most diminutive scale.

A similar example among plants exists in the lycopodiaceæ; which during the carboniferous period, formed trees from 40 to 60 feet high. But above that period, they rarely appear; and their only remaining representatives on earth at the present time, are obscure plants a few inches in height.

Much more information of this sort may be obtained by a few moments inspection of this chart; which will prevent the necessity of details. As this however is the first effort that has been made to give such a representation of the leading facts in palæontology, I shall expect that defects and imperfections will be discovered in it.*

9. *Comparison of Fossil and Living Species.*

Prin. It is a moderate estimate to reckon the species of organic remains hitherto described in the rocks below the tertiary strata, at 10,000. Yet scarcely none of this number have

---

* Since the above was in type, (in the first edition,) I have received the *Lethæa Geognostica* of Professor Bronn, published at Stuttgart in 1837 and 1838, where I find a chart constructed on essentially the same principles. The wonder with me is, not that I have been anticipated, but that so simple a plan to exhibit the leading facts of palæontology has not been employed by writers in the English language.

thus far been identified with any now living on the globe. In many cases they differ even generically.

Rem. On another page of this section I have given approximate estimates of the number of living and fossil species hitherto discovered on the globe. I add another estimate, made by Prof. Charles B. Adams, and given in his *Second Annual Report on the Geology of Vermont*, p. 81.

	Living Species.	*Fossil Species.*
Mammalia,	1300	275
Birds,	8000	20
Reptiles,		120
Fishes,	8000	900
Insects,	100.000	250
Articulata,		500
Mollusca,	12.000	6000
Polypi,		900
Echinodermata,		200 ?
Animalcula,	800	100
Plants,	100.000	800
	230.100	10.165

Rem. The actual number of species of animals on the globe is estimated by some authors as high as two millions; and of plants, as high as 134.000, (*Am. Jour. Sci. Vol.* I. *Second Series, p.* 132,) and Agassiz supposes the actual number of species of fossil fish in the rocks to be 30.000.

Prin. The deeper we descend into the earth, that is, the older the rock, the more unlike in general are its organic remains to existing species. As we ascend, we find a nearer and nearer approximation to existing species in each successive formation.

Descr. In 1833, the number of shells in the tertiary strata, that had been discovered and described by M. Deshayes in Europe, amounted to 3036 : of these, 568 were identical with species found in our present seas. They were distributed however, very unequally through the different groups of these strata, as follows.

In the Eocene or oldest Group,	1238	species : Living analogues,		42
In the Miocene,	1021	do	do	176
In the Pliocene,	777	do	do	350

Lyell's Elem. Geol. Vol. 1. *p.* 281.

Prin. The organic remains in the northern parts of the globe, correspond more nearly to existing tropical plants and animals, than to those now living in the same latitudes.

Proof. It is well know that the Fauna and Flora of tropical regions are so different from those in higher latitudes as to strike every observer. Now any one who is acquainted with these peculiar features of tropical organic life, even as they are exhibited in books, will be struck with their resemblance to the organic remains in the fossiliferous strata. The following examples may serve for illustration: beginning with the highest of the strata, viz. diluvium. 1. Along the shores of the Arctic Ocean in the banks of the great rivers, such as the Oby, the Yenesi, and the Lena, are found immense quantities of the bones of the extinct species of elephant called the mammoth. The region in which these remains occur, is almost as large as the whole of Europe. Now although the fact that these animals were covered with hair, proves that the climate where they lived was colder than that where naked elephants now live, yet it must have been much warmer than the present temperature of Siberia, in order to produce vegetables for their sustenance. The rhinoceros found fossil in the same country confirms this conclusion. 2. The bones of extinct species of elephant, rhinoceros, hippopotamus, lion, tiger, hyæna, &c.—genera confined almost exclusively within the tropics at this day, are found scattered through the diluvium of almost every part of Europe. 3. When shells are found in the tertiary strata in northern countries, identical with those in existing seas, their analogues are almost universally found in tropical seas: and when the same species occurs in the Mediterranean, for instance, as is found fossil upon its shores, the latter is much larger than the former: and it is a well known fact that the same species in tropical regions attains a greater size than in colder climates. 4. The great size, both of the animals and plants found in the secondary strata, compared with that of living organic beings of a similar kind, shows a state of climate during their growth very favorable to their developement: such a climate, in fact, as exists in tropical countries. 5. The great number of chambered shells, such as ammonite, orthocera, &c. found in the secondary rocks, confirms this proposition, since the few representatives of these shells still found alive, occur in warm latitudes. 6. But perhaps the most striking evidence of a warm climate, during the deposition of the secondary rocks, exists in the fossil flora of the coal formation. This is filled with gigantic plants of genera mostly found within the tropics, such as equiseta, lycopodiaceæ, tree ferns, palms, &c.: and a person who is familiar with these remains, is struck, on going to a tropical country, with their resemblance to the vegetation around him; as he is with their want of resemblance to the flora of high latitudes. These tropical plants have been found in the rocks around Baffin's Bay, and even as far north as Melville Island, in 75° north latitude. 7. Numerous organic remains in the secondary rocks even in the oldest fossiliferous strata, appear to have once constituted coral reefs, such as are now found only in tropical seas. Such relics as these, also, have been found in the rocks of Melville Island.

Rem. 1. Agassiz, Lyell, and Smith of Jordan Hill, seem to have proved that the climate of northern regions, immediately preceding the alluvial period, was lower than it is at present: and it is the opinion of Agassiz, that a similar fall of temperature took place near the close of each great geological period. But admitting all this to be true, it cannot affect the preceding arguments concerning the general temperature during those periods, except as to the glacio-aqueous, which is of little consequence in this respect. *Edinburgh New Philosophical Journal, April,* 1838. *Dr. Buckland's Anniversary Address before the London Geological Society in* 1841. *An. and Mag. of Nat. History for February,* 1841.

Rem. 2. Those devoted to fossil botany say that the land plants found in the older strata, correspond more nearly with those now growing upon the low islands of the Pacific Ocean, between the tropics: and hence they infer that when these flourished, the land was but little elevated above the waters; and that the climate was constantly very warm and moist. *Amer. Jour. Sci. Vol.* 34. *p.* 324.

Prin. It is probable that during the deposition of the older fossiliferous rocks, the climate was ultra-tropical; that is, warmer than at present exists on the globe.

Proof. Tropical species of equiseta, lycopodiaceæ, tree ferns, &c. are much larger than those found growing without the tropics. But those found fossil are much larger than any now living. Equiseta, for instance, in the ancient world, were sometimes 10 feet high; tree ferns, from 40 to 50 feet, and arborescent lycopodiaceæ, 60 or 70 feet high. Recent equiseta are rarely more than half an inch in diameter; whereas the fossil calamites, a very similar plant, is sometimes 7 and even 14 inches in diameter, and no living lycopodiaceæ are more than 3 feet high. This extraordinary developement, which is found also in other species of plants and animals, can be explained only by a higher temperature: though Adolphe Brongniart suggests that in those early times, when perhaps no land animals existed, the atmosphere might have been more highly charged with carbonic acid than at present. *Histoire des vegetaux fossiles, par M. Adolphe Brongniart: 2d Livraison. p.* 113. *Buckland's Bridgewater Treatise, Vol* 1. *p.* 450. *Phillips's Edinburgh Treatise on Geology, p.* 118.

Prin. The temperature of the climate seems to have gradually sunk during the successive deposition of the different groups of fossiliferous rocks.

Proof. While the whole number of species of ferns, now growing upon the globe, is 1500; only 144 are found in the northern, temperate, and frigid zones; and 140 in the southern, frigid, and temperate zones; while the remaining 1200 are found within the tropics. Now the number of fossil ferns diminishes in nearly the same ratio, in ascending from the oldest secondary rocks, as it does in going north or south from the equator. Hence it is inferred that a similar decrease of temperature is in both cases the cause. 2. This is the most rational mode of explaining the gradual approach of organic remains to existing species, as we come near the surface; so that during the tertiary periods the climate could not have been much different from that around the Mediterranean. *Buckland's Bridgewater Treatise, Vol. 1. p. 471.* 3. If the former high temperature of the globe be admitted, we should expect this gradual reduction of temperature by radiation. *Phillips's Edin. Treatise on Geology, p. 96.*

10. *Description of individual and peculiar Species of Organic Remains.*

PLANTS.

Descr. The number of fossil plants hitherto described, according to Professor Unger, in his *Synopsis Plantarum Fossilium*, amounts to 1643; which are thus distributed in the several *classes* of Endlicher. *Am. Jour. Sci., Vol. 2, Second Series, p.* 136.

Algæ,	119	Iuliflorœ, (Amentaceae, &c.)	93
Characeæ,	6	Oleraceæ,	1
Lichenes,	1	Thymeleæ,	17
Fungi,	9	Contortæ,	11
Musci,	2	Nuculiferæ,	1
Calamariæ, (Equisetaceæ and Calamitæ,)	109	Petalanthæ,	1
		Discanthæ,	1
Filices,	444	Polycarpieæ,	1
Hydropterides, (Sphenophyllum)	11	Nelumbia,	2
Selagines (Lycopodiaceæ, Lepidodendreæ, &c.)	207	Peponiferæ,	1
		Columniferæ,	14
Zamiaæ,	100	Hesperides,	2
Glumaceæ,	11	Acera,	19
Enantisblastæ,	2	Frangulaceæ,	4
Coronariæ, (Lilia)	13	Tricoccæ,	1
Scitamineæ, (Musaceæ)	14	Terebinthineæ,	20
Fluviales,	21	Calycifloræ,	5
Spadicifloræ, (Pandanocarpum, &c.)	18	Myrtifloræ,	2
Principes, (Palmae)	43	Rosifloræ,	2
Coniferæ,	141	Leguminosæ,	45
Aquaticæ,	1	Plantae Incartæ Sedis,	118

Algæ or Sea Weeds.

Descr. Existing submarine vegetation, amounting to more than 500 species, may be arranged in three divisions; dependent for their characters upon climate: the first group occupying the frigid, the second the temperate, and the third the torrid zone. A similar distribution of the fossil marine plants, will bring all those below the top of the new red sandstone into the class of tropical plants; while those higher in the series, approximate more and more to those now existing on the globe. *Adolphe Brongniart's Histoire des Vegetaux Fossiles, Livraison,* 1: *p.* 41.

Descr. In the lowest rocks most of the plants are marine. Fig. 38, represents a species of fucoides.

Fig. 38.

Fucoides.

Musci and Filices: or Mosses and Ferns.

Descr. On account of their delicate structure, mosses are rarely preserved in the rocks. But ferns are very abundant; especially in the more ancient strata, where they are found of a size at least equal to those now growing in the torrid zone, which are often from 40 to 50 feet high. Fig. 39, is a sketch of some of these tree ferns, now growing in tropical climates.

Fig. 39.

Tree Ferns

Descr. Europe at the present time does not contain more than 30 or 40 species of ferns, and these of diminutive size; whereas more than 200 species have been found in the coal formation of the same quarter of the globe. *Adolphe Brongniart in American Journal of Science, Vol.* 34. *p.* 319.

Lycopodiaceæ, or Club Mosses.

Descr. The *Lycopodiaceæ* are a tribe of plants intermediate between ferns and coniferæ on the one hand, and ferns and mosses on the other. *Lindley's Natural System of Botany, p.* 313.

Lepidodendron. This fossil plant approximates in its character to the lycopodiaceæ: or rather, it seems to be intermediate between the club moss tribe, and the coniferæ or pine tribe. It is abundant in the coal formation, where it is some times found from 20 to 45 feet long; and M. Ad. Brongniar has described 34 species. The genus is wholly extinct. Fig 40, will convey some idea of these plants.

Fig. 40.

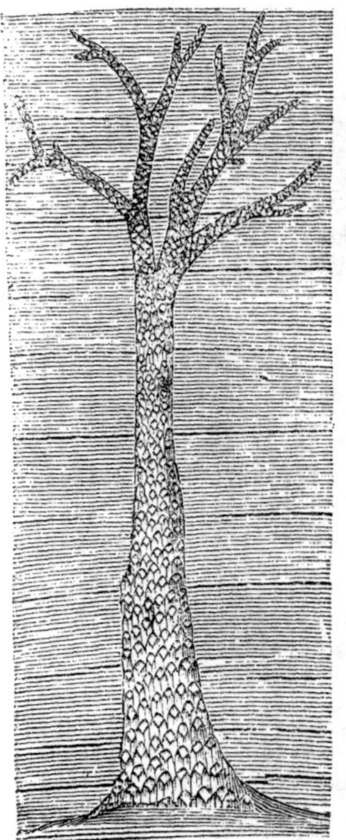

Lepidodendron.

Rem. The lepidodendra fill up a chasm in the existing series of plants, between flowering and flowerless plants, better than any living genus. Similar blanks in the existing organization are filled by other extinct genera of organic remains. *Lindley and Hutton's Fossil Flora, Vol.* 2, p. 53.

Equisetaceæ.

Descr. Living plants of this tribe are called *horsetails, cat-tails, scouring rushes,* &c.: and although of frequent occurrence in all climates (the most frequent in the temperate zones) they are of diminutive size, even in the torrid zone, compared to those found fossil. The latter are divided into two genera, *Equisetum* and *Calamites:* the former corresponding very nearly to living equiseta, but the latter differing a good deal in structure and size; being much larger than the equiseta. Fig. 41, is a calamites destitute of leaves.

Fig. 41.

Calamites.

Descr. The Cycadeæ are a remarkable family of plants, occupying an intermediate place between palms, ferns, and coniferæ; filling up an important link between dictyledonous, monocotyledonous, and acotyledonous vegetation. Only two genera and 22 species are known as now living upon the globe. But during the deposition of the rocks above the coal, they formed a large part of the vegetation. For out of 70 species of land plants found fossil, during this period, 29 species are cycadeæ, referable to 4 genera. They have lately been found also in the coal formation. The living species mostly grow in tropical climates. Fig. 42, represents a living species of these plants.

Fig. 42.

Cyclas Revoluta.

Descr. Large trunks of trees, from half a foot to three feet in diameter, and from 50 to 60 feet long, covered with flutings and scars, have long been known in the coal formation unde the name of *Sigillaria*, from the resemblance of the scars t(the impression of a seal. It was not till recently that geolo gists have been able to refer these trunks to any known family of vegetables. They now suppose them to have been exogen. ous plants and gymnosperms; but like the zamia above des. cribed, combining also the vascular system of ferns and lyco. podiaceæ. And they probably, also, were a principal source of the beds of coal in the coal measures. Fig 43 exhibits one of these Sigillaria.

Fig. 43.

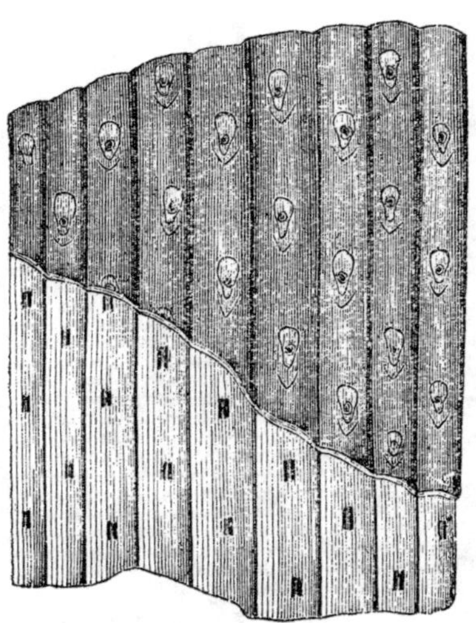

Sigillaria.

Descr. Another fossil plant still more common in the "underclay" beneath the coal beds, is called Stigmaria. It consists (Fig. 44) of a dome shaped centre, three or four feet in diameter, from which proceed branches, 30 or 40 feet long, covered with tubercles, or rootlets. These Stigmariæ were pro bably the roots of the Sigillaria; although till recently they have been thought to be dome shaped aquatic plants.

Fig. 44.

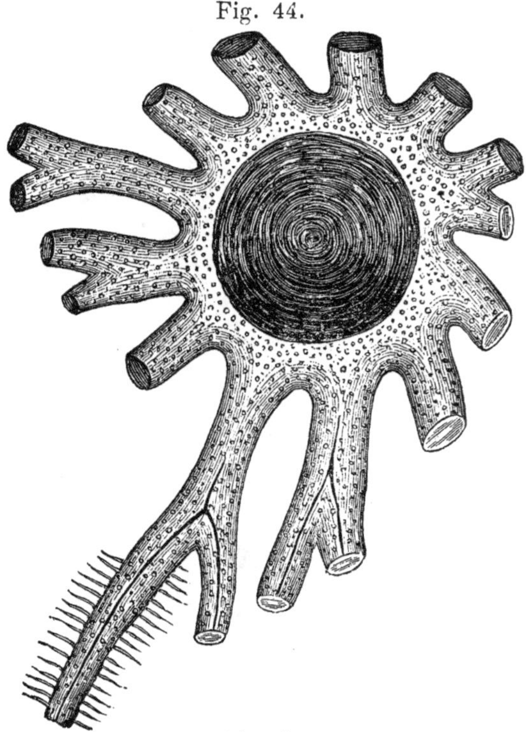

Stigmaria.

Descr. Several other extinct genera, with scars similar to those on the Sigillaria, occur in the same rocks, and are probably Coniferæ. Fig. 45 shows a portion of one called ***Ulo dendron.***

Fig. 45.

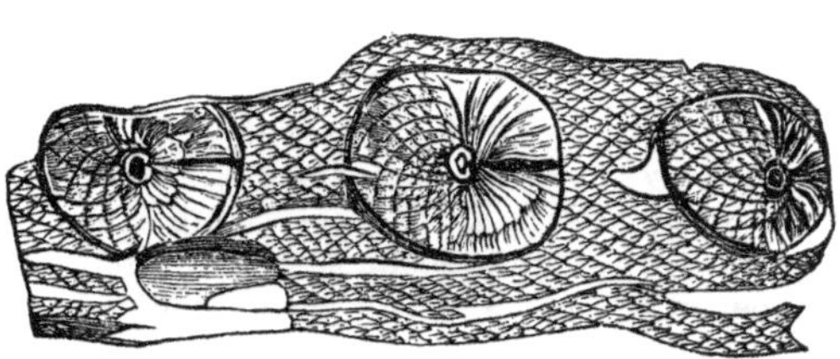

Ulodendron.

Descr. Another remarkable and beautiful tribe of plant not unfrequent in the coal formation, has whorled leaves lik the flower of the Aster: hence one genus is called Asterophy lites. Fig. 46 shows one of these from the coal mine in Man: field, Massachusetts.

Fig. 46.

Annularia.

Coniferæ and Cycadeæ.

Descr. Coniferæ and Cycadeæ are the only two families of fossil plants whose seeds are originally naked. Hence they are called *Gymnospermous Phanerogamiæ.* The coniferæ under the name of pines, araucarias, &c. constitute a large and important part of the existing trees of all climates: and they occur in the rocks of all ages. More than 20 species have been found in the tertiary strata, 13 species in the oolite and lias, 4 (of voltzia) in the new red sandstone, and several in the carboniferous formation. Mr. Witham has figured the trunk of an araucaria, 47 feet long, from Cragleith quarry in the carboniferous limestone near Edinburgh. *Witham's description of a Fossil Tree, &c. Edinburgh,* 1833. Araucarias are found fossil in Great Britain alone; but genuine pines occur in the coal formation in Nova Scotia and New Holland. The four living species of araucaria that have been described, occur in tropical climates south of the equator.

Descr. Sometimes the trunks of those gigantic trees, as well as of some of the other plants that have been described, are found standing erect, rarely in the very place where they grew; but generally they appear to have been transported, and to have assumed an upright position by the greater specific gravity of the roots. Fig. 47 shows the stumps of an ancient forest of coniferæ, with the roots imbedded in the black vegetable mould in which they grew; the whole being now converted into stone. The section was taken in the Isle of Portland, Eng.

Fig. 47.

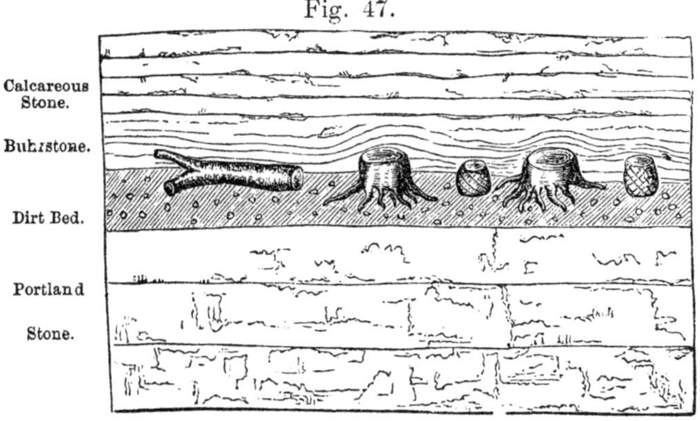

Subterranean Forest: Isle of Portland.

Prin. It is probable that dicotyledonous plants, as well as the frailer kinds of flowerless ones, such as fungi, and mosses, may have been more abundant in the earlier periods of geological history, than the specimens of these plants found fossil would lead us to infer.

Proof. Most organic remains must have been preserved in water, or at least in wet sand, or mud. Now Prof. Lindley, having immersed in a tank of fresh water 177 species of living plants for more than two years, arrives at the following conclusions.

"1. That the leaves and bark of most dicotyledonous plants are wholly decomposed in two years, and that of those which do resist it, the greater part are Coniferæ and Cycadæ."

"2. That Monocotyledons are more capable of resisting the action of water, particularly Palms and Scitamineous plants; but that grasses and sedges perish."

"3. That Fungi, Mosses, and all the lowest forms of vegetation disappear."

"4. That Ferns have a great power of resisting water if in

a green state, not one of those submitted to the experiment having disappeared; but that their fructification perished." *Buckland's Bridgewater Treatise, Vol. 1. p.* 480.

Rem. 1. It is interesting to observe that by this experiment those plants were most enduring in water, which we find most abundant in a fossil state. Yet some circumstances prevent us from inferring with certainty that all the more frail and the dicotyledonous species perished in the process of petrifaction. If a corresponding experiment had been made with these plants in wet mud, or sand, another in salt water, or salt mud, these results might have been somewhat modified, and probably in nearly every case where plants are carried to the bottom of water, they are covered by mud in a shorter time than two years; and most of those preserved in the rocks were fossilized beneath salt water. Other substances, as iron, or lime, in solution in the water, might essentially modify the experiment. After all, however, the experiment does show us that we must not place too much dependence on the relative numbers of different classes of fossil plants, as hitherto discovered.

Rem. 2. Other peculiar and interesting plants occur in a fossil state; as the pandaneæ, palms, &c., but the limits of this treatise do not permit their introduction.

Rem. 3. The great size of many fossil plants and the vast accumulations of carbonaceous matter in the coal formation render it probable that the vegetation of the early periods of the globe was far more abundant than at the present day. Yet as the trees were mostly without flowers, and unenlivened by the presence and voices of any vertebral animals, the landscape must have presented a very uniform and sombre though imposing aspect: better adapted to a state of preparation for the higher orders of animals, than for their actual existence: better adapted to prepare fuel for man, than for his happy dwelling.

Rem. 4. A recent discovery has enabled geologists to ascertain with greater certainty and ease than before, the nature of fossil plants. The different families of living plants are distinguished not merely by external characters, which mostly disappear when petrified, but by a corresponding anatomical structure; chiefly by the form of the minute vessels, of which they are composed. Now it is found that these vessels retain their form when petrified. Hence by cutting a fragment of fossil wood very thin, and polishing it, a microscope will show these vessels, and thus enable the enquirer to determine the nature of the plant. For this discovery we are indebted to Mr. Witham, who has given directions for preparing fossil wood for such an examination. *Witham's Observations on Fossil Vegetables, &c. Edinburgh,* 1831, *Quarto.*

ANIMALS.

1. RADIATED ANIMALS.

Descr. This extensive division of animals are the most simple of any in their organization, and the most removed from common observation in general. They are distinguished by their radiated structure: though in some of the first order, the Echinodermata, it has been lately shown that they possess somewhat of a "bilateral symmetry," like the higher orders of animals. The whole class are frequently denominated *zoophytes.*

Descr. The number of zoophytes in a fossil state is very large; and, in almost every case, they differ specifically, and frequently generically, from existing species. I shall notice those chiefly that are most unlike such as now live on the globe.

Crinoideans, or Encrinites.

Descr. These animals have long attracted attention from their peculiar structure and the immense quantity of their remains in some limestones called *entrochal* or *encrinal marble.* They belong to the first order of Radiata, or the Echinodermata. They are exceedingly rare among living animals, but two species, the *Pentacrinus caput Medusaæ,* and the *Comatula fimbriata,* that have been discovered in the ocean, have thrown much light upon those that are fossil. Mr. Miller, who has written an excellent work, entitled *the Natural History of the Crinoidea,* has divided them into nine genera. The two genera that have attracted most attention are the *Encrinites moniliformis, or lily encrinite, or stone lily,* and the *Pentacrinus Briareus.* The former consists of a vast number of little joints, or bones, forming a column, (which may be called the vertebral column, although these animals are invertebral,) for the support of a cup-like body, containing the viscera, and from whose margin proceed five articulated arms, dividing into tentaculated fingers, more or less numerous, surrounding the mouth. The animal was fixed at the bottom of the ocean, or to a piece of wood, and merely moved as far as it could reach by bending its very flexible column, which was admirably fitted for this purpose. The number of little bones, or joints, composing the head alone of this species, is estimated at 26.000. These bones are perforated and are used sometimes for rosaries. This animal relic is shown in Fig. 48.

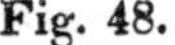

Fig. 48.

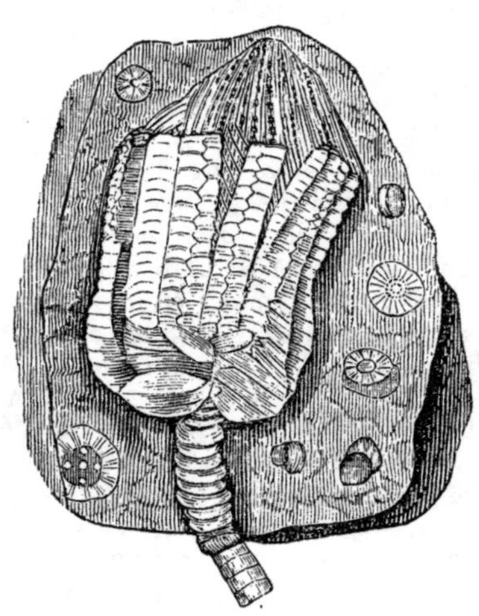

Lily Encrinite.

Descr. The stem of the encrinite is circular, but that of *pentacrinite*, pentagonal. The latter, also, had usually a greater number of side arms and of joints. One of the most remarkable of them was the *briarean pentacrinite*, (Pentacrinus Briareus) so called on account of the great number of its hands or tentacula. The bones in its fingers and tentacula, amount at least to 100.000: and those of the side arms, to at least 50.000 more. And since each bone must have had two set of muscular fibres for contraction and expansion, the bundles of muscular fibres in the whole animal must have been as many as 300.000. This vastly exceeds the muscular apparatus in any other animal. What a contrast to man, whose bones are only 241, with 232 pairs of muscles!

Fig. 49, shows another genus of this family, the *apiocrinites* or *pear encrinite.* It is represented as restored, and situated as if in the water.

Fig. 49.

Pear Encrinite.

Polyparia, or Corals.

Descr. *Polypi*, *Polypifers*, or *Polyparia*, are those minute radiated animals that have the power of secreting carbonate of lime, and thus of building up large stony structures from the bottom to the surface of the ocean. They swarm in immense numbers in the seas of tropical climates, and form coral reefs which sometimes extend hundreds of miles. They seem to have existed in all ages, and to have formed similar deposits, which are now ranked among the limestones. Figs. 50, 51, and 52, show several living species of these animals as they are attached to their stony habitations.

Fig. 50.

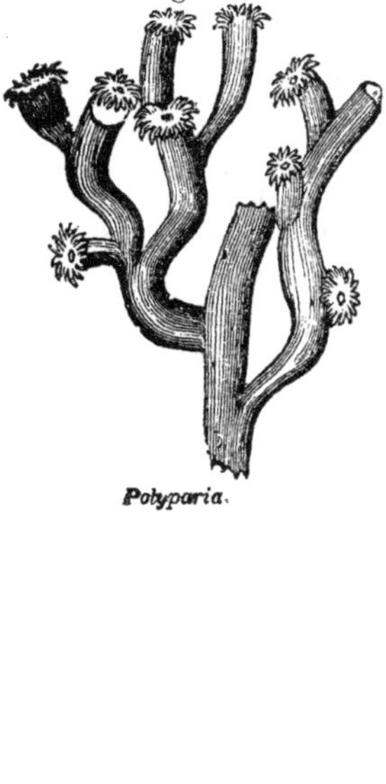

Polyparia.

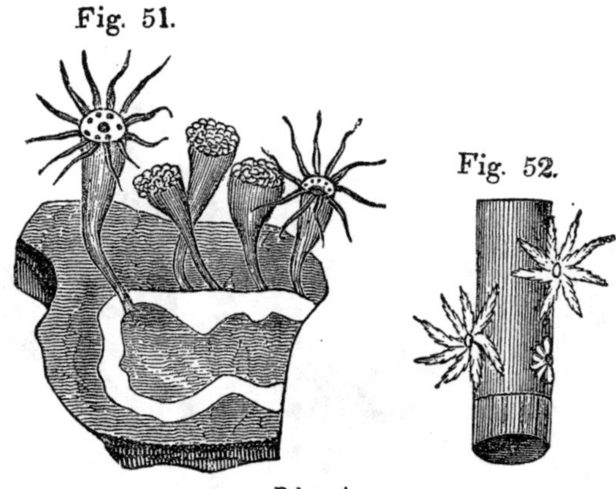
Fig. 51. Fig. 52.

Polyparia.

Descr. The tentacula of these animals are provided with cilia or minute hairs on their margins, which are capable of being rapidly moved, so as to keep currents of water in motion, that food may be conveyed to their mouths. Immense numbers of the polyparia unite in building up a single habitation, and they do this as if influenced by one instinct: so that the structure rises with the most symmetrical proportions. Hence it is still a question, whether all the animals upon each structure are not to be regarded rather as one compound animal. In the *Flustra carbasea* each polype has usually 22 tentacula; and on these, 2200 cilia. An ordinary specimen of this species will contain 18.000 polypi; and of consequence, 396.000 tentacula, and 39.600.000 cilia. On the *Flustra foliacea*, Dr. Grant estimates 400.000.000 cilia. *Roget's Bridgewater Treatise. Vol.* 1. *p.* 122

Descr These polyparia mostly multiply by buds, called gemmules, which grow like the buds of plants from the parent, and after a time fall off and become distinct animals. A single polype in this mode may produce a million of young in a month. They may also be multiplied by division, when each separate part becomes in a short time a whole animal. Different parts may also be made to grow together, and monsters of every form be produced. The Hydra is one of the genera of polypi; and by taking the heads of several individuals, and grafting them to one body, a Hydra with seven, or any other number of heads, may be produced.

Descr. Small as these animals are, they have nevertheless effected important geological changes on the globe; for some of the most extensive rock formations appear to have resulted from their labors. But the next tribe of animals, which is to be described, furnishes still more striking evidence how powerful an agency the minutest of all beings are able to exert upon our globe.

Infusoria.

Descr. These animals are not discernible, with a few exceptions, but by powerful microscopes: and as they usually occur in some sort of infusion, they have been called *Infusoria;* though they generally go by the name of *animalcula.* The recent astonishing discoveries of Ehrenberg, a Prussian naturalist, have given a new aspect to this department of animated nature, even in a geological point of view. He has described 722 living species, which swarm almost everywhere, even in the fluids of living and healthy animals, in countless numbers.

Descr. Formerly they were thought to be the most simple of all animals in their organization: to be in fact little more than mere particles of matter endowed with vitality; but he has discovered in them mouths, teeth, stomachs, muscles, nerves, glands, eyes, and organs of reproduction. Some of the smallest animalcula are not more than the 24.000th of an inch in diameter; and the thickness of the skin of their stomachs, not more than the 50.000.000th part of an inch. In their mode of reproduction they are viviparous, oviparous, and gemmiparous. An individual of the *Hydatina senta* increased in ten days to 1.000.000: on the eleventh day, to 4.000.000; and on the twelfth day, to 16.000.000. In another case Ehrenberg says that one individual is capable of becoming in 4 days, 170 billions! *Am. Journal of Science, Vol.* 35. *p.* 372.

Descr. Leuwenhoeck calculated that 1.000.000.000 animalcula, such as occur in common water, would not altogether make a mass so large as a grain of sand; Ehrenberg estimates that 500.000.000 of them do actually sometimes exist in a single drop of water.

Descr. In the Alps there is sometimes found a snow of a red color; and it has been recently ascertained by M. Shuttleworth, that the coloring matter is composed chiefly of infusoria, with some plants of the tribe of Algæ. And what is most singular, is, that when the snow has been melted for a short time, so as to become a little warmer than the freezing point, the animals die, *because they cannot endure so much heat!* *Bi-*

bliotheque Universelle de Geneve, for February, 1840; as quoted in *Etudes sur les Glaciers, par Agassiz, p.* 62. *Neuchatel,* 1840. A specimen of *meteoric paper,* which fell from the sky in Courland in 1686, has been examined by Ehrenberg, and found to consist, like the red snow, of *Conferva and Infusoria.* Of the latter he found 29 species. *Ann. and Mag. Nat. Hist. No.* 16, *p.* 185.

Descr. Surprising as these facts are, it will perhaps seem still more incredible, that the skeletons of these animals should be found in a fossil state, and actually constitute nearly the whole mass of soils and rocks several feet in thickness, and extending over areas of many acres. Yet this too has been ascertained by the same acute Prussian naturalist. The following formations, he says, are of this description.

	1. Bog Iron Ochre. 2. *Kieselguhr,* a siliceous incrustation, from hot springs.	Alluvial.
	3. *Polierschiefer,* Polishing Slate, a variety of Tripoli, or rotten stone. 4. The Semi-opal of the Polierscheifer.	Tertiary.
Probably of the same nature.	5. Semi-opal of the Dolerite, 6. Precious Opal of the Porphyry, 7. Flint of the Chalk.	Secondary and Primary.

Descr. Some of the above substances occur in large quantity. The polishing slate, for example, at Bilin in Germany, forms a bed 14 feet thick, and the eatable infusorial earth near Lunebourg, a bed above 20 feet thick. Yet it would take 41.000 millions of these skeletons to make a cubic inch; their weight being only 220 grains! A single shield or skeleton weighs about the 187 millionth part of a grain!

Descr. Professor Bailey calculates that a cubic inch of the infusorial deposit from Maidstone in Vermont, (composed mainly of the minute *Gaillionella varians,*) contains 15.625.000.000 skeletons! *Adams's Second Annual Report on the Geology of Vermont, p.* 152.

Descr. The chalk of northern Europe is more than half inorganic matter, the rest animalcula. Forty species are found in it. *Dr. Buckland, as quoted in the Edinburgh New Philos. Journ. Jan. to April,* 1841, *p.* 441.

Descr. The animalcula differ from all other animals, in having their softer parts protected by a shield, or skeleton, which may consist either of silica, lime, or oxide of iron. These shields, of course, will not be altered by the strongest heat; and although some of the rocks above named have been subject to heat, the skeletons often remain very entire, and their organic structure is very obvious through a powerful glass.

Descr. In New England and New York, the siliceous marl (*Bergmehl*, or *Mountain meal*,) already described as occurring beneath peat in swamps, has been recently shown by Prof. J. W. Bailey, of West Point, to be almost entirely made up of the fossil skeletons of infusoria belonging to the family of Bacillariæ: some of which appear to be identical with those found by Ehrenberg in Germany. Fig. 53 shows a group of these fossil skeletons, sketched by Prof. Bailey, as they appear when diffused in water, under the microscope. *American Journal of Science, Vol.* 35. *p.* 118. Deposits of this siliceous marl are very common in Massachusetts; and all hitherto examined, contain vast numbers of these relics; indeed, they constitute nearly the whole of the deposits. I have examined specimens from Spencer, Pelham, Barre, Manchester, Fitchburg, Wrentham, North Bridgewater, and Andover. Professor Bailey has given a detailed account of the species from these localities, with drawings, in my *Final Report on the Geology of Massachusetts, Vol.* 2. *p.* 310.

Fig. 53.

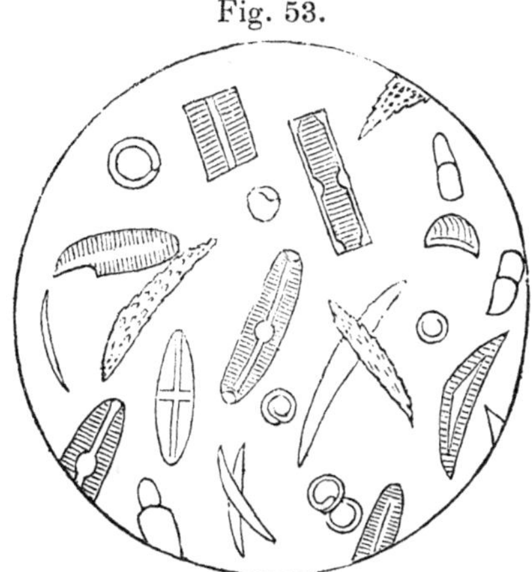

Fossil Infusoria.

Descr. Of 80 species of fossil infusoria discovered by Ehrenberg, nearly one half belong to extinct species. Those in the recent strata are all fresh water animals; but those in the chalk are marine. *London and Edinburgh Philosophical Magazine, for May*, 1839, *p.* 277.

Descr. "The fossil animalcule from iron ochre is only the one twenty-first part of the thickness of a human hair; and one cubic inch of this ochre must contain one billion of the skeletons of living beings." *Wonders of Geology, Vol.* 2. *p.* 689.

Descr. The ferruginous scum that appears upon the water of some springs, as well as the red deposit at their bottom, is composed partly of the remains of animalculæ, and partly of inorganic oxide of iron. *Edinburgh New Philos. Journ. Jan. to April*, 1841, *p.* 441.

2. ARTICULATED ANIMALS.

Descr. The animals of this class are distinguished by having envelopes connected by annulated plates or rings.

Exam. The earth worm, blood sucker, lobster, crab, horse shoe, spiders, scorpions, insects.

Rem. Excepting the insects, of which more than 100.000 living species are already known, the animals of this class are not numerous; and but few of its tribes are found in the rocks.

Insects.

Descr. Until recently, no insects had been discovered lower in the rocks than the oolite: but two species of Coleoptera, and one of Corydalis, have of late been disinterred in the coal formation. *Buckland's Bridgewater Treatise, Vol.* 1. *p.* 409. Not less than 25 species occur in the oolite, several in the Wealden, and 244 species in a fresh water formation of the tertiary group. *Bronn's Lethæa Geognostica, p.* 811. *Annals and Mag. Nat. Hist. Feb.* 1841, *p.* 495.

Rem. If there is any probability that insects were numerous in early times, and no sufficient reason can be given to show that they were not, it may seem strange that their remains so rarely occur. But in the first place, a large part of these animals are too frail to be preserved in a fossil state. Secondly, only one or two species of insects are found in salt water, which is the principal medium by which organic remains have been preserved: and thirdly, they are so light as to sink with difficulty in the waters, while a great number of insectivorous animals are watching to devour them, as they float along on the surface.

Arachnidans, or Spiders and Scorpions.

Descr. The scorpion has recently been found in the coal formation in Bohemia, by Count Sternberg; and is the first example of this animal in a fossil state. Spiders have not been found lower than the oolitic series, where only two species are recognized; but in the fresh water tertiary several species occur.

Crustaceans.

Descr. Crustaceous animals are not common in the rocks: yet the King Crab (*Limulus,*) so abundant on the coast of New England, has been found in the coal formation, and also in the oolite, where other animals of this family occur. But the most remarkable animal of this class is extinct, viz.

The Trilobite.

Descr. This singular animal, which is found in the older fossiliferous rocks, in all the northern parts of Europe, in North and South America, and at the Cape of Good Hope, was long confounded with insects. But it was at length ascertained that it corresponded most nearly to the living genera of crustaceans the Serolis, Limulus, and Branchipus. Figs. 54, and 55, represent two genera of trilobites, out of the many genera and 224 species that are known. It will be seen that this animal is composed of a shield covering the anterior part of the body, while the abdomen has numerous segments which fold over one another like those on a lobster's tail. By this arrangement some of the species had the power to roll themselves up like the wood louse, or the armadillo, and thus of defending themselves against enemies. These animals are usually from one to six inches long, and are divided by longitudinal furrows into three lobes; and hence their name. They seem to have been destitute of antennæ, and their legs, which were soft, and which answered the purpose of legs and wings, have disappeared.

Descr. By far the largest trilobite yet described, is the *Isotelus maximus*, found in Ohio, and figured by Professor Locke, in the Second Annual Report on the Geology of that State, p. 248. Its entire length is 19½ inches!

Fig. 54.

Fig. 55.

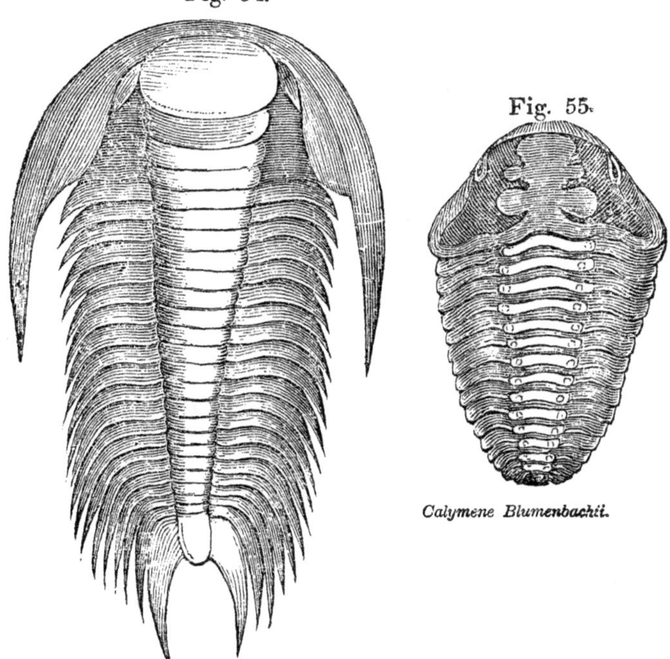

Restoration of Paradoxoides Tessini of Brong.

Calymene Blumenbachii.

Descr. Trilobites abounded among the earlier inhabitants of the globe, being most common in the graywacke. A few species occur in the carboniferous strata; above which, not a trace of them has been discovered. In the carboniferous strata they are accompanied by the Limulus. This latter also occurs in the oolitic group, with other crustaceans of a higher order.

Descr. Perhaps the most curious fact respecting the trilobites, is the discovery of their eyes, which are sometimes perfectly preserved. It is well known that the eyes of crustaceous animals, like those of insects, are made up of a vast number of facets, or lenses, placed at the end of tubes, which are arranged side by side, so as to produce a radiating mass of eyes, which being generally of a hemispherical or conical form, and sometimes elevated from the head on a stem, enable the animal to see in every direction; although their eyes have no motion. In some insects the number of these lenses in both eyes, as in the house fly, is 14.000: in other cases, (the dragon fly,) 25.000: in others, (the butterfly,) 35.000: in others (the Mordella,) 50.000. But in the trilobite they amount only to about 800. The whole mass is of a conical shape as is shown in Fig. 56.

Fig. 56.

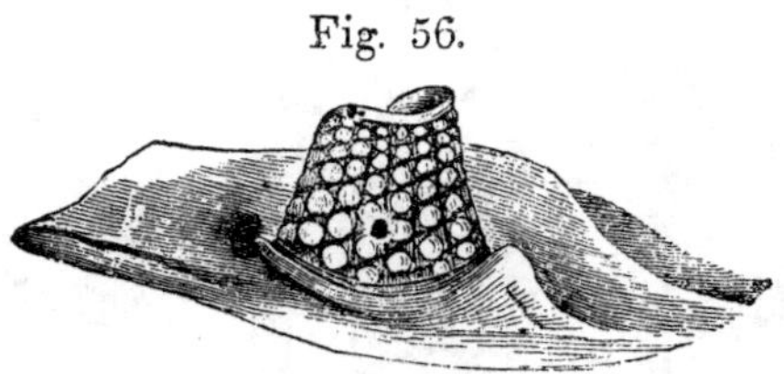

Eyes of the Trilobite.

3. MOLLUSCA.

Descr. The Molluscous animals compose the third great division of those that are invertebral, reckoning from the least to the most complex. This division embraces those animals that are destitute of a spinal marrow, or articulated skeleton; but whose muscles are attached to a calcareous covering, called a shell; or to a soft skin externally, and to a hard body within, analogous to shells. They are most abundantly diffused among living animals; and the great number of their remains in the rocks, proves, either that they were more numerous than other animals in early times, or that they were more readily preserved. Perhaps we must call in both circumstances to explain the fact.

Rem. The science that describes molluscous animals is called *conchology*. and from this science geology derives the greatest aid. Even its fundamental principles cannot be described in this treatise: nor, indeed, but a few of the vast number of shells that are found in a fossil state. Some of the most remarkable will be selected.

Chambered Shells.

Descr. These are univalve shells, which are divided into numerous compartments, or chambers, by cross partitions; as is shown in Fig. 57, which is a section of the common *Nautilus pompilius.* These partitions are all perforated by what is called the *siphuncle,* which consists mostly of a membrane, having the form of a tube, and being so firmly fastened to the partitions that no air can pass by into the chambers. The animal resides in the outer chamber, and is connected with the others only by the siphuncle. Around the heart of the animal is a sac, which may contain fluid enough to fill the siphuncle. Now the object of this structure is to enable the animal to rise or sink at pleasure in the water. When the sac around the heart is filled with fluid, the siphuncle is empty, and the air in the posterior chambers expands, so as to cause the shell to rise and float in the water; but when the animal withdraws its arms into the shell, the fluid in the sac is compressed, and forced into the siphuncle, which condenses the air, and thus the animal is made heavier than the water, and sinks. In short, he rises and sinks in exactly the same manner as a *water balloon.*

Fig. 57.

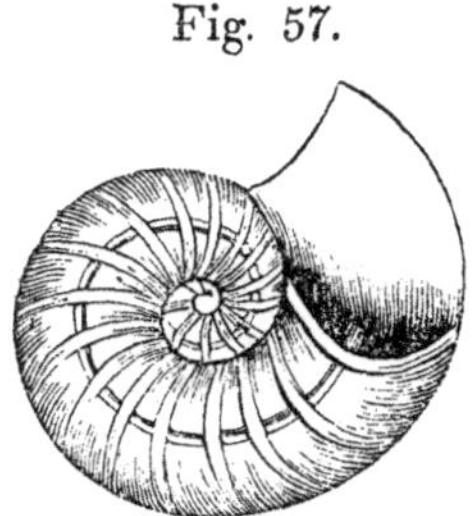

Nautilus.

Rem. Although the nautilus has attracted great attention from the earliest times, it is only within two or three years that Dr. Buckland first discovered the true explanation of the manner in which it could rise or sink at pleasure in the water. *Bridgewater Treatise, Vol. 1. p.* 325. Mr. Owen, however, doubts the correctness of this explanation.

Ammonites.

Descr. With the exception of one or two species of nautilus, all the larger species of multilocular or chambered shells have disappeared from the earth, although in early times they were very numerous and widely diffused, and often of enormous size. They resembled the nautilus in general form and structure, although generically different; and they are sometimes found more than four feet in diameter. Figs. 58, 59, represent two species of ammonites.

Fig. 58. Fig. 59.

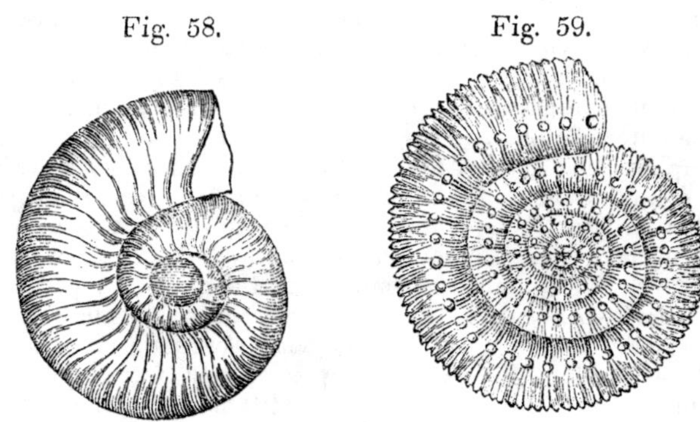

Ammonites.

Details. Brochant enumerates 270 species of ammonites: Phillips mentions 274, which he distributes as follows: In graywacke, 17: in the carboniferous system, 33: in the saliferous system, 3: in the oolitic system, 164: in the cretaceous system, 57: in the tertiary strata, 0? *Treatise on Geology, Vol.* 1. *p.* 83.

Descr. It is well ascertained that in some chambered testacea, the shell is contained within the animal; as in the *Spirula Peronii*, Fig. 60. This appears to have been the case with several extinct genera, as the Orthoceratite, Lituite, Baculite Hamite, Scaphite, Turrilite, Nummulite, and Belemnite.

Fig. 60.

Spirula Peronii.

Orthoceratite.

Descr. As its name implies, this was a straight shell divided by transverse septa into chambers, of which nearly 70 have sometimes been counted. It has been found a yard in length, and half a foot in diameter; forming a *float*, which would have been sufficient for an animal far larger than any existing cephalopod. Fig. 61, shows the shell of an orthocera with one of the septa.

Fig. 61.

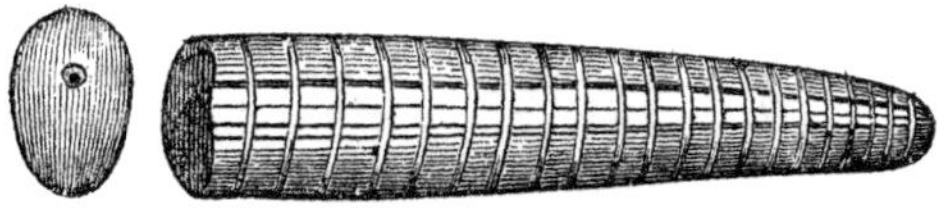

Orthocera.

Inf. The great size of these shells, as well as of the ammonites, confirms the views already presented of the existence of a very warm climate when they were alive in northern seas.

Rem. Not less than 20 species of orthoceratite are found in the graywacke, and 28 in the carboniferous group. They have not been found in any later rock.

Belemnite.

Descr. This internal shell resembled a conical arrow, with a cavity of similar shape, in which was a thin horny sheath, and within this, a thin conical chambered shell, or *alveolus.* It was provided, also, with an ink bag, like the living Sepia, or cuttle fish, as a defence against enemies, or rather, as a means of making good its retreat. These shells are found only in the oolite and the chalk, and 83 species have been described.

Fig. 62, shows an imaginary restoration of the Belemnosepia, as made by Dr Buckland, (*Bridgewater Treatise, Vol.* 2. *plate* 44. *fig.* 1.) exhibiting the animal with the internal shell and ink bag

Fig. 62.

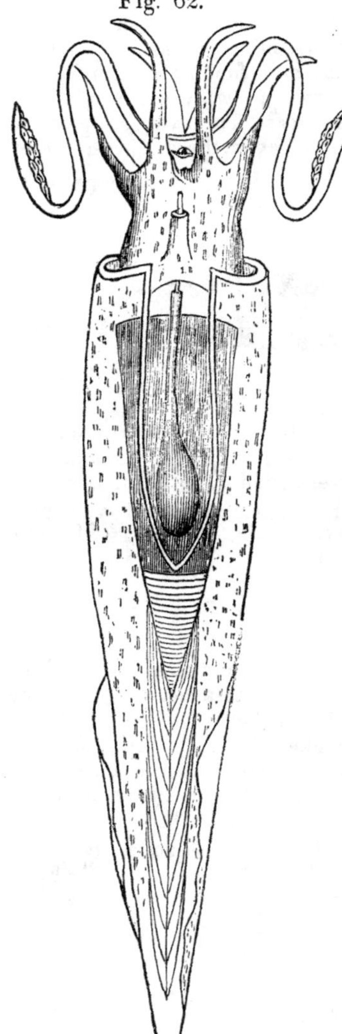

Nummulites.

Descr. These extinct shells are so called from their resemblance to money. They are generally of very diminutive size, and belong to wha are called the *foraminatea polythalamous,* or many chambered shells. They are chiefly remarkable for their vast numbers, constituting often almost the entire mass of whole mountains, in the tertiary and newer secondary limestones. The Sphinx and some of the pyramids of Egypt are composed of nummulitic limestone. Fig. 63, exhibits a species of nummulite.

Fig. 63.

Nummulite.

Loligo or Cuttle Fish.

Descr. It is well known that the cuttle fish, (Sepia or Loligo,) is provided with a bag of ink within its body, from which the sepia used in painting is obtained; and also with an internal bone, or in some species, a mere thin cartilaginous substance like horn, that resembles a quill. Both the ink and the pen of the Loligo have recently been discovered in a fossil state in England. *Buckland's Bridgewater Treatise, Vol. 1. p. 303.*

Rem. It is a very curious fact, that a substance so easily destroyed as ink should have been so perfectly preserved in the lias limestone of Lyme Regis, that after thousands and perhaps ten thousands of ages, it can be extracted, and the paint formed from it cannot be distinguished from the best which artists now prepare! It is also interesting to learn that for this discovery we are indebted to the industry and skill of a lady (Miss Mary Anning,) who, with others of her sex that might be named, is doing much for the science of geology in England.

Bivalve Shells, mostly extinct.

Descr. Of the Terebratula (Fig. 64) 50 species have been found in the Silurian rocks; 49 in the Devonian; 44 in the Carboniferous; 14 in the Saliferous; 49 in the Oolitic; 57 in the Cretaceous; 18 in the Tertiary; and quite a number among living molluscs.

Descr. Of the Producti, (Fig. 65,) 22 species (including the Leptænæ,) occur in the Silurian series; 14 in the Devonian; 76 in the Carboniferous; and seven in the Lower New Red, or Permian system.

Descr. Of the Spirifer, (Fig. 66, which shows the spiral appendage within the shell,) 29 species have been found in the Silurian series; 50 in the Devonian; 74 in the Carboniferous; 5 in the New Red; and 6 in the Oolitic.

Descr. Of the Orthis, another brachiopodous shell, that became extinct early, 64 species have been found in the Silurian series; 35 in the Devonian and 15 in the Carboniferous.

Fig. 64.

Terebratula.

Fig. 65. Fig. 66.

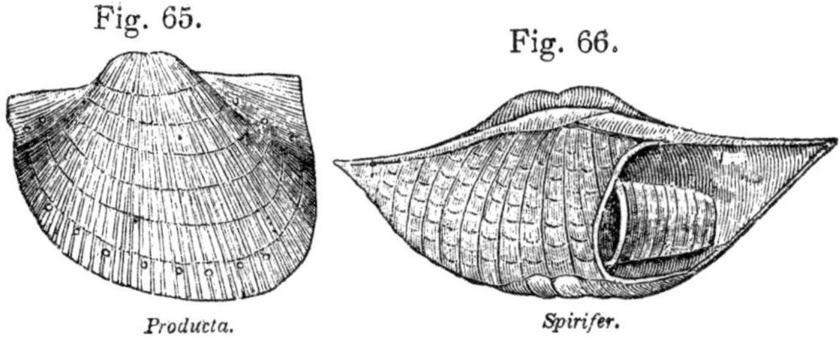

Producta. *Spirifer.*

VERTEBRAL ANIMALS.

Descr. This extensive division of the animal kingdom embraces those animals whose organization is the most perfect, with man at their head. A cranium and vertebral column, which enclose the principal part of the nervous system, and a regular skeleton, covered by muscles, constitute the principal anatomical distinctions between this division and the three that have already been considered. It is divided into four well marked classes: 1, Mammalia. 2, Birds. 3, Reptiles. 4, Fishes.

Fishes.

Rem. Ichthyology, or the history of fishes, has received such great improvements from the labors of Professor Agassiz, as developed in his great work, entitled *Recherches sur les Poissons Fossiles, par L'Agassiz, &c.* that it may almost be regarded as a new science. Especially is this the case in respect to fossil ichthyology.

Descr. The number of living species of fishes now known, amounts to about 8000; and the number of fossil species to more than 1700. *Buckland's Bridgewater Treatise, Vol.* 1. *p.* 267. *Ed. Phil. Journ. Oct.* 1845, *p.* 301. Agassiz estimates the number of species that have lived during all the geological periods, at 30.000.

Descr. Fishes are found in all the great rock formations, from the Silurian upwards:—a fact which is not true of any other class of vertebral animals; and therefore the history of fossil fishes becomes of great importance.

Descr. Agassiz divides fishes into four orders; deriving their characters from the scales: 1. The *Placoidians,* or those whose skin is covered irregularly with plates of enamel (from πλαξ, a *broad plate.*) 2. The *Ganoidians;* (from γανος, *splendor,*) or those having angular scales of horny or bony plates, covered with a thick plate of enamel. 3. The *Ctenoidians,* (κτεις, *a comb,*) or those having jagged or pectinated scales. 4. The *Cycloidians,* (κυκλος, *a circle,*) or those having scales smooth and simple at their margin.

Descr. Three fourths of the existing species of fish belong to the two last orders, whose existence has not been ascertained below the chalk: the remaining fourth belongs to the two first orders: which are not at all numerous now, but existed alone in all the periods during which the fossiliferous rocks below the chalk were deposited. *Agassiz in the American Journal of Science, Vol.* 30. *p.* 39.

Prin. Not one species of fish has yet been found that is common to any two of the great geological formations; or is now living in the ocean.

Prin. Fossil fishes do not change insensibly as we pass vertically from one formation to another, but abruptly; and these changes occur simultaneously with those in other classes of organic remains.

Inf. Hence the conclusions that have been made from the history of other organic remains, are confirmed by this new branch of palæontology.

Descr. Below the chalk not even any genus is found that embraces any living species: those of the carboniferous strata disappear with the deposition of the new red sandstone: those in the oolite, introduced after the epoch of the new red sandstone, suddenly vanish with the appearance of the chalk. Two thirds of those in the chalk and one third of those in the lower tertiary strata, belong to genera no longer existing. *American Journal of Science, Vol.* 30. *p.* 40, 41.

Rem. This constancy in the character of fossil fishes has enabled M. Agassiz to determine the true situation of several groups of rocks in the geological scale, that geologists had been unable to fix: or had put them in the wrong place.

Desc. In the old red sandstone is found a genus of fishes, of singular form. Fig. 67. exhibits a species of this kind.

Fig. 67.

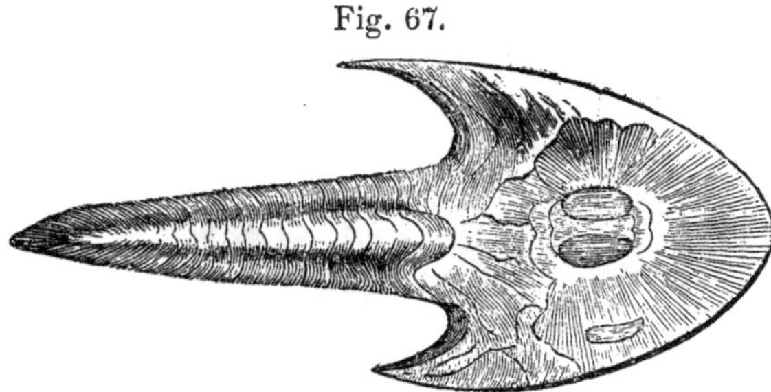

Cephalaspis Lyellii.

Descr. Agassiz has described 17 genera of sauroid fishes, found in all the formations from the carboniferous upward, except the tertiary; but only two genera remain among living fishes: viz. the *Lepidosteus osseus*, or bony pike of North America; of which there exists five species; and the *Calypterus* of two species. Some of these sauroid fishes in the rocks were of enormous size; their teeth being larger than those of the living crocodile.

Sharks.

Descr. These fishes occur in a living state all over the globe; and there seems to have been no period in geological history in which they did not prevail. More than 150 species have been found fossilized. Fig. 68, shows a shark's tooth of enormous size, found in Lenoir County, North Carolina, by Mr. John Limber. Estimating the size of the animal by the rules given by zoologists, its length must have been between 70 and 100 feet.

Fig. 68.

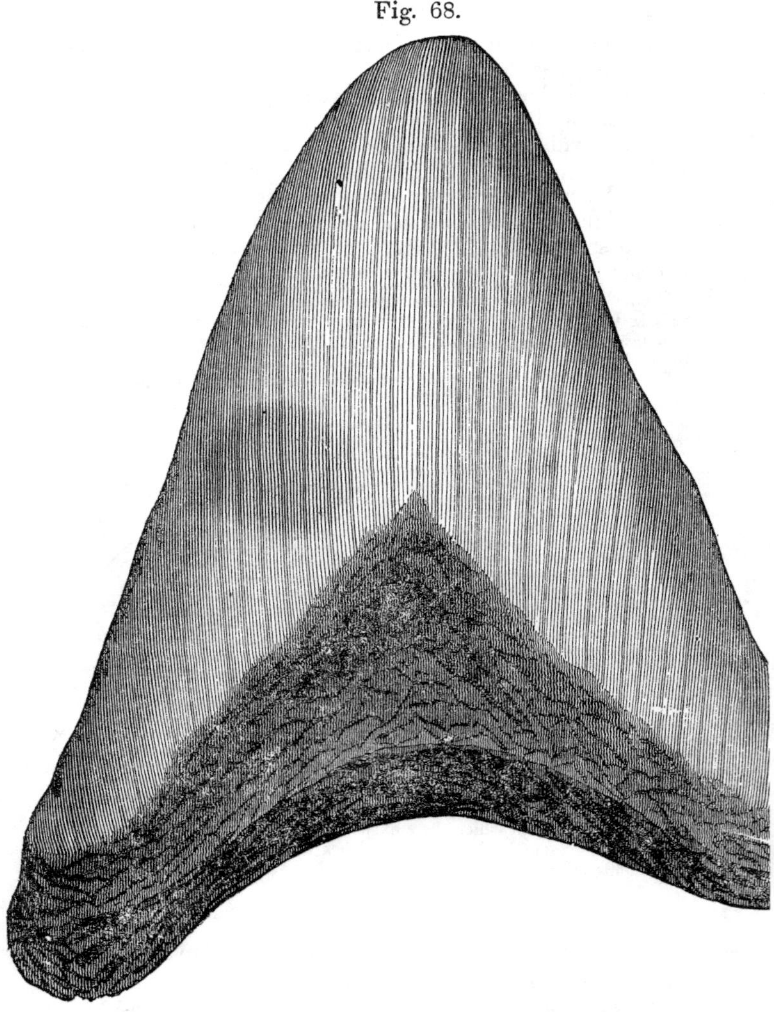

Shark's Tooth, natural size · North Carolina.

Descr. Another singular variety of fish is found in all the strata below the lias, distinguished by their heterocercal or unsymmetrical tails; that is, by tails whose upper lobe extends much the farthest, by the prolongation of the vertebral column. Fig. 69, represents one of these fishes of the genus *Palæoniscus.* Most living fishes have homocercal or equally bilobate tails.

Fig. 69.

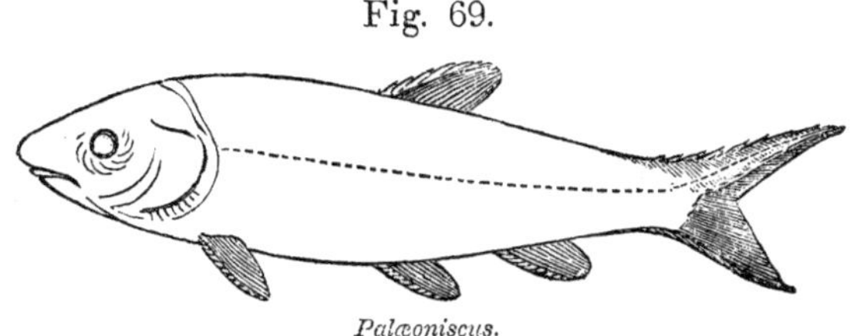

Palæoniscus.

Coprolites.

Descr. The fecal remains of fishes and other animals which have been found frequently in several formations, have been called *Coprolites.* Sometimes they are found detached and sometimes in the body of the animal; and the information derived from them, as to the food, habits, &c. of the animals to which they belong, is of great importance.

Illus. The coprolites of birds have recently been found, in connection with their tracks, in Springfield, Massachusetts. From a careful analysis by Dr. S. L. Dana, they were found to contain uric acid, and in such a proportion as shows not only that they are the droppings of birds, but of a particular kind of birds, viz. the omnivorous tribes, such as produce the guano.

Reptiles.

Descr. Prof. Owen has recently divided the reptiles into eight orders, all of which are found more or less in a fossil state, and three of which, (No. 1, 2, and 5) are exclusively fossil. *Order* 1. *Dinosauria,* or Land Saurians; embracing only three genera of gigantic size: the Megalosaurus, Hylæosaurus, and Iguanodon, found in the wealden and oolite formations. 2. *Enaliosauria,* or Marine Saurians; embracing three genera also, viz. Plesiosaurus, Ichthyosaurus, and Pliosaurus; which, however, contain 28 species even in Great Britain. Found in muschelkalk, lias, oolite, wealden, and chalk. 3. *Crocodilia,* (Crocodiles, &c.) embracing in England eight genera and 14 species, and found in the lias, oolite, wealden, chalk, and tertiary. Among these the largest was the *Cetiosaurus,* which was 60 feet long, and rivalled the modern whale in size. 4. *Lacertilia,* or lizards. Found in magnesian limestone of the Permian system, and the new red sandstone of the Triassic system. It embraces in England eight fossil genera and ten species. The most remarkable of these was the Rhynchosaurus, whose structure approached that of birds. It is doubtful even whether it had teeth, and it probably had a hind toe, extending backwards. 5. *Pterosauria*, or Flying Saurians; embracing only one fossil and no living genus, viz. the Pterodactyle, which occurs in the lias, oolite, and wealden. 6. *Chelonia*, or Tortoises. Found fossil in new red sandstone, oolite, wealden, chalk, and tertiary; and embracing in England six genera and 20 species. 7. *Ophidia*, or Serpents. Found fossil only in the tertiary, where in England only one species has been discovered, which was sometimes 20 feet long. 8. *Batrachia*, or Frogs. Found fossil in the new red sandstone and the tertiary; where occur in England, five species of the Labyrinthodon. The whole number of fossil reptiles found in Great Britain alone, and described in the admirable Report of Professor Owen on British Fossil Reptiles, is 85. *See Reports of the British Association for* 1839 *and* 1841. *Also Ansted's Geology, Vol.* 1. *p.* 303. Professor Bronn, in his Lethea Geognostica, described in 1837 one genus and one species of reptiles in the Permian system; eight genera and 10 species in the Triassic system; 18 genera and 24 species in the oolite; 4 genera and 5 species in the chalk, and 20 genera and 35 species in the tertiary: in all, 51 genera and 75 species: but how many of these are included in Prof. Owen's Report, I have not ascertained.

Rem. I shall now describe more particularly a few of the most remarkable of the fossil reptiles.

Ichthyosaurus.

Descr. This animal, sometimes more than 30 feet long, and of which 10 species are known, had the snout of a porpoise, the teeth of a crocodile, (sometimes amounting to 180,) the head of a lizard, the vertebræ of a fish, the sternum of an ornithorhynchus, and the paddles of a whale: uniting in itself a combination of mechanical contrivances which are now found among three distinct classes of the animal kingdom. One of its paddles was sometimes composed of more than 100 bones; which gave it great elasticity and power, and enabled the animal to urge its way through the water with a rapid motion. Its vertebræ were more than 100. Its eye was enormously large: in one species, the orbital cavity being 14 inches in its longest direction. This eye also, had a peculiar construction to make it operate both like a telescope and a microscope: thus enabling the animal to descry its prey in the night as well as day, and great depths in the water. The length of the jaws was some- ıes more than 6 feet. Its skin was naked, some of it having en found fossil: its habits were carnivorous, its food, fishes d the young of its own species; some of which it must have 'allowed several feet in length. This fish-like lizard was an habitant of the ocean. Fig. 70, exhibits a restored ichthyo- urus, by Mr. Hawkins. *Memoirs of Ichthyosauri and Plesio- ıuri, extinct Monsters of the ancient earth. By Thomas Haw ns, Esq. Folio, with 28 plates.*

Fig. 70.

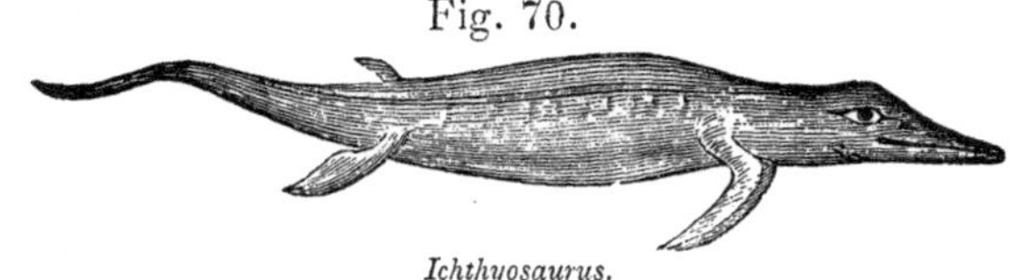

Ichthyosaurus.

Plesiosaurus.

Descr. This animal, of which 16 species have been found, ıas the general structure of the ichthyosaurus. Its most re- narkable difference is the great length of the neck, which has from 20 to 40 vertebræ; a larger number than in any known animal: those of living reptiles varying from 3 to 6, and those of birds from 9 to 23.

The largest perfect specimen yet found is 11 feet long; with about 90 vertebræ. Its paddles were proportionally larger than in the ichthyosauri. It was carnivorous; an inhabitant of the ocean, or rather of bays and estuaries: where it probably used its long neck for seizing fish beneath, and perhaps flying reptiles, above the waters. Fig. 71, exhibits a restoration of one of the most remarkable species, the *P. dolichodeirus,* by Mr. Hawkins.

Fig. 71.

Plesiosaurus.

Rem. According to Dr. Harlan's Medical and Physical Researches, ichthyosauri and plesiosauri have been found in the secondary rocks of the United States. *Mining Review for January,* 1839, *p.* 10.

Mosasaurus.

Descr. Up to the time of the deposition of the chalk, the ichthyosaurus and plesiosaurus appear to have ruled in the ocean: but then they disappeared, and the mosasaurus to their place, to keep the multiplication of the species of ot animals within proper limits. It was most nearly related in structure to the monitor, a species of lizard now living. Wh the head of the largest monitor does not exceed five inches length, that of the mosasaurus is 4 feet long; and the wh animal 25 feet in length: while the monitor is only 5 feet length. It had paddles instead of legs; and the number of vertebræ was 133.

Megalosaurus.

Descr. This name (meaning a *great saurian,*) has be given by Dr. Buckland to a gigantic terrestrial reptile, 30 fe long, allied to the crocodile and monitor in structure, and four in the oolite. The animal was carnivorous; and in the stru ture of its teeth are combined the knife, the saw, and the sabr Its principal food was probably crocodiles and tortoises. It ha a Dinosaurian companion, called the Hylæosaurus, about fiftee feet long.

Iguanodon.

Rem. For our knowledge of this most gigantic of all the reptiles of former world, we are indebted to the industry and scientific acumen of D Mantell, who found its bones along with those of the megalosaurus, hylæ saurus, plesiosaurus, crocodile, &c. in the wealden rocks in England; a fres water, or rather an estuary formation, extending over more than 1000 squar miles. It must once have formed the delta of a large river, which has dis appeared, as well as the country from which it originated. *Wonders of Geo logy, Vol.* 1. *p.* 368.

Descr. This animal approaches nearest in its structure, es pecially that of the teeth, to the living iguana: a reptile of the warmer parts of this continent; and hence its name; signify ing an animal with teeth like the iguana. Its average length was about 30 feet: circumference of the body, 14.5 feet: length of the hind foot, 6.5 feet: circumference of the thigh, more than 7 feet! The form of the teeth shows it to have been herbivorous, like the living iguana. It had a horn 4 inches long upon the snout, like some species of iguana. Fig. 72, will give some idea of the iguanodon.*

* In looking at Mr. Martyn's thrilling picture, that forms the Frontispiece to Dr. Mantell's Wonders of Geology, entitled *The country of the Iguanodon Restored,* one cannot but be reminded of Milton's graphic description of Satan:

"With head uplift above the waves and eyes
That sparkling blazed, his other parts besides
Prone on the flood, extended long and large
Lay floating many a rood, in bulk as huge
As whom the fables name of monstrous size.

Fig. 72.

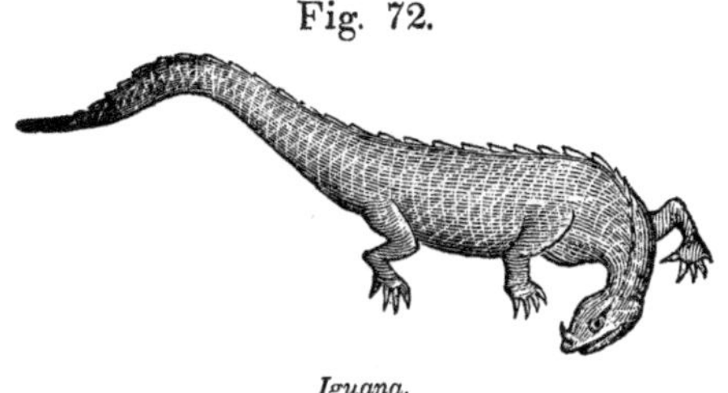

Iguana.

Pterodactyle.

Rem. Of all the fossil animals probably this is the most heteroclitic, and at first view, monstrous. Hence anatomists were unable for a long time to refer it to its true place among animals; some pronouncing it a bird, some a reptile, and some a bat. But Cuvier at last developed its true characters, and proved it to be a beautiful example of the wisdom that adapts creatures to peculiar and varied modes of existence.

Descr. This animal had the head and neck of a bird, the mouth of a reptile, the wings of a bat, and the body and tail of a mammifer. Its teeth, as well as other parts of its structure, show that it could not have been a bird; and its osteological characters separate it from the tribe of bats. But in many respects it had the characters of a reptile. The outer toe of its fore feet was enormously elongated, to furnish support, it is probable, for a membranous wing. By this means these animals were doubtless able to fly like the bat; while the fingers with claws projecting from their wings enabled them to creep or climb. When their wings were folded, they could probably walk on two feet; and it is most likely, also, they could swim. Their eyes were enormously large; so that they could seek their prey in the night. They probably fed on insects chiefly; though perhaps, also, they had the power of diving for fish. Eight species, from the size of a snipe to that of a cormorant, have been found in the oolite and lias in England, and on the continent of Europe at Solenhofen. Fig. 73, shows several of these animals restored.

Titanian, or earth born, that warred on Jove.
Briareos or Typhon, whom the den
By ancient Tarsus held, or that sea beast
Leviathan, which God of all his works.
Created hugest that swim the ocean stream.

Paradise Lost, *Book* 1. *line* 198.

Fig. 73

Pterodactyle.

Rem. "Thus," says Dr. Buckland, "like Milton's fiend, all qualified fo all services, and all elements, the pterodactyle was a fit companion fo. th kindred reptiles that swarmed in the seas, or crawled on the shores of a tur bulent planet.

"The Fiend,
O'er bog, or steep, through straight, rough, dense, or rare,
With head, hands, wings, or feet, pursues his way,
And swims, or sinks, or wades, or creeps, or flies."
Paradise Lost, Book 2. *line* 947.

"With flocks of such-like creatures flying in the air, and shoals of no less monstrous ichthyosauri and plesiosauri swarming in the ocean, and gigantic crocodiles and tortoises crawling on the shores of the primeval lakes and rivers: air, sea, and land must have been strangely tenanted in these early periods of our infant world." *Bridgewater Treatise, Vol.* 1. *p.* 224.

Crocodiles and Tortoises.

Descr. Of twelve species of the crocodile family now living, three are alligators, eight true crocodiles, with a short and broad snout, and one, the gavial, with a long and narrow beak, adapted for feeding on fish. Of the six or more fossil species, all those in the secondary rocks approach in structure to the gavial; probably because few other animals then existed but fish on which they could feed. These gavial saurians are called *teleosauri and steneosauri.* Of the former, one has been found 18 feet long, with 140 teeth. The fossil crocodiles with short snouts, differ so little from existing species, as to need no description.

Descr. Tortoises, both marine, fresh water, and terrestrial, have been found fossil in all the formations above the coal formation. One of these in the muschelkalk, a sea turtle, was 8 feet long; another in India was 20 feet across. *Ansted's Geology, Vol.* 2. *p.* 99. Numerous tracks of tortoises have also been found on the new red sandstone in Scotland: but these will be described in a subsequent part of this section.

Birds.

Descr. Twenty species of birds have been found in diluvium; 28 in the tertiary strata, one in the chalk, and recently, one species of Grallæ, or waders, in the Wealden formation; and these are all the fossil relics of this order of animals yet

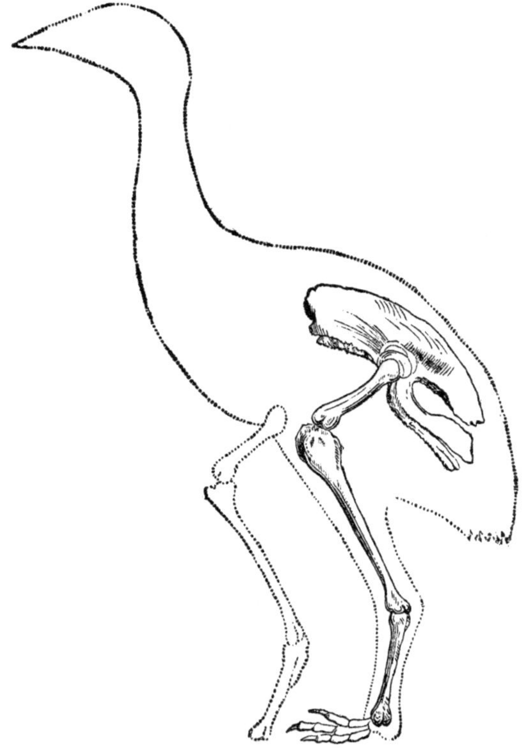

Dinornis Giganteus.

discovered. The tracks of more than 30 different species animals have, however, recently been brought to light in tl valley of Connecticut river, in Massachusetts and Connectic on what is thought to be the equivalent of European new r sandstone; and a large part of these resemble the impressio of bird's feet more than those of any other known animal But these will be more particularly noticed farther on.

Descr. The most remarkable discovery in zoology durin the present century, is that by Prof. Owen in relation to th Dinornis, an enormous species of bird that once inhabited Ne Zealand. Having received bones, collected by missionarie from alluvium, he has been enabled to describe six species th must have lived within a few hundred years in that islan They varied in size from that of the great bustard, three o four feet high, to a bird ten or twelve feet high, called the *Din ornis giganteus*; an outline sketch of which is given on p. 143 as restored by Professor Owen. *Owen's Paper on the Dinorni in Zoological Transact., Vol.* 35. *Part* 3. *p.* 235.

MAMMALIA.

Descr. There is reason to believe that the Marsupial Mam malia appeared earlier on the globe than any other animals of this class. For Dr. Buckland has found two undoubted species of the marsupials in the oolite of England.

Descr. With the above exception, all the other fossil mammalia occur in the tertiary strata and drift. In the eocene tertiary, as many as 50 species have come to light. Cuvier described 78 species in his great work, the *Ossemens Fossiles;* 49 of which are extinct: and since that time the number has been increased to nearly 200.

Descr. Among these fossil animals, are many of existing genera, and so nearly related to existing species, that a particular description will be unnecessary. Such are the fossil species of the rhinoceros, hippopotamus, hog, cat, glutton, horse, ox, deer, bear, hyena, weasel, hare, rabbit, rat, mouse, &c. In general, however, the fossil species are of a larger size than those now living; indicating a warmer climate when they were upon the globe.

Descr. The marine fossil mammalia, such as the whale, the dolphin, the seal, the walrus, and lamantin, occur as we should expect, in tertiary strata deposited in the ocean; and some of the terrestrial mammalia are found mixed with marine animals in salt water formations; having been drifted into the ocean by rivers. Other terrestrial mammalia are found in fresh water rmations, deposited at the bottom of lakes and ponds during e tertiary epoch. Others occur in caverns and fissures, which cisted in the dry land during the same period: and finally, milar remains are found dispersed through the diluvium, hich is spread over formations of every age.

Descr. At the meeting of the Association of American Geologists in 1841, r. Harlan stated that the continuous vertebræ of a cetaceous animal, had een found in Alabama, as much as 100 feet long. Subsequently this animal as named *Zeuglodon cetoides* by Prof. Owen, who describes it "as one of e most extraordinary of mammalia, which the revolutions of the globe have lotted out of the number of existing beings." A supposititious specimen of is animal has been exhibited of late in the cities of this country, 114 feet ng, and made up of bones collected at different localities. *Amer. Quart. ourn. Agriculture, April,* 1846, *p.* 223.

Descr. The history of bone caverns and fissures, as decribed by Dr. Buckland in his splendid work entitled *Reliquiæ Diluvianæ,* deserves a more extended notice than can here be given. From a careful examination of these osseous caverns, by Dr. Buckland, it appears that some of them, as that of Kirkdale, Kent's Hole, &c. in England, were the residence of hyenas, for a long time previous to the historic period; and that these animals dragged in thither great quantities of bones of other animals on which they fed. This is proved by the broken and gnawed state of the bones, and by the great quantity of coprolites belonging to the hyena found in the caverns. Other caverns appear to have been the abodes of bears for a long period, as those of the limestones of central Germany. In one of these, the cave of Kuhlok, more than 500 cubic feet of black animal dust, impressively denominated the *dust of death,* were found resulting from the decomposition of bears; which must have required at least 2500 of these animals! The bones of the osseous breccia, found in fissures in Somersetshire in England, and on the northern shores of the Mediterranean, belong mostly to animals that fell into fissures that were afterwards filled with detritus. *Reliquiæ Diluvianæ, p.* 137, *and* 148. *London,* 1823. *De la Beche's Manual of Geology, p.* 181.

Descr. Sometimes ossiferous caverns have been used by man as a place of habitation, or more frequently as a place of sepulture. And hence his bones, as well as fragments of pottery, and other relics of a rude people, sometimes are found so mixed with the remains of extinct animals, as to lead to the inference that they were deposited during the same period. Indeed, in some of these caverns in the south of France, it is still believed by some geologists, that the remains of men were of contemporaneous deposition with those of the extinct mammalia. The English geologists appear to be decidedly of tl opposite opinion.

Descr. Osseous breccias, have been found in Australasia containing bones of the kangaroo, wombat, dasyurus, koal halmaturus, elephant, hypsiprimnus, &c. most of which are an mals living in that country.

Descr. In the United States a few ossiferous caverns have been foun though none of them exactly corresponding with the European cavern Some of them appear to have been filled with fragments of rocks and bone which were partially cemented by stalagmite, during the alluvial or historic period. An example of this kind has been communicated to me by the lat Prof. L. F. Clarke, of the college of East Tennessee, in a letter dated Aug 23d, 1838. The cavern is situated in the town of Chittenden, Vt. from 200 to 3000 feet above Otter Creek, and in limestone interstratified with mic slate, having an easterly dip of about 65°. The bottom of the cavern has westerly slope, as shown in Fig. 74, which is an imperfect sketch fron memory.

Fig. 74.

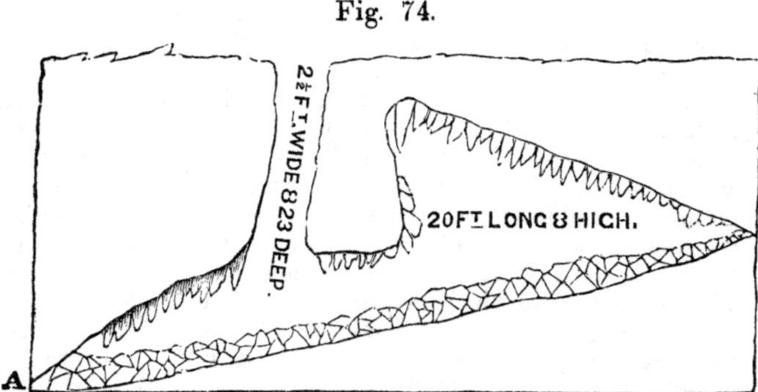

Bone Cavern: Chittenden, Vermont.

The roof and sides of the cavern are covered with large stalactites, and the bottom with loose fragments of rocks, in many places to the depth of at least 5 feet. The fragments at the surface are mostly cemented together by stalagmite, but rarely at much depth. The bones are scattered among the fragments, and are much broken, and many of them gnawed by the teeth of some small animal. These bones, so far as they have been anatomically examined, appear to belong to existing animals; among which those of the bat were very abundant; and it is supposed that some of them belong to the wolf, racoon, &c. They appear very fresh, still retaining most of their animal matter, much more than the bones from the English caverns. From all the facts I infer, that this cavern was produced by running water which probably had an outlet at A, and that the fragments of stone were thus accumulated, while the stalagmite was forming and the bones were being collected, both by carnivorous animals, and by the death of others that might have made this cavern their habitation, during the alluvial or historical period.

Early Pachydermata.

Descr. In the older tertiary strata around the city of Paris, Baron Cuvier has brought to light more than 40 extinct quadrupeds, many of them allied to the modern pachydermata, or thick-skinned animals. Fig. 75, exhibits the form of the *Anoplotherium gracile*, which was of the size and form of the gazelle, living like the deer and the antelope. Fig. 76 is the *Anoplotherium commune;* an animal about the size of the wild boar, with the means of swimming with facility. Four other species have been described. Fig. 77, is the *Palæotherium magnum*, of the size of the horse, but more thick and clumsy. Probably it had a trunk. Fig. 78, is the *Palæotherium minus*, of the size of the roebuck. Ten other species have been described

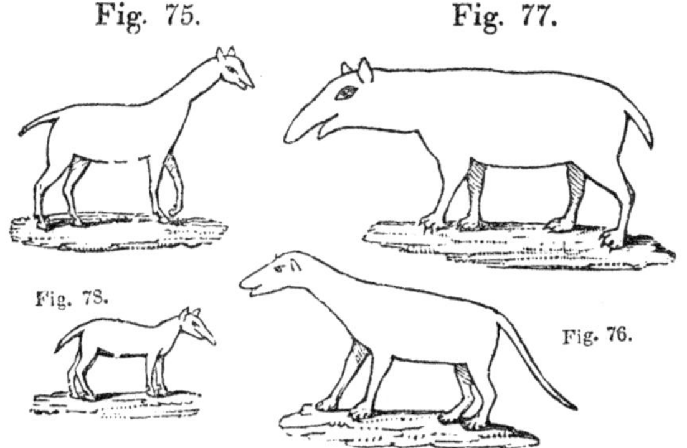

75 *Anoplotherium gracile:* 76 *Anoplotherium commune:* 77 *Palæotherium magnum:* 78 *P. minus.*

Mammoth, or Fossil Elephant.

Descr. Two species of living elephant are known; the Asiatic, or Indian, which extends only to the 31st degree of north latitude: and the African, which occurs as far south as the Cape of Good Hope. A third species is found in a fossil state; especially in the northern parts of Asia and Europe; as well as America. It is this extinct species that goes by the name of *mammoth*,—an Arabic word (*behemoth*,) signifying elephant. Fig. 79, exhibits the skeleton of the remarkable specimen found encased in frozen mud on the shores of the arctic ocean, in Siberia, with its softer parts preserved, as has been already described. This skeleton is now deposited in the Museum of Natural History in St. Petersburgh. Its length is 16 feet; and its height 9 feet. Its hair, of which 30 pounds were preserved, consisted of black bristles, 15 inches long, mixed with wool of a reddish brown color.

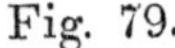

Fig. 79.

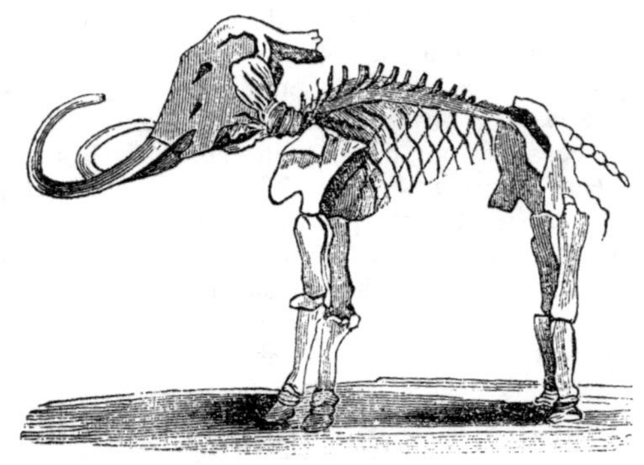

Mammoth.

Mastodon.

Descr. Although the mastodon is frequently called the mammoth in this country, where the remains of the largest species are abundant, yet it differs generically from the elephant in the form of its teeth; as may be seen in Figs. 80 and 81, below Fig. 80 is a side view of the grinder of the mastodon, and Fig 81, represents the flat surface of the mammoth's grinder.

Fig. 80. Fig. 81.

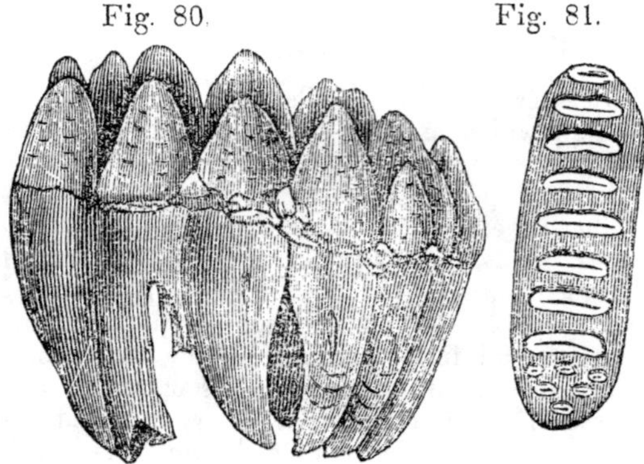

Descr. No less than seven species of mastodon have been discovered in a fossil state; (13 species according to Prof. Grant: *Murchison's An. Ad. for* 1843, *p.* 110;) viz. three in Europe, two in South America, one in the United States, and one in India. The largest species, *M. maximus* has been found in almost every part of the United States; though most abundantly in the *salt licks* of Kentucky, Ohio, &c. The most easterly point where the bones of these animals have been found, is Berlin in Connecticut. An almost entire skeleton has been put up in the Museum of Mr. Peale in Philadelphia, which is 15 feet long, and eleven feet high. This was found in Orange County, New York. The most remarkable locality in this country is at the Big Bone Lick in Kentucky, where a vast number of bones of various extinct animals are imbedded in dark colored mud and gravel, which appears to have been formerly the bottom of a marsh. This spot has been examined by William Cooper, Esq., with his usual discrimination and accuracy; and he is of opinion that the deposit containing the bones is to be regarded as drift. He estimates that the bones of 100 mastodons, 20 elephants, 2 oxen, 2 deer, and one megalonyx, have already been carried from this spot.

Descr. In 1845 the skeleton of a mastodon, almost entire, was dug up in New Jersey, with skulls and other bones of four or five others. These have been purchased for Harvard College. A more gigantic and perfect skeleton was found the same year in Orange County, New York, in a peat bog, with marl beneath; where it stood in an erect position, as if the animal lost its life by getting *mired* in search of food. In the place where its stomach and intestines lay, was found a large mass of fragments of twigs and grass, hardly fossilized at all. The sketch below represents this skeleton, which weighed 2000 pounds, and is perhaps the most perfect and gigantic ever found. It is now owned by Dr. John C. Warren of Boston. *Am. Quart. Jour. of Agricul. & Sci. Vol.* 2, *No.* 2, *July*, 1845, *p.* 203.

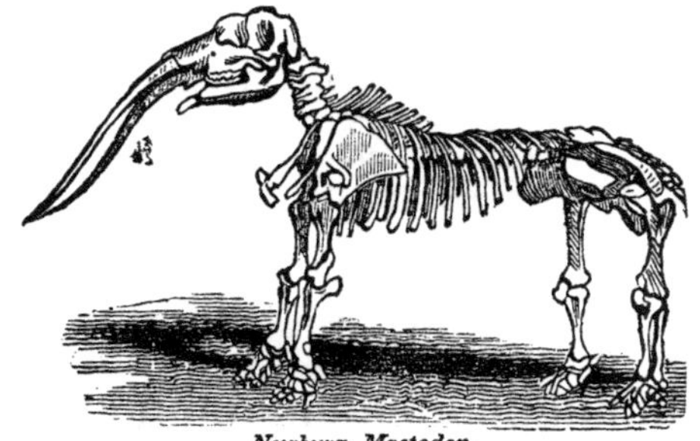

Newburg Mastodon.

Inf. This skeleton was found, as indeed most others probably have been, in an alluvial formation, and the question whether it was antediluvian or postdiluvian is a natural one, but not easily answered. Those familiar with tertiary fossils, know that wood and leaves are sometimes preserved almost unchanged, even as low down as the Eocene; so that the apparent freshness of the bones and the contents of the stomach, does not settle the question. The grand objection to the animals having lived within a few hundred years, is the difficulty of conceiving how so enormous a creature could have sustained himself through our winters. Yet it ought to be recollected, that the Moose,—often nearly as large as a horse,—does thus live by browsing in the forests of Maine. It would seem, however, more probable, that the Mastodon lived when the climate was warmer; since the allied huge animals do live in warmer latitudes. Perhaps the change of climate destroyed the Mastodon.

Rhinoceros, Hippopotamus, Hyena, Horse, Ox, Deer, Sivatherium, Monkey, Camel, &c.

Descr. Most of these animals in their fossil state, differ so little from the existing species, that they need not be particularly described in this work. They are generally, however, of larger size than the living species. The rhinoceros found undecayed in the frozen gravel of Siberia, has already been noticed; and several other species of this animal occur in Europe and in India, associated with the bones of the elephant, also with several species of hippopotamus, and one or two of oxen, aurochs and deer. The horns of the fossil ox are sometimes very large; in one example 31 inches long. So also the famous fossil elk of Ireland, (*Cervus giganteus*,) had horns that measured 10 feet 10 inches between their tips. *Jameson's Cuvier's Theory of the Earth, p.* 246. The most interesting remains of the hyena are those found in caverns. (See *Reliquiæ Diluvianæ.*) The *sivatherium* is an extinct animal recently found in India, in concretionary drift, larger than the rhinoceros, furnished with four horns and a proboscis, and forming an intermediate link between the ruminantia and pachydermata. In the same deposit were found the remains of a gigantic species of monkey and of a camel. Another species of monkey has also been discovered in tertiary deposits in France: so that the important fact seems now well established, that the animals approaching nearest to man in their structure, have been found in a fossil state. *Wonders of Geology, Vol.* 1. *p.* 138 *and* 226.

Megatherium.

Descr. Sloth and Armadillo. Because the Sloth (*Bradypus*

tardigradus) is not adapted for walking on the ground, some writers, and even some naturalists, (Buffon, &c.) have ridiculed its structure, as if it indicated a want of wisdom. But it now appears that the animal was intended to live continually upon trees; and that its long forelegs, with long and very crooked claws, are admirably adapted for this purpose. The armadillo, as is well known, is covered with a bony armor for defence against enemies, dust, &c. The few living species of this animal are small and confined chiefly to South America, where they burrow like the woodchuck.

Descr. The Megatherium is an enormous extinct animal, which was once abundant in the vast plains or pampas of the same continent. They have been lately found by Mr. Darwin over an extent of 600 miles, accompanied with bones and teeth of five other quadrupeds, some of them of a similar construction. Dr. Mitchell and Mr. Cooper have also described bones of this animal from the island of Skiddaway, on the coast of Georgia. *Buckland's Bridgewater Treatise, Vol.* 2. *p.* 20. *Annals of N. Y. Lyceum, May,* 1824. It was larger than the rhinoceros, and its proportions were perfectly colossal. With a head and neck like those of the sloth, its legs and feet exhibit the character of an armadillo, and the ant-eater. Its body was 12 feet long, and 8 feet high. Its forefeet were a yard in length, and more than 12 inches wide; terminated by gigantic claws. Across its haunches it measured five feet; and its thigh bone was nearly three times as thick as that of the elephant. Its spinal marrow must have been a foot in diameter; and its tail, at the part nearest the body, twice as large, or six feet in circumference! Its teeth were admirably adapted for cutting vegetable substances; and its general structure and strength seem intended to fit it for digging in the ground for roots, on which it principally fed. Fig. 82, exhibits the entire skeleton of this animal, which exists in the museum at Madrid, in Spain. *Buckland's Bridgewater Treatise, Vol.* 1. *p.* 139.

Rem. 1. It has been generally supposed that the megatherium was covered with a bony armor like the armadillo: but Mr. Owen, the distinguished comparative anatomist, is of opinion, after a careful examination, that such was not the case; and that this animal approached more nearly in character to the ant-eaters and sloths, than to the armadillo. *Lond. and Edin. Philos. Mag. July,* 1839. He has lately ascertained that this bony armor belonged to another extinct animal called the *Glyptodon. An. and Mag. Nat. Hist. Feb.* 1841, *p.* 492.

Rem. 2. Dr. Harlan has recently described the remains of a huge extinct animal, allied to the above, from Missouri, under the name of *Orycterotherium Missouriensis:* and Dr. Henry C. Perkins has described, in the Boston Journal of Natural History, another species from Oregon Territory, as the *Oryct. Oregonensis.*

Fig. 82.

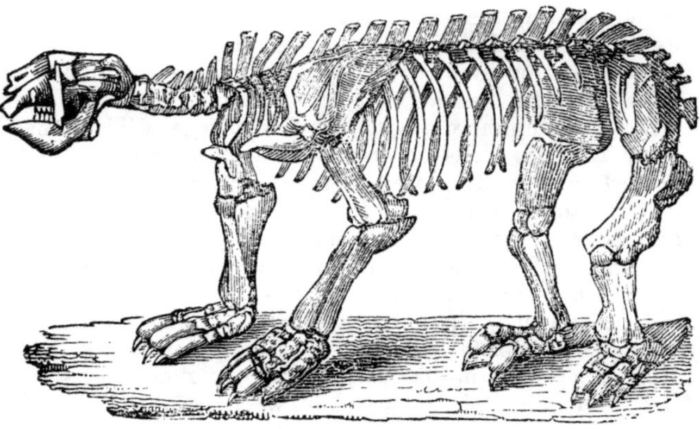

Megatherium.

Descr. In the superficial deposits of South America, several interesting extinct animals have been found, belonging mostly to the Pachydermata, or thick-skinned, and the Edentata. The Toxodon, which had a skull 28 inches in length, approximates in its structure to several families of animals, viz. the Rodentia, the Ruminantia, and Cetacea; although, in fact, a Pachyderm. The Macrauchenia greatly resembled the llama, and had a neck almost as long as that of the giraffe, with a body nearly as large as that of the rhinoceros. This also was a Pachyderm. The Mylodon, an Edentate animal, was of massive and singular proportions. Its body was shorter than that of the hippopotamus, but was terminated by a pelvis as broad as that of the elephant and deeper, resting on two massive but short hind legs, with feet as long as the thigh bones. The tail, as long as the legs, and very thick and strong, was probably used like that of the kangaroo, to support the body when the animal raised up its anterior extremities. It is supposed by Mr. Owen, that the peculiar structure of this animal adapted it, first for digging around trees, and then, resting upon the tripod base of its hind legs and tail, it seized the trunk with its fore legs, and rocked it to and fro, until it was prostrated, and its leaves furnished food for several days, perhaps. The following sketch will give a good idea of this animal.

Mylodon robustus.

Descr. The Scelidotherium was an analogous animal not larger than some of the existing ant-eaters of South America, but with excessively large hind legs. The Glyptodon was covered with armor like the armadillo, and was of great size. These animals are all called Megatheroids, because they resemble the Megatherium.

Megalonyx.

Descr. This animal was first described by Mr. Jefferson, who mistook its characters. It was found in the nitre caverns of Virginia and Kentucky, and has since been discovered in other places. It was of the size of the ox, and appears to have been nearly related to the sloth.

Dinotherium.

Descr. Until recently the mammoth and the mastodon have been supposed the largest of all the terrestial mammalia that ever inhabited the earth; but they must give place to the Dinotherium described by Cuvier as a gigantic tapir, but recently by Professor Kaup, a distinguished German naturalist, as a new genus between the tapir and the mastodon; and adapted to that lacustrine condition of the earth which seems to have been so common during the deposition of the tertiary strata. Its remains have been found in tertiary strata, in the south of France, in Austria, Bavaria, and especially in Hesse Darmstadt. Its length must have been as much as 18 feet. One of its most remarkable peculiarities consisted in two enormous tusks, at the anterior extremity of the lower jaw, which curved downwards, like those of the walrus. Its general structure seems to have been adapted to digging in the ground; and for this purpose its feet as well as tusks, projecting a foot or two beyond the jaws which were four feet long, were intended. It lived principally in the water like the hippopotamus; and it probably used its tusks for tearing up the roots of aquatic vegetables, which, as is shown by its teeth, constituted its food. Dr. Buckland suggests also, that these tusks might have been useful as an anchor fastened into the bank of a river, while the body of the animal floated in the water and slept. They might have been useful also, to aid in dragging the body out of the water and for defence.

Fig. 83, is a sketch of the *Dinotherium giganteum* as restored by Prof. Kaup. *Buckland's Bridgewater Treatise, Vol.* 1. *p.* 135. Another species the D. Bavaricum, has been discovered.

Fig. 83.

Dinotherium.

Ichnolithology: or the History of Fossil Footmarks: (*Ichnology, Buckland.*)

Descr. This singular branch of palæontology has but lately begun to attract the attention of geologists; since it is only within a few years that genuine examples of the tracks of animals in stone have been found. It is, indeed, and long has been the common belief, that such impressions are frequent; but the geologist usually finds that they are merely the effects of disintegration or aqueous action, by which the softer parts of the rock are more worn away than the harder parts. The following are all the well authenticated examples of fossil foot marks that have been discovered up to the present time.

Exam. 1. In the Transactions of the Royal Society of Edinburgh for 1828, Dr. Duncan has given an account, with drawings, of the tracks of a quadruped on new red sandstone in the quarry of Corn Cockle Muir, in Dumfries-shire, Scotland. These tracks have been found there in great abundance, on many successive layers of the stone, to the depth of 45 feet; or as low as the quarry had been opened. They occur also in another quarry, 10 miles south of Corn Cockle Muir, where one series of tracks extended from 20 to 30 feet. Dr. Buckland seems to have satisfactorily shown that they were made by tortoises. *Bridgewater Treatise, Vol.* 1. *p.* 259.

Exam. 2. In 1831, Mr. G. P. Scrope found numerous footmarks of small animals on the layers of forest marble, north of Bath in England. They occur along with ripple marks, and were probably made by crustacea, crawling along the bottom of an estuary. The impression of the tail or the stomach is sometimes visible between the rows of Tracks. *Journal of the Royal Institution of London,* 1831, *p.* 538.

Exam. 3. In 1834, an account was published in Europe of some remarkable fossil footmarks on the *gres bigarre* (new red sandstone,) at Hessberg, near Hildberghausen in Saxony. Accounts of these impressions have been given by Drs. Hohnbaum and Sickler, Professor Kaup, M. Link, and Baron Humboldt. The largest track appears to have been made by a marsupial animal, whose hind foot was 8 inches long. This animal Prof. Kaup has named *chirotherium*, from the resemblance of its track to a human hand. Some of the tracks appear to have been made by tortoises; and M. Link, who has made out four distinct species from these tracks, suggests that some of them may have been made by gigantic batrachians, (frogs, salamanders, &c.) *Buckland's Bridgewater Treatise, Vol.* 1. *p.* 263, *and Vol.* 2. *p.* 36. *Wonders of Geology, Vol.* 2. *p.* 423. *Am. Journal of Science, Vol. XXX. p.* 191. Fig. 84, shows a few tracks of the chirotherium on a slab of sandstone from Hessberg. (*See specimens in the Cabinet at Amherst College.*)

Fig. 84.

Rem. Mr. Owen, with great sagacity, has rendered it probable that the tracks referred to the chirotherium, were made by a gigantic batrachian, or frog, whose hind feet were much larger than his fore feet. He has given a sketch of the animal restored, (the bones of the head only have been discovered,) and of the manner in which the tracks might have been produced. Fig. 85 is a copy

Fig. 88.

Labyrinthodon pachygnathus, Owen.

Descr. 4. In 1835, a few very distinct tracks resembling those of birds, were discovered in the red sandstone of the Connecticut valley, at Montague in Massachusetts. Subsequent examination discovered similar impressions at several quarries in the same valley and state: and an account of seven species was given in the American Journal of Science for January, 1836, under the name of *Ornithichnites*, or *stony bird tracks*. Some of them are very small, the toes not being more than half an inch long; and the length of the step not more than 3 or 4 inches: while others are of enormous size: the foot being 17 inches long, including a claw of 2 inches; and the length of the step from 4 to 6 feet. In another species, if we include a singular impression behind the toes, that appears to have been made by a large heel, the whole length of the track is 2 feet, and of the step, 6 feet.

Descr. Since 1836 great progress has been made in the knowledge of fossil footmarks. In 1838, and 1840, tracks of chirotheria, tortoises, and saurian reptiles, to the amount of 6 or 7 species, were found in the new red sandstone near Liverpool; also in that city, and near Tarpoly. In 1839, Dr. O. Ward described tracks from Grinshill Hill, Shropshire, which had three toes, with claws, directed forward, and one behind; which Prof. Owen supposes were made by a singular reptile called the Rhynchosaurus, whose bones were found in the same quarries. In 1842, Mr. Hawkshaw found tracks of chirotheria and crustaceans, with others resembling those of birds, at Lymm, in Cheshire. *Reports of Brit. Assoc. for* 1840, *p.* 99; *for* 1839, *p.* 75; and *for* 1842, *p.* 57. In 1839 Dr. Cotta described some tracks in new red sandstone in Saxony, that are two toed, or rather like a horse shoe: and M. Feldman has described others near Jena, that are tridigitate, or three toed, and disposed parallel to one another. *Geologist for Jan.* 1845, *p.* 18. In 1841, Dr. Buckland described the tracks of deer and large oxen upon mud, beneath a bed of peat, in Pembrokeshire, which must have been made centuries ago, and which show how tracks may be preserved on mud unconsolidated for a period indefinitely long. *Buckland's An. Add. before Lond. Geol. Soc.* 1840, p. 44. The same gentleman, in 1843, gave an account of certain "*Ichthyopòdolites*, or petrified trackways of ambulatory fishes upon sandstone of the coal formation" in Flintshire, England. *Phil. Mag. March*, 1844, *p.* 230. About the same period, Mr. Logan and Mr. Dawson found tracks, some with three and some with five toes, analogous to those of birds and reptiles, in the lower carboniferous series in Nova Scotia. *Lyell's Travels in N. America, Vol.* 2, *p.* 178. At a session of the Geological Society of France, Nov. 1843, M. Degenharett announced the existence of impressions of the feet of great birds, in a red sandstone 5000 feet above the ocean, in a hill called Cuchilla de las pesunnas del venado, (literally, *the hill of deer's tracks,*) in Colombia, S. America. *Bulletin of the Society, p.* 13. In 1846, Mr. Cunningham detected the tracks of what he considers small Grallæ, in the Storeton quarries, near Liverpool. *Jour. Geol. Soc Vol.* 2. *p.* 410. In 1844, Dr. Alfred T. King discovered in the lower part of the carboniferous series of western Pennsylvania, 800 feet below the top of the coal formation, the tracks, probably, of a large Batrachian. *Amer. Jour. of Sci., April*, 1845, *p.* 348. *See specimen in Amherst College.*

Descr. Since the author of this work first published an account of the Ornithichnites in 1836, numerous other facts have been discovered in relation to the footmarks along Connecticut river. He is now acquainted with more than 40 species which he supposes were made by as many species of animals. In his Report on the Geology of Massachusetts, in 1841, he figured and described 27 species, which he divided into *Ornithoidichnites*, or tracks resembling those of birds; and *Sauroidichnites*, or tracks resembling those of Saurians. Nine species more have since been described in the Transactions of the Association of American Geologists, and in the American Journal of Science. Several species have more recently been discovered; and on a careful revision of the whole subject for this work, more than 40 species appear to be well defined and established. Five of these are quadrupeds, and probably more; and the rest bipeds, with the exception of two species of Annelids, which have left only a trackway or furrow, and three species of a very anomalous character. Regarding the names Ornithichnites, Ornithoidichnite, &c., as provisional merely, I now propose names for the animals that made the tracks, instead of the tracks. For since we are certain of their existence, and that they have never been described, why should they not be named like other extinct animals? This has been done in Europe in respect to the *Chirotherium*, by Prof. Kaup; and in respect to the *Testudo Duncani*, by Prof. Owen; who had no other evidence of the existence of such animals except their tracks.

Descr. I have divided these animals into as many groups, or families, as seemed to be indicated by their tracks; and have sometimes added the known living families to which they probably belong; not, however, with any great confidence. I have also embraced the animals described by Dr. Alfred T. King, whose footmarks are found in Pennsylvania. Ere long I hope to give a systematic description of the animals here named, so far as their tracks in the Connecticut valley will enable me to do it,—through the transactions of some scientific society.

Names of the Animals that made the Fossil Footmarks found in the United States.

GROUP 1. (*Struthionidæ?*) 1 *Brontozoum*, (βροντης and ζωον) 1 giganteum, 2 Sillimanium, 3 expansum, 4 gracillimum, 5 parallelum. SUB-GROUP, (Wing-footed) 2 *Aethiopus*, (αιθυια and πȣς) 1 Lyellianus, 2 minor.

GROUP 2. 2 *Leviathan* Moodii; *Moody's Leviathan:* from an enormous track with four thick toes, distinctly jointed; the whole 20 inches long; (!) recently discovered (January, 1847) in South Hadley, Mass., by Pliny Moody, Esq., who more than 40 years ago, dug up the first fossil footmark ever noticed in this valley.

GROUP 3. 4 *Steropozoum*, (ϛεροπης and ζωον) 1 ingens, 2 elegans, 3 elegantior; 5 *Argozoum*, (αργης and ζωον) 1 Redfieldianum, 2 robustum, 3 dispari-digitatum, 4 pari-digitatum, 5 minimum; 5 *Platypterna*, (πλατυς and πτερνα) 1 Deanianus, 2 tenuis; 3 delicatulus.

GROUP 4. 7 *Ornithopus*, (ορνις and πȣς) 1 Adamsanus, 2 gallinaceus, 3 gracilior, 4 loripes.

GROUP 5. 8 *Polemarchus*, (πολεμαρχος) 1 gigas; 9 *Plectropus*, (πληκτρον and πȣς) 1 minitans, 2 longipes, 3 tenui-digitatus; 10 *Triaenopus*, (τριαινα and πȣς) 1 Baileyanus, 2 Emmonsianus.

GROUP 6. 11 *Palamopus*, (παλαμη and πȣς) 1 anomalus, 2 Dananus; 12 *Typopus*, (τυπος and πȣς) 1 abnormis.

GROUP 7. (Quadrupedal,) (Lacertilia? or Chelonia?) 13 *Anomœpus*, (ανομοιος and πȣς) 1 scambus, 2 Barrattii; 14 *Harpedactylus*, (ἁρπη and δακτυλος) tenuissimus

Group 8. (Chelonia?) 15 *Ancyropus*, (αγκυρα and πυς) 1 heteroclitus; 16 *Helcura*, (ελκω and υρα) 1 litoralis.

Group 9. (Marsupialia?) 17 *Anisophus*, (ανισος and πυς) 1 Deweyanus, 2 gracilis; 18 *Macropterna*, (μακρος and πτερνα) 1 Rhynchosauroideus.

Group 10. (Batrachia?) 19 *Thenaropus*, King, (θεναρ and πυς) 1 heterodactylus, in Pennsylvania.

Group 11. (Annelata, *Owen*, *Annelida*,) 20 *Herpystezoum*, (ερπυστης and ζωον) 1 Marshii, 2 minutum.

Group 12. (Obscure) 21 *Harpagopus*, (αρπαγη and πυς) 1 giganteus, 2 Hudsonius, 3 dubius.

Descr. The following sketch will give a very good idea of the different varieties of tracks in the Connecticut valley, and the manner in which they occur together. It is copied from a slab in the cabinet of Amherst College, obtained at Turner's Falls. It exhibits four rows of a biped, or bird; one row of a quadruped, whose hind foot was the largest, and two rows of what is supposed to be the track and the trail of a tortoise. The whole slab is covered also with rain drops, which fell after the tortoise had passed over the mud, but before the passage of the other animals. It will be seen that the joints of the toes are distinctly manifest in the bird tracks.

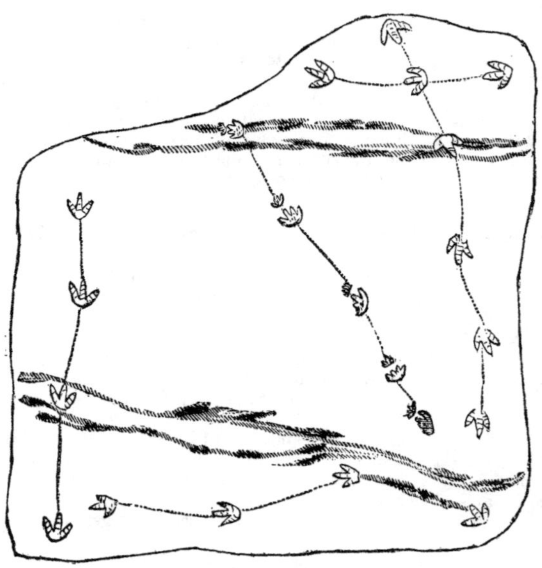

Footmarks in Gill, Mass.

Descr. The figure following shows another slab from the same locality, obtained by Mr. Dexter Marsh, of Greenfield, and now in his cabinet. This exhibits seven rows of the bird tracks, two of unequal-footed quadrupeds, and two of a tortoise. This slab is about seven feet by four and a half.

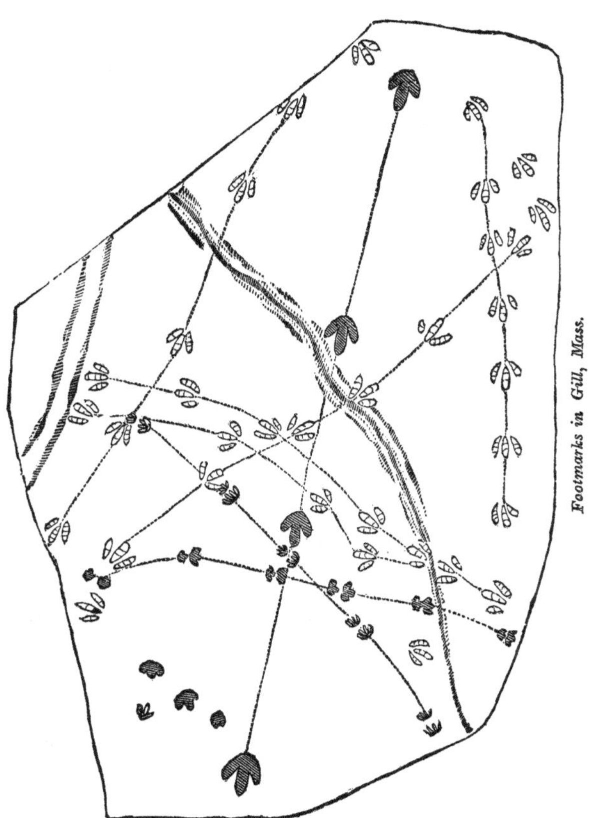

Footmarks in Gill, Mass.

Descr. The next figure will convey an idea of the footmarks found in the coal formation of Western Pennsylvania, as figured by Dr. A. T. King. They were made probably by a Batrachian, or a Chelonian, whose feet were semi-palmate.

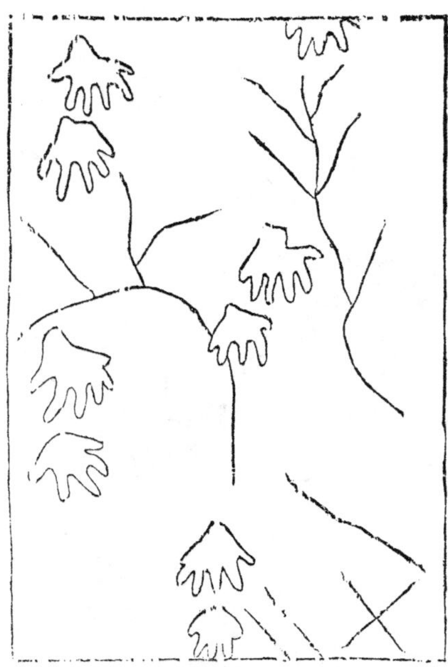

Footmarks in Pennsylvania.

Inf. These footmarks were probably made in most cases along the shores of an estuary, either beneath the water, or between high and low tide mark. That many of them were impressed on dry land, is obvious from the numerous impressions of rain drops (to be described below) found on the same surface. In such cases the returning tide or river floods, brought in silt and sand enough to preserve the impressions; and there they might have been kept thousands of ages, uninjured, before the mud was hardened into stone. For the arguments to show that some of these tracks were made by birds, and others by quadrupeds, I must refer to my *Report on the Geology of Massachusetts, Vol. 2. p.* 508. *Trans. Amer. Geol. Assoc. p.* 254, and *Amer. Jour. Sci., commencing with Jan.* 1836.

Rem. 1. It will be seen that I do not adduce as an example of fossil footmarks, the impressions of the human foot, which were found on limestone on the banks of the river Mississippi, in front of the city of St. Louis, in Missouri, and which have been described by Mr. Schoolcraft in the 5th volume of the American Journal of Science. I omit them because I do not think there is sufficient evidence that they are natural impressions. They occur in a blue limestone, containing an abundance of encrinites and other analogous remains, and which is probably the carboniferous limestone of Europe. *Featherstonhaugh's Report*, 1835, *p.* 28, *&c.*

Rem. 2. If any are in doubt whether these footmarks are the work of art, let him consult a paper by D. D. Owen on the subject in the *Am. Jour. Sci., Vol.* 43, *p.* 14; and he will be satisfied that they were excavated by man, to indicate the low water mark of the Mississippi.

Descr. It seems that the aborigines of this country were in the habit of cutting figures on soft rock, representing the feet of birds and quadrupeds. These, after long exposure, sometimes appear so natural that they have been mistaken for real tracks, especially in Pennsylvania. But the fact that on the same rock are sometimes seen representations of arrows, human bodies, &c., having the same appearance as the tracks, (and I am assured by gentlemen who have seen them that such is the case,) settles the point that they are of artificial origin. *Amer. Journ. Sci., Vol. 2. p. 25, New Series.*

Impressions of Rain Drops.

Descr. In the same formation of red sandstone that contains the footmarks in England, are found most distinct impressions of what appear to have been drops of rain. In the Storeton quarry, where are found the tracks of the chirotherium, "the under surface of two strata at the depth of 32 or 35 feet from the top of the quarry, presents a remarkably blistered or watery appearance, being densely covered by minute hemispheres of the same substance as the sandstone. These projections are casts in relief of indentations in the upper surface of a thin subjacent bed of clay, and due in Mr. Cunningham's opinion to drops of rain." *Lond. and Edin. Philos. Mag.* 1839. Sometimes these impressions are perfect hemispheres: in other cases they are irregular and are elongated in a particular direction as if the drops struck the surface obliquely: appearing in fact as if a wind had accompanied the rain.

Descr. Specimens corresponding to the above description exactly, are very common and cover large surfaces at Wethersfield, in Connecticut; West Springfield, Northampton, South Hadley, and Turner's Falls, in Massachusetts. They are abundant, also, New Jersey; and Mr. Lyell describes them on the hardened red mud of Nova Scotia. *Travels in North America, Vol. 2. p.* 140. Furthermore, I have produced precisely the same indentations upon clay made plastic by sprinkling water upon it, and the same impressions may be seen at almost every brick kiln.

Rem. It is a most interesting thought, that while millions of men, who have striven hard to transmit some trace of their existence to future generations, have sunk into utter oblivion, the simple footsteps of animals, that existed thousands, nay, tens of thousands of years ago, should remain as fresh and distinct as if yesterday impressed; even though nearly every other vestige of their existence has vanished: Nay, still more strange is it, that even the pattering of a shower at that distant period, should have left marks equally distinct, and registered with infallible certainty, the direction of the wind!

Addenda.

Descr. The following list of all the fossil plants found in the rocks up to the year 1845, with their distribution in the several formations, by M. Gœppert, has just met my eye, and is too important to be omitted.

	Families.	*Species.*
Graywacke, or rocks older than the carboniferous,	8	52
Carboniferous System,	21	819
Lower New Red or Permian System,	7	58
Triassic System,	18	86
Lias,	12	75

	Families.	*Species.*
Oolite,	9	159
Wealden,	8	16
Cretaceous System,	16	62
Tertiary System,	66	454
Unknown Geological Position,	4	11
Quart. Jour. Geol. Soc., Vol. 1. *p.* 566.	169	1792

General Inferences.

Rem. I have passed over several important inferences derived chiefly from palæontology, because they were not deducible from any one statement that has been made, and I thought it best to present them in the conclusion of this section with a summary of the proof.

Inf. 1. The present continents of the globe, (except perhaps some high mountains,) have for a long period constituted the bottom of the ocean, and have been subsequently elevated.

Proof. 1. Two thirds at least of these continents are covered with rocks, often several thousands feet thick, abounding in marine organic remains; which must have been quietly deposited, along with the sand, mud, and calcareous or ferruginous matter, in which they are enveloped, and which could have accumulated but slowly. 2. The secondary and primary stratified rocks are almost universally fractured and raised up at various angles, just as they would have been if lifted from the bottom of the ocean by a force acting beneath them. 3. Anticlinal ridges are so frequently found with a nucleus of unstratified rocks, as to point us to a sufficient cause, viz. volcanic agency, for the elevations that appear to have taken place.

Rem. This inference is to be regarded as probably the most important principle in geology, and as established on an immovable foundation.

Inf. 2. Different continents, and different parts of the same continent, appear to have been elevated at different epochs.

Proof. Let A, B, Fig. 87, represent a mountain ridge, with an axis *a* of unstratified rock. Let the three systems of strata, *bb*, *cc*, and *dd*, rest upon the axis *a*, and upon one another, unconformably, and dip at different angles, except *dd*, which suppose horizontal. Now it is obvious that the formations *cc*, and *bb*, must have been elevated previous to the deposition of *dd;* otherwise the latter would have partaken of the upward movement. And if there be no regular member of the series of rocks wanting, between *d*, and *c*, it is obvious that we thus ascertain the geological though not the chronological epoch, when *cc* was elevated. *cc*, however, is unconformable to *bb;* and therefore *bb*, was partially elevated before the deposition of *cc:* in other words, *bb* has experienced at least two vertical movements. Now this is a just representation of the actual state of things in the earth's crust; and hence by ascertaining the dip of the formations that are in juxtaposition, we ascertain the different epochs of elevation.

Fig. 87.

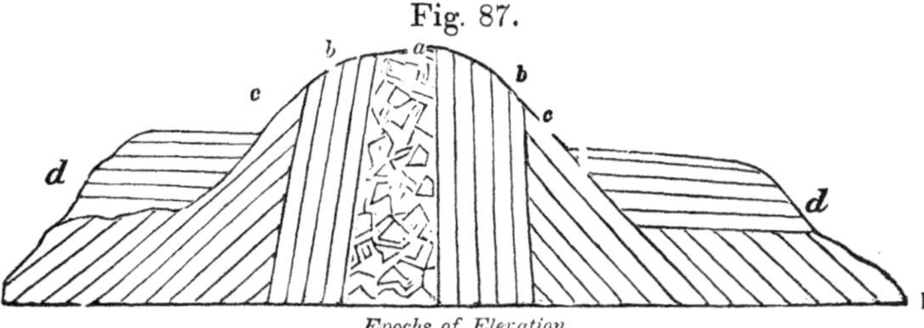

Epochs of Elevation.

Facts. By the application of these principles, it is found that the mountains of Europe have been elevated at no less than twelve different epochs; the oldest of which dates as far back as the time when the slates of Westmoreland were tilted up: and the most recent, (the principal chain of the Alps,) is said to be subsequent to the deposition of the tertiary strata. So far as the subject has been examined in this country, it appears that five or six systems or epochs of elevation can be traced in our mountains: though since the deposition of our secondary rocks, scarcely any movement has taken place: and though Elie de Beaumont suggests that the elevation of the Andes was so recent that it may have produced the historical deluge, yet the eastern side of our continent is probably of an older date than most of Europe.

Prin. According to Elie de Beaumont, to whom we are indebted for the first extensive generalizations on this subject, although Von Buch and other geologists had previously made local applications of the same principles, chains of contemporaneous elevation are parallel to one another, and "to the semi-circumference of the earth's surface:" and hence he infers that mountain chains were elevated at the same epoch, even though on different continents, if they are parallel: that the chain of the Alleganies, for instance, belongs to a system of elevation which includes the Pyrenees, part of the Appenines, the mountains of the Morea, a part of the Hartz mountains, Mount Atlas, and other ridges in Africa, the Carpathian mountains, Mount Carmel, and Sinai, a part of the Caucasian chain, and the Ghauts. In the present state of our knowledge of these mountains, this conclusion is unwarranted; for it is admitted by Beaumont, that systems elevated at different epochs may be parallel; and until the geology of these mountains is better understood, it cannot be known but that these belong to different parallel systems. *See Lyell's Prin. Geol. Vol.* 1. *p.* 304. Yet it seems capable of mathematical demonstration, that fissures produced by an elevating force acting beneath a portion of the earth's crust, will be parallel when no two consecutive fissures are remote from each other; and if another set of fissures be produced, by the force acting unequally at different points, these will be parallel to one another, also; and approximately perpendicular to the first set. Hence Beaumont's principle may prove true over large areas; but much more extensive observations will be necessary before it can be applied, except in limited districts. *Phillips's Geology, p.* 259.

Inf. 3. The convulsive movements by which systems of strata were elevated, appear to have been in most instances short compared with the intervening periods of repose, during which successive formations were deposited.

Proof. 1. The deposits appear generally not to have been disturbed by any elevating force while in a state of formation, as this would have changed the character of the organic remains; *De la Beche's Theoretical Geology, Chap.* 12. and the period of deposition must have been in most cases very protracted. 2. Had the elevating force been going on slowly during the deposition, the lower beds of the formation ought to have a greater dip than the upper beds; which is rarely found to be the case. 3. Paroxysmal convulsions are sufficient to account for the appearances in most cases of the elevation of the strata. 4. In most cases there is no evidence of any long interval between the deposition of two rocks whose position is unconformable. 5. Some single local dislocations are of enormous size, amounting to 3000 or 4000 feet; as in the Penine region of the north of England: and it is difficult to conceive how such faults could have resulted from a succession of minor forces acting through long intervals. 6. The doctrines of internal heat, if admitted, furnish a sufficient force to elevate the highest mountains by a single effort. *Phillips's Geology, p.* 260.

Rem. The interval between the deposition of consecutive strata may sometimes have been very great, if measured chronologically; especially where one member or more of the series of rocks, is wanting. In such cases the principles above explained do not enable us to determine during what part of this interval the elevation of the strata took place. It is chiefly on this ground that Mr. Lyell attempts to overthrow the whole of Beaumont's reasonings and conclusions on this subject. But it seems to me that he has shown only that it is difficult to fix upon that point in the interval between two consecutive rocks, when the convulsive movements took place; and that the fundamental principle of Beaumont's theory remains unaffected. And in respect to the exact geological time when the elevation occurred, it is, to say the least, very probable that it took place just at the termination of the period during which the elevated rock was in a course of formation; for such a convulsion furnishes, in many instances at least, the only known reason why its deposition was brought to a close. *Lyell's Principles of Geology, Vol.* 1. *p.* 304. Other writers have adduced various objections to the views of Beaumont; as Dr. Boue, for instance, in the *Journal de Geologie, Tome* 3. *p.* 338.

Inf. 4. It is maintained by Beaumont, that the changes in the zoological and botanical characters of the formations, correspond in general to the epochs of elevation: that is, the period of elevation seems to have been the time for the destruction of one group of organic races and the introduction of new species. But though this may be generally, it is not usually true. No great change, for instance, appears to have taken place in the organic characters of rocks below the zechstein inclusive. But this may in part be explained by the fact that all or nearly all the animals before that period were marine, and consequently might very probably survive the up-burst of a continent; since violent agitation of the waters would be the principal effect.

Inf. 5. In many instances the rocks appear to have suffered one alternation or more of elevation and subsidence.

Proof. 1. The phenomena of what is called the *Dirt Bed*, of the oolite formation in the isle of Portland, in G. Britain (*See Fig.* 46.) are perhaps the best example in proof of this proposition that occurs. In a bed of black mould, lying between the Portland stone beneath, and the Purbeck stone above, (both of them oolitic limestones,) there exist large prostrate trunks, and erect stumps of cycadeæ, or tropical trees, which must have grown on the spot where the stumps now stand. The following conclusions seem to be fairly inferrible from the facts detailed. 1. That the limestone beneath this dirt bed was deposited at the bottom of the ocean. 2. That this bottom must have been elevated above the waters long enough for the accumulation of the soil and the growth of the trees. 3. That the surface was next submerged beneath the waters of a freshwater lake; next beneath an estuary; and next beneath a deep sea, long enough for the deposition of strata 2000 feet thick. 4. That these strata have been subsequently elevated into their present terrestrial state in England. *Buckland's Bridgewater Treatise, Vol.* 1. *p.* 495.

2. The coal formation may be mentioned as another example in point. In these formations there is sometimes an alternation of marine and fresh water remains, and always an alternation of coal with shales and sandstones. Hence some geologists are of opinion that the land, where the vegetation grew that formed the coal, must have sunk and risen again, every time these alternations occur. Others, however, suppose that the coal plants grew on low islands of tropical archipelagos, and were transported into the bottom of the ocean, or of estuaries, when they were covered by deposits of sand, clay, and limestone, and again by other beds of vegetables, until a great thickness of these interstratified layers had accumulated. Freshwater and marine relics sometimes alternate in successive strata of the coal measures, because in an estuary the salt water would occasionally prevail over the fresh water from the river which emptied into it. *See De la Beche's Manual, p.* 444. *Also his Theoretical Geology, p.* 264. *Macculloch's system of Geology, Vol.* 2. *p.* 312. *Phillips's Geology, p.* 116.

3. A third example is the alternating strata of freshwater and marine deposits in the tertiary series. But in this case, also, it is very possible to conceive how these alternations might have been effected by the successive predominance of rivers

and the sea in an estuary as explained above. *Phillips' Geology, p.* 164.

Rem. Examples of more recent elevation and subsidence will be given in Section VIII: where will also be found the various theories proposed for the explanation of the phenomena of elevation and subsidence.

Inf. 6. From the phenomena of organic remains, it appears that the species of animals and plants now existing on the globe, could not, with a few exceptions, have been cotemporaries with those found in the rocks.

Proof. 1. If they had been cotemporaries, no reason can be given why the remains of the living species do not occur in the rocks; which, with the exception of a few hundred species in the more recent tertiary strata, is well known to be the case. 2. Comparative anatomists decide from the structure of the extinct animals and plants, that they were intended for a climate and other physical circumstances so different from those now existing, that the organic beings adapted to one state, could not have endured the other. The period of the tertiary strata is the only exception: and even then, the climate appears to have been in high northern latitudes nearly as warm as at present between the tropics, until near the close of the period.

Inf. 7. Hence too we learn the mistake of those, who are in the habit of pronouncing very confidently that certain organic remains are petrifactions of existing animals and plants. For if they are obtained from the secondary rocks, the presumption amounts almost to certainty, that they cannot be the representatives of existing species.

Exam. Fossil trees are called oak, maple, hemlock, &c. fibrous tremolite and some varieties of mica and talcose slates, are called petrified wood: encrinites are called snakes: coal plants are called rattle-snakes: favosites and certain fossil shells are called butternuts and walnuts: some varieties of ancient polyparia are regarded as the horns of deer, others as petrified pork: and even petrified squaws, pappooses, and buffaloes have been announced as existing in the far west. It is often amusing to see with how much confidence a man ignorant of zoology and botany, will pronounce upon these supposed cases of identicalness.

Inf. 8. It appears that there have been upon the globe several distinct periods of organized existence, in which particular groups of animals and plants, exactly adapted to the varying physical condition of the globe, have been created and have successively passed away.

Proof. If we take only those larger groups of animals and plants, whose almost entire distinctness from one another has been established beyond all doubt, we shall still find at least five nearly complete organic revolutions on the globe: viz. 1. The existing species. 2. Those in the tertiary strata. 3. Those in the cretaceous and oolitic systems. 4. Those in the upper new red sandstone group. 5. Those below the new red sandstone. Comparative anatomy teaches us that the animals and plants in these different groups could not have lived in the same physical circumstances.

Rem. 1. The palæontological chart, as well as the palæontological classification described at the end of Sect. 1st, exhibits seven distinct groups. But some of these are not entirely independent of one another; that is, some species are common to two groups. This fact, however, does not militate against the general principle of the above inference; since a few species may have survived the causes by which particular groups were destroyed. Nay, if with some geologists we admit that the destruction of old, and introduction of new species was gradual, the proposition would still remain true. If only the greater part of the species have been changed several times, it establishes the inference. And these have been essentially changed, certainly as many as six times, as shown on the palæontological chart; and therefore, I prefer to speak of seven distinct periods of organic life on the globe, rather than five. But from the next remark, it appears that five *entirely distinct* periods can be made out.

Rem. 2. It appears from the following quotation by Dr. Smith, (*Scripture and Geology, p.* 514, *second Ed.*) that Deshayes, an eminent palæontologist, is able to make out five distinct periods of organized existence, besides that now passing, on the globe. "M. Deshayes has lately announced that he had discovered, in surveying the entire series of fossil animal remains, *five great groups, so completely independent, that no species whatever is found in more than one of them.* The first of those groups is that to which trilobites give the character; the three succeeding belong to the system of the large saurians, and the fifth includes the systems in which I have pointed out the palæotheria, the mastodons, and the elephants." *D'Omalius d'Halloy, Elem. de Geol. p.* 127; *Paris*, 1839.

Obj. Perhaps the deposits containing these different organic groups, may have been going on at the same time in different countries, or in different parts of the same country.

Answer. Although all the rocks composing these different systems are not found piled upon one another in any one place, they are all found so connected at different points, as to prove that they were formed successively. Yet where any are wanting in the series, as the Wealden, for instance, in North America, the interval during which these were forming in particular localities, may have been occupied by a prolonged deposition of the next older, or by an earlier commencement of the next newer rock. Most probably however, the same formation was begun and completed in different places about the same period: otherwise the climate would have varied so much as to produce a marked change in the organic remains.

Rem. The *Palæontological Chart* appended will aid in impressing upon the mind the origin, expansion, and termination of the organic beings that have lived on the globe. In order to make it more impressive, it ought to be more extensive.

Inf. 9. It appears that amidst all the diversities of organic life that have existed on the globe, the same general system has always prevailed.

Illus. 1. All the leading forms of organization that now exist on the globe, have existed from the beginning: for instance, all the four great classes of animals, the Vertebral,

Molluscous, Articulated, and Radiated, and the two great classes of plants, the Vasculares and the Cellulares. The relative number of species, however, in these different classes has varied very much at different periods. 2. Carnivorous races have always existed to keep down the excessive multiplication of the herbivorous races. Thus, when the sauroid fishes of the earliest rocks disappeared, their place was supplied in the more recent secondary strata, by the voracious marine saurians; and when these became extinct, sharks and other predaceous fishes, more like those now existing, made their appearance. So among the molluscs. During the deposition of all the secondary rocks, carnivorous cephalopods abounded; such as the nautilus, ammonite, &c.: but in the tertiary strata, and in our present seas, these are rare, and their place is taken by carnivorous trachelipods, which were not common at an earlier date. *Buckland's Bridgewater Treatise, Vol. 1. p.* 298. 3. From the eyes of trilobites and the orbits of other animals found in the rocks, we learn that the same relations of animals to light always existed as at present.

Rem. It has been observed in another place, (p. 93,) that no plants have been found below the upper part of the Silurian rocks; yet it seems certain that they must have existed as early as animals. It is also true, that no vertebral animals have been found in the lower Silurian group. Hence a late anonymous writer very strenuously maintains the doctrine of the creation and gradual development of animals by law, without any special creating agency on the part of the Deity. *Vestiges of the Natural History of the Creation, and a Sequel to the Same: New York,* 1844 *and* 1846. But the facts in the case show us merely that the different animals and plants were introduced at the periods best adapted to their existence, and not that they were gradually developed from monads. In the whole records of geology, there is not a single fact to make such a metamorphosis probable; but, on the other hand, a multitude of facts to show that the Deity introduced the different races just at the right time. That he did this according to certain laws, though not by their inherent force,—for laws have no such force—may be admitted; as may be done in respect to all his operations: but this does not prove them any the less special or miraculous. To find as we do, vertebral animals as low down as the upper Silurian strata, is nearly equivalent to finding them at the very bottom of the series; since the upper and lower Silurian periods appear to have been but parts of one period. It is also true, that almost every year brings to light some fact proving the earlier existence of the vertebral animals. The fossil footmarks found in Nova Scotia and Pennsylvania, for instance, in the lower part of the carboniferous rocks, demonstrate the existence of reptiles at that early period. In short, we may consider it as proved that all the great classes of animals and plants have been represented on the globe so near the commencement of organic life, that no geologist will doubt that it was so from the very beginning. Certainly no geologist will imagine that his science furnishes any evidence of the development theory.

Inf. **10.** It does not appear that any of the ancient forms of animal or vegetable life can be properly regarded as monstrous: or, when compared with the proper standard, even heteroclitic.

Proof. When compared with existing races, they sometimes seem monsters in size, and anomalies in character. But their great size resulted from a climate better adapted, than that now upon the globe, for the development of organic life; and their peculiar construction adapted them most admirably for the peculiar situation which they occupied. So that what seems heteroclitic at this day, was exact and harmonious adaptation then.

Inf. 11. The whole period occupied in the deposition of the fossiliferous rocks must have been immensely long.

Proof. 1. There must have been time enough for water to make depositions more than 10 miles in thickness, by materials worn from previous rocks, and more or less comminuted. 2. Time enough, also, to allow of hundreds of changes in the materials deposited: such changes as now require a long period for the production of one of them. 3. Time enough to allow of the growth and dissolution of animals and plants, often of microscopic littleness, sufficient to constitute almost entire mountains by their remains. 4. Time enough to produce by an extremely slow change of climate, the destruction of several nearly entire groups of organic beings. For although sudden catastrophes may have sometimes been the immediate cause of their extinction, there is reason to believe that those catastrophes did not usually happen, till such a change had taken place in the physical condition of the globe, as to render it no longer a comfortable habitation for beings of their organization. 5. Time enough for erosions to have taken place in the rocks, in an extremely slow manner, by aqueous and atmospheric agencies, on so vast a scale that the deep cut through which Niagara River runs, between Niagara Falls and Lake Ontario, is but a moderate example of them. We must judge of the time requisite for these deposits by similar operations now in progress; and these are in general extremely slow. The lakes of Scotland, for instance, do not shoal at the rate of more than 6 inches in a century. *Macculloch's Geology, Vol.* 1. *p.* 507. See also a full view of the arguments on this subject in *Dr. J. Pye Smith's Lectures on Scripture and Geology, p.* 394. *Second London Edition,* 1840.

Obj. 1. The rapid manner in which some deposits are formed at the present day: ex. gr. in the lake of Geneva; where, within the last 800 years, the Rhone has formed a delta two miles long and 600 feet in thickness. *Lyell's Prin. Geol., Vol.* 1. *p.* 423.

Ans. Such examples are merely exceptions to the general law, that rivers, lakes, and the ocean, are filling up with extreme slowness. Hence such cases show only that in ancient times, rocks might have been deposited over limited areas, in a rapid manner; but they do not show that such was generally the case.

Obj. 2. Large trunks of trees, from 20 to 60 feet long, have sometimes been found in the rocks, penetrating the strata perpendicularly, or obliquely; and standing apparently where they originally grew. Now we know that wood cannot resist decomposition for a great length of time, and therefore, the strata around these trunks must have accumulated very rapidly; and hence the strata generally may have been rapidly formed.

Ans. Admitting that the strata enclosing these trunks were rapidly deposited it might have been only such a case as is described in the first objection. But sometimes these trunks may have been drifted into a lake or pond, where a deep deposit of mud had been slowly accumulating, which remained so soft, that the heaviest part of the trunks, that is, their lower extremity, sunk to the bottom by their gravity, and thus brought the trunks into an erect position. Or suppose a forest of trees sunk by some convulsion, in the manner described by Rev. Mr. Parker in the Columbia River: (See Section VIII. *Exploring Tour beyond the Rocky Mountains, p.* 132, *first Ed.*,) how rapidly might deposits be accumulated around them, were the river a turbulent one, proceeding from a mountainous region.

Obj. 3. The vast accumulations that have been made of the shields of animalcula since the commencement of the historic period, show that similar deposits of other animal remains might have been made of much greater thickness in ancient times, in a comparatively short period.

Ans. If it can be shown that the larger animals, found fossil, have a power of increase that will compare at all with the astonishing multiplication of animalcula, the objection will be valid: but not till then: and this can never be shown.

Obj. 4. All the causes producing rocks may have operated in ancient times with vastly more intensity than at present.

Ans. This, if admitted, might explain the mere accumulation of materials to form rocks. But it would not account for the vast number of changes which took place in their mineral and organic characters; which could have taken place, without a miracle, only during vast periods of time.

Obj. 5. The fossiliferous rocks might have been created, just as we now find them, by the fiat of the Almighty, in a moment of time.

Ans. The possibility of such an event is admitted: but the probability is denied. If we admit that organic remains, from the unchanged elephants and rhinoceros of Siberia, to the perfectly petrified trilobites and terebratulæ of the transition strata, were never living animals, we give up the whole groundwork of analogical reasoning; and the whole of physical science falls to the ground. But it is useless formally to reply to an objection which would never be advanced by any man, who had ever examined even a cabinet collection of organic remains.

Inf. 12. There is reason to suppose that immense numbers of the softer species of animals, which have no solid parts, may have lived and died, during the deposition of the older fossiliferous rocks, without leaving in the rocks a vestige of their existence.

Proof. Limestone, (that is, the carbonate of lime,) being then less common than at a later period, it is probable that animals not needing it to form a covering or a skeleton, would have been created; since we find that in all periods living beings had natures exactly adapted to their condition. Again, we find many of the older secondary limestones highly bituminous; and the decomposition of soft and gelatinous animals would have produced a large amount of bitumen. *De la Beche's Manual, p.* 476.

Inf. 13. The greater part of the accessible crust of the globe may once have constituted portions of the animal frame.

Proof. In respect to limestone, which has been thought to constitute about one seventh of the earth's crust, the presumption in favor of its animal origin, seems quite probable: that is, animals have the power of separating lime from its other combinations and converting it into the carbonate. *Macculloch's Syst. Geol. Vol.* 1. *p.* 219. *Lyell's Prin. Geol. Vol.* 3. *p.* 401. The recent discoveries of Ehrenberg, respecting fossil animalcula, already detailed, make it probable that a large amount of silica and oxide of iron may have thus originated. At a late meeting of the British Association for the Advancement of Science, this naturalist exhibited "a large glass full of artificial siliceous earth," which he had prepared from existing infusoria; and he says that "pounds and tons of this earth may be easily prepared." *Am. Jour. Sci., Vol.* 35. *p.* 372. I am not prepared, however, to go so far on this subject, as a recent able and elegant writer, who says, that "probably there is not an atom of the solid materials of the globe, which has not passed through the complex and wonderful laboratory of life." *Mantell's Wond. Geol., Vol.* 2. *p.* 670.

Inf. 14. It appears that every successive general change, that has taken place on the earth's surface, has been an improvement of its condition.

Proof. Animals and plants of a higher organization have been multiplied with every change, until at last the earth was prepared for the existing races; with man at their head, the most generally perfect of all.

SECTION VI.

OPERATION OF AQUEOUS AND ATMOSPERIC AGENCIES IN PRODUCING GEOLOGICAL CHANGES.

Prin. The basis of nearly all correct reasoning in geology, is the analogy between the phenomena of nature in all periods of the world's history: in other words, similar effects are supposed to be the result of similar causes at all times.

Illus. and Proof. This principle is founded on a belief in the constancy of nature: or that natural operations are the result of only one general system, which is regulated by invariable laws. Every other branch of physical science, equally with geology, depends upon this principle: and if it be given up, all reasoning in respect to past natural phenomena, is at end.

Rem. It does not follow from this principle that the causes of geological change have always operated with equal intensity, nor with entire uniformity. How great has been the irregularity of their action, is a subject of debate among geologists.

Inf. We see from the preceding principle, how important it is to ascertain the *true dynamics* of existing causes of geological change: that is, the amount of change which they are now producing. For until this is done, we cannot determine whether these causes are sufficient to account for all the changes which the earth has undergone.

Rem. Heat, cold, and water in its manifold states, liquid, solid and vaporous, so often act conjointly upon rocks, that their separate agency cannot be pointed out. But the results of their combined action are numerous.

Glaciers, Avalanches, Icebergs, and Landslips.

Descr. Glaciers are masses of ice which are enclosed in Alpine valleys, or are suspended upon the flanks of the mountains which rise into the regions of eternal frost. Being of a white color, they appear at a distance like vast streams of snow issuing from lofty summits, and extending into the lower valleys. In the Alps, where the large glaciers have received distinct names, as Bossons, Montanvert, Aletsch, Viesch, &c., they terminate sometimes as high as 7000 or 8000 feet; but some descend to 3000 feet. They are sometimes three miles wide and fifteen miles long: and their thickness at the lower end varies from 80 to 100 feet, and at the upper end, from 120 to 180 feet. *Etudes sur les Glaciers par Agassiz, Neuchatel*, 1840. From this most interesting work, I derive nearly all the statements that follow respecting glaciers.

Descr. Glaciers are composed of snow that has been more or less melted and again frozen. The lower part becomes pure solid ice: the upper part is composed of a sort of granular snow, called in French, *neve*, and in German, *firn*. A new layer is added at least each year, so that the mass is stratified. The upper surface is rough and sometimes covered by pointed masses of ice, called *Aiguilles or Needles.* Fissures across the glacier, 20 or 30, and sometimes even 100 feet wide, are very common, produced by the unequal temperature of different parts of the mass. The slope of glaciers is frequently quite moderate. The lower end of the glacier of Aar, which is 15 miles long, is 3000 feet below the upper end. The slope of that of Aletsch, is 2° 58: that of the mer de glace of Chaumoni, is 3° 15.

Descr. The crests and higher parts of the Alps, which are frequently vast *plateaux*, or table land, containing 100, 200, or even 300 square miles, are covered with thick continuous masses of ice, through which the lofty peaks sometimes rise as volcanic mountains from the ocean. These *plateaux* are denominated *Mers de Glace*, or seas of ice: and from them the glaciers originate. The *mers de glace* by expansion, send down the sides of the mountain, especially through the valleys, enormous streams of ice, which continue to descend until they are melted. The common opinion that the glaciers slide down the mountains by their weight, is found to be incorrect; as they are not detached from the *mer de glace;* and it is only the lower part, where the ground is thawed beneath, that slides over the surface, and that by the expansion of the ice. In the winter they do not slide at all on their bottom.

Illus. To give a more accurate idea of glaciers, I have reduced a few of the splendid drawings of the *Etudes sur les Glaciers*, and had them executed on wood. They have been so well executed by the artist, as to give almost as good an idea of the phenomena of glaciers, as the originals. Fig. 88, is a view of the upper part of the glacier of Viesch, as it proceeds from the distant *mer de glace*, and winds through the long valley.

Descr. When the slope down which the glacier descends is very steep, or it is crowded to the edge of a precipice, huge masses sometimes are broken off by gravity, and tumbling down the mountain, produce great havoc. In the Alps these are called *avalanches;* and so large are they sometimes, that from one to five villages, with thousands of inhabitants, have been destroyed in a moment.

Descr. Landslips are a somewhat similar occurrence, happening in countries where perpetual frost does not exist. They frequently occur in the spring, when the frost leaves the soil, and the great weight of snow and ice drags along with it trees, soil and rocks, down the mountain's side. Sometimes these slides take place in the summer, after powerful rains; as that in the white Mountains in 1826, by which a family were destroyed. Marks of ancient slides are visible on the sides of the Hopper on Saddle Mountain in Massachusetts.

Fig. 88.

View of the Glacier of Viesch.—(Upper part.)

Descr. When avalanches occur on the steep shores of the ocean in high latitudes, the mass of ice, loaded perhaps with detritus, is precipitated into the sea and constitutes an *iceberg*. These icebergs are drifted often to a great distance by oceanic currents. Those from the northern ocean are seen as far south as 40° N. latitude. They are sometimes a mile in diameter, and rise 250 or 300 feet above the surface, and consequently must sink more than 2000 feet below; as every cubic foot above, implies 8 cubic feet below the surface. They were exceedingly numerous and large in the Atlantic Ocean in W. longitude about 50, reaching probably from N. latitude 40° to 43°, in the spring of 1841. *N. York Mercury, June* 10, 1841. *Capt. Hosken's Statement.*

Illus. Fig. 90, shows the lower extremity of the glacier of Aletsch in the Alps, where it enters the lake of Aletsch, which it formerly caused to overflow with great devastation. Masses of ice are frequently broken off and float about in the lake, as shown in the figure. This is one of the largest glaciers in the Alps.

Descr. As glaciers advance, they break off masses of rock from the sides and bottoms of the valleys, and crowd along whatever is movable, so as to form large accumulations of detritus in front and along their sides. When the glacier melts away, these ridges remain, and are called *moraines*. Agassiz describes three kinds: 1. The *terminal moraine*, or that at the extremity of the glacier: 2. *Lateral moraines*, or those ridges of detritus formed along the flanks of the glacier. They are thickest at their lower part, and decrease upwards. 3. *Medial moraines*, or those longitudinal trains of blocks which sometimes accumulate upon the top of the glacier, especially where glaciers unite from two valleys, and crowd the detritus between them upon their tops. The moraines are sometimes 30 or 40 feet high.

Descr. At the lower extremity of the glacier, there is a vault from which issues, especially in the summer, a stream of water, which ramifies upward beneath the ice like rivers in general. This stream, continuing from generation to generation, wears out a channel in the rocks as it descends from the glacier.

Illus. Fig. 89, shows the lower extremity of the glacier of Viesch, with a distinct terminal moraine, which, at the sides, is connected with lateral moraines. A stream of water is seen issuing from the glacier, which has worn a channel in the rocks. On Fig. 88, are shown both lateral and medial moraines: the latter considerably scattered.

Descr. The progress of glaciers down their slope, varies with their situation and the temperature and hygrometric state of the season: for they move only in the summer.

Fig. 89.

Glacier of Viesch.—(*Lower part.*)

Fig. 90.

Glacier of Aletsch.

In a single case the advance was 4400 feet in nine years. Sometimes they advance lower and lower for several years in consequence of the low temperature, and then they retreat by being melted and evaporated when the seasons are warmer. But these changes have no regular period, nor are they dependent upon any general changes in the temperature of the globe. As the glaciers advance and retreat they produce and leave successive moraines, especially terminal ones.

Rem. Water in freezing enlarges about one ninth of its volume. *Turner's Chemistry, p.* 19. *Fifth American Edition.*

Descr. Although the inferior surface of the glacier is pure smooth ice, yet it is usually thickly set with fragments of rock, pebbles and coarse sand, firmly frozen into it, which make it a huge rasp; and when it moves forward, these projecting masses, pressed down by the enormous weight of the glacier, wear down and scratch the solid rocks; or when the materials in the ice are very fine, they smooth and even polish the surface beneath. The movable materials beneath the ice are crushed and rounded, and often worn into sand or mud. The rocks in place, against which the glacier presses, are also smoothed and striated upon their sides. These striæ, wherever found, are perfectly parallel to one another, because the materials producing them are fixed in the bottom of the ice. But as the glacier advances and retreats, new sets of scratches will be produced, which sometimes cross those previously made, at a small angle.

Illus. Fig. 91, shows a specimen of schistose serpentine smoothed and scratched beneath a glacier in the Alps.

Fig. 91.

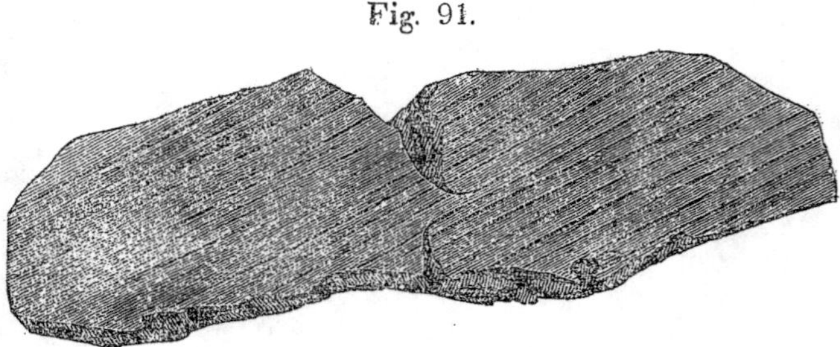

Rocks striated by Glaciers.

Fig. 92. is another similar specimen, exhibiting two sets of striæ crossing one another at a considerable angle.

Fig. 92.

Rocks striated by Glaciers.

Descr. When the ledges beneath the glaciers are uneven, and exhibit many angular projections, the angles are worn off and the surfaces assume that peculiar rounded and undulating appearance denominated by Saussure, *roches moutonnees;* some imperfect examples of which are shown on Figs. 88 and 89, where a considerable surface near the lower left hand corner of the drawing, shows the effect of former glaciers.

Descr. Currents of water sometimes conspire with the movements of the glacier, and form grooves or troughs of considerable depth and width on the top of precipitous rocks, to which currents of water could have no access were not the space around them filled with ice. Such furrows are called in Switzerland, *lapiaz*, or *lapiz.*

Rem. A different theory to explain the progress of glaciers than that by expansion, has more recently been advocated with great ability by Prof. James D. Forbes, and is generally adopted. "A glacier," says he, "is an imperfect fluid, or a viscous body, which is urged down slopes of a certain inclination by the mutual pressure of its parts." Gravity is, of course, the impelling agency. *Forbes's Travels in the Alps of Savoy, &c.*

Degradation of Rocks and Soil by Frost and Rains.

Descr. Water acts upon rocks and soils both chemically and mechanically: chemically, it dissolves some of the substances which they contain, and thus renders the mass loose and porous: mechanically, it gets between the particles and forces them asunder; so that they are more easily worn away when a current passes over them. Congelation still more effectually separates the fragments and grains, and thus renders it easy for rains and gravity to remove them to a lower level. In a single year the influence of these causes may be feeble: but as they are repeated from year to year, they become in fact some of the most powerful agencies in operation to level the surface of continents.

Detritus or Debris of Ledges.

Descr. It is chiefly by the action of frost and gravity, that those extensive accumulations of angular fragments of rocks are made, that often form a *talus*, or slope, at the foot of naked ledges, and even high up their faces. In some cases, though not generally, this detritus has reached the top of the ledge, and no farther additions are made to the broken spoils, which usually slope at an angle not far from 40°. Examples of this detritus are usually most striking along the mural faces of trap rocks; as for instance in the valley of Connecticut river in New England.

Inf. From these facts it appears that the earth cannot have existed in its present state an immense period of time: otherwise these slopes of *debris* would in every instance have extended to the top of the ledge: that is, the work of degradation would have been finished. We cannot, indeed, determine from this geological chronometer, the chronological epoch when this work of degradation commenced: but we are at least made sure, that the present state of the earth had a beginning.

Rivers.

Descr. Rivers produce geological changes in four modes: 1. By excavating some parts of their beds. 2. By filling up other parts. 3. By forming deposits along their banks. 4. By forming deposits, called *deltas*, at their mouths.

Exam. Most of the larger rivers, especially where they flow through a level country, are filling up their channels: but where smaller streams pass through a mountainous region, the power of excavation is still going on: and it is accomplished in a good measure by means of *ice freshets.* It is impossible for one who has not witnessed the breaking up of one of these streams in the spring, when for many miles the whole channel becomes literally choked with ice, to form an adequate idea of the immense excavating force which it exerts.

Descr. The deposit formed in the lake of Geneva by the waters of the Rhone, has been already mentioned. Another is formed at the mouth of this river, on the shores of the Mediterranean, and is said to be mostly solid calcareous and even crystalline rock. *Lyell's Prin. Geol. Vol.* 1. *p.* 433. The delta of the Mississippi has advanced several leagues since New Orleans was built. The delta of the Ganges commences 220 miles from the sea, and has a base 200 miles long, and the waters of the ocean at its mouth are muddy 60 miles from the shore. Since the year 1243 the delta of the Nile has advanced a mile at Damietta; and the same at Foah since the 15th century. In 2000 years the gain of the land at the mouth of the Po, has been 18 miles, for 100 miles along the coast. The delta of the Niger extends into the interior 170 miles, and along the coast 300 miles, so as to form an area of 25.000 square miles.

Descr. An immense alluvial deposit is forming at the mouth of the river Amazon and Oronoco; most of which is swept northerly by the Gulf Stream. The waters of the Amazon are not entirely mixed with those of the ocean at the distance of 300 miles from the coast. The quantity of sediment annually brought down by the Ganges, amounts to 6.368.077.440 tons; or 60 times more than the weight of the great pyramid in Egypt. The quantity of matter chemically and mechanically suspended in the waters of Merrimac river, that run past the city of Lowell in Massachusetts, in 1838, according to the very accurate experiments of Dr. Samuel L. Dana, amounted to 1.678. 343.810, pounds avoirdupois. The annual amount of anthracite coal used in the Merrimac Print Works in Lowell, is 5000 tons: and Dr. Dana estimates, that if the above amount of sediment were coal, it would supply those works 167 years. The quantity of water discharged by the Merrimac in 1838, was, 219.598.840.800. cubic feet.

Inf. 1. The extensive deposits thus forming daily by rivers, need only consolidation to become rocks of the same character as the shales, sandstones, and conglomerates of the secondary series.

Inf. 2. Rivers in general have not excavated their own beds: but run in valleys formed for the most part by other causes.

Proof. 1. In a majority of instances they are filling up their beds. 2. Transverse valleys frequently cross the course of rivers in such a way, that the water must have originally passed through them instead of excavating their present channels.

Illus. Fig. 94, shows Connecticut River, crossing Massachusetts and Connecticut, and emptying into Long Island Sound. If it had been left at first to find its own way to the ocean, and the passage between Holyoke and Tom (which are in fact but one ridge,) had not been formed, it must have passed through the valley A, to the sound: since no part of that valley, (through which the Farmington canal passes,) is more than 134 feet above the present bed of the river, where it runs between Holyoke and Tom. Or if the bed of the river had not existed through the mountains below Middletown in Connecticut, the river, instead of forming it, would have passed to the Sound through the valley B, through which the Hartford and New Haven Rail Road now runs, and no part of which can be more than 20 or 30 feet above the present level of the river at Hartford.

Descr. *Terraced valleys*, (of one of which a cross section is given in Fig. 93, at A,) sometimes exist in alluvial or tertiary regions, with the terraces on each side of equal height: and these appear to have been formed by the excavating operations of the rivers themselves.

Mode of Formation. Some suppose them formed by the sudden and successive bursting away of the barriers by which the river has been restrained. But it is wholly unnecessary to suppose any such bursting of barriers. The slow and uniform operation of the stream upon alluvial soil, will explain all the phenomena. But the explanation cannot be given here for want of room. It will be found in my *Final Report on the Geology of Massachusetts, Vol. 2. p.* 332.

Dr. Buckland has recently suggested that these parallel terraces in Scotland are probably "the effects of lakes produced by glacial action. *Proceedings of the Geological Society, Nov. 18th.* 1840, *p.* 333.

Fig. 93.

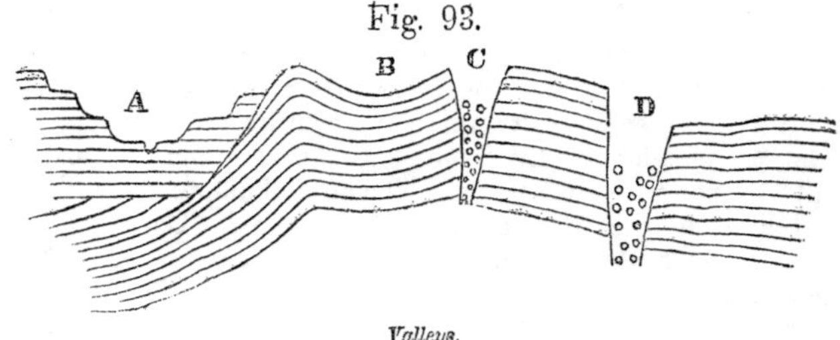

Valleys.

Other Valleys.

Descr. The terraced valleys above described are denominated *valleys of denudation*, when produced by the denuding force of water: and numerous valleys of other shapes, having been formed in the same manner, are thus called. Indeed, scarcely a valley exists that has not been more or less modified by this cause. But the greater part of the larger valleys that furrow the earth's surface, had a different origin; viz. the elevations, fractures, and dislocations which the strata have experienced.

Fig. 94.

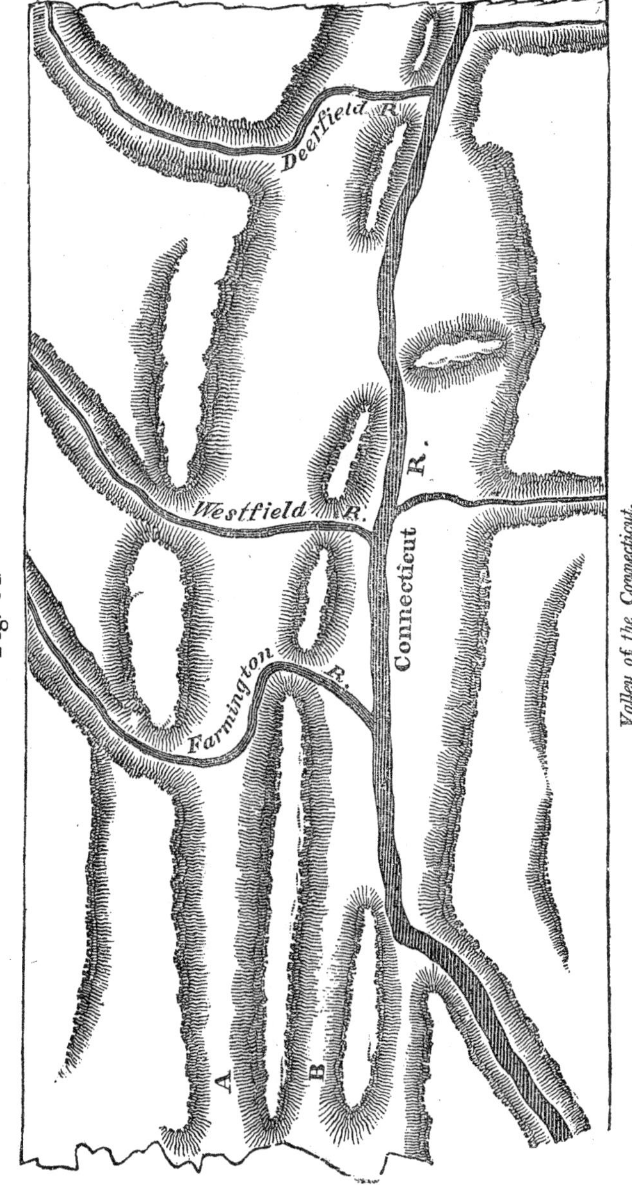

Valley of the Connecticut.

Proof. The phenomena of longitudinal and transverse valleys prove that all of them cannot have been the result of running water. In Fig. 94, which is a sketch of the valley of Connecticut river, with a portion of the mountainous region on both sides, it will be seen that the general direction of the mountain ridges, and of course of the valleys, is nearly north and south. Nevertheless, it will be seen that the tributaries of the Connecticut, Farmington, Agawam, and Deerfield rivers, and also the Connecticut itself, pass across these ridges and longitudinal valleys, in transverse valleys, which must be deeper than the others, else the water would flow out laterally into the longitudinal valleys. Now it is obvious that both sets of these valleys could not have been excavated by water. For if the longitudinal valleys were thus formed, how could the water afterwards be raised to the requisite level, for cutting valleys through the longitudinal ridges? We must therefore, suppose that one and often both of these sets of valleys originated in the fractures and foldings of the strata at the time of their elevation; and that water has only rounded their outlines and covered their inequalities with detritus.

Def. When a valley is produced by the sinking of the strata; or which is the same thing, by their elevation along two parallel anticlinal lines, it forms, what is called a *valley of subsidence*, as B. Fig. 93. When by the elevation of the strata they are made to separate at their highest point, a valley is produced, called a *valley of elevation;* as C. When a fracture has taken place in the strata, so as to leave the sides very steep and the valley narrow, a ravine is produced; as at D. In such a case the lower part of the fissures is usually filled up with detritus.

Prin. In some instances it is very difficult to decide, whether a particular gorge or ravine has been excavated by existing streams or by earlier agency, or was in part formed by some original dislocation of the strata, or by all these causes combined.

Exam. The deep ravine, seven miles long, between the falls of Niagara and Lake Ontario, has occasioned much discussion among geologists, respecting its origin. The falls are about 150 feet in height; and 670.000 tons of water are precipitated over them every minute. The upper stratum of rock is limestone; beneath which is shale, which wears away faster than the limestone, so as to cause the latter occasionally to break off in large masses; and the falls have been said in this way to recede 50 yards in 40 years. At this rate it would have required 10,000 years for them to have reached their present situation, if they commenced at Lake Ontario; and 30.000 years longer will be necessary to reach Lake Erie. But the rate of retrocession is yet unsettled; and probably it would be very different at different periods; so that the time already occupied in forming this ravine, and the time requisite to carry it to Ontario, must be regarded as determined only conjecturally. The geologist, however, who witnesses the extensive excavations made by other streams in our country in the solid rocks, will not be disposed to reject the above calculations simply because they require vast periods of time.

Bursting of Lakes.

Descr. A few examples have occurred in which a lake, or a large body of water long confined, has broken through its barrier and inundated the adjacent country. An interesting example of this kind occurred in the town of Glover in Vermont; in which two lakes, one of them a mile and a half long and three fourths of a mile wide, and in some places 150 feet deep; and the other, three fourths of a mile long, and half a mile wide, were let out by human labor, and being drained in a few minutes, the waters urged their way down the channel of Barton river, at least 20 miles to Lake Memphramagog, mostly through a forest, cutting a ravine from 20 to 40 rods wide, and from 50 to 60 feet deep; inundating the low lands, and depositing thereon vast quantities of timber. *American Journal of Science, Vol.* 11. *p.* 39. In 1818, the waters of the Dranse in Switzerland, having been long obstructed by ice, burst their barrier and produced still greater desolation, because the country was more thickly settled than the borders of the lake above named. *De la Beche's Manual, p.* 56.

Rem. It has been supposed, that should the falls of Niagara ever recede to lake Erie, a terrible inundation of the region eastward would be the result: but De la Beche has proved satisfactorily, that the only effect would be a gradual draining of lake Erie, with only a slight increase of Niagara river *Theoretical Geology, p.* 154.

Agency of the Ocean.

Descr. The ocean produces geological changes in three modes: 1. By its waves: 2. By its tides: 3. By its currents. Their effect is twofold: 1. To wear away the land: 2. To accumulate detritus so as to form new land.

Descr. The action of waves or breakers upon abrupt coasts, composed of rather soft materials, is very powerful in wearing them down, and preparing the detritus to be carried into the ocean by tides and currents. During storms, masses of rocks weighing from 10 to 30 tons, are torn from the ledges, and driven several rods inland, even up a surface sloping with a considerable dip towards the ocean.

Exam. In the 13th century, a strait half as wide as the channel between England and France, was excavated in 100 years in the north part of Holland: but its width afterwards did not increase. The English Channel also, has been supposed to have been formed in a similar manner. In England, several villages have entirely disappeared by the encroachments of the sea. At Cape May, on the north side of Delaware Bay, the sea has advanced upon the land at the rate of about 9 feet in a year; and at Sullivan's Island, near Charleston, S. Carolina, it advanced a quarter of a mile in three years. But perhaps the coast of Nova Scotia and New England, exhibits the most striking examples of the powerful wasting agency of the waves, whose force there is often tremendous, especially during violent northeast storms. Where the coast is rocky, insulated masses of rocks (in Scotland called *Drongs*,) are left on the shore, giving a wild and picturesque effect to the scenery, as in the following sketch, Fig. 95, which was taken upon Jewell's Island in Casco Bay.

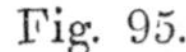

Fig. 95.

Drongs on Jewell's Island: Casco Bay.

Descr. It is difficult to examine the coast of Nova Scotia and New England;—to witness the great amount of naked battered rocks, and to see harbors and indentations chiefly where the rocks are rather soft, while the capes and islands are chiefly of the hardest varieties,—without being convinced that most of the harbors and bays have been produced by this agency. In Boston Harbor, the outer islands are composed of naked rock, and farther within the harbor, the outer borders of the islands are being swept of their loose soil. Here we see the steady progress of this encroaching process.

Fig. 96, shows the rocks and light houses (there is scarcely anything else to be seen,) on the island near the extremity of Cape Ann, in Massachusetts, which is peculiarly exposed to N. E. storms.

Fig. 97, will give an idea of the appearance of the islands in Boston Harbor, as it is entered from the north-east.

Purgatories.

Descr. When the rocks exposed to the waves are divided by fissures, running perpendicular to the coast, the mass between two fissures is sometimes removed by the water, thus leaving a chasm, often several rods long and very deep, into which the waves rush during a storm with great noise and violence. Such caverns have received in New England, the sin gular appellation of *purgatories.* Very good examples occur in the vicinity of Newport, Rhode Island.

Rem. Similar fissures with the same name occur in the interior; as in Sutton and Great Barrington in Massachusetts.

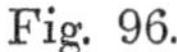
Fig. 96.

Extremity of Cape Ann.

Fig. 97.

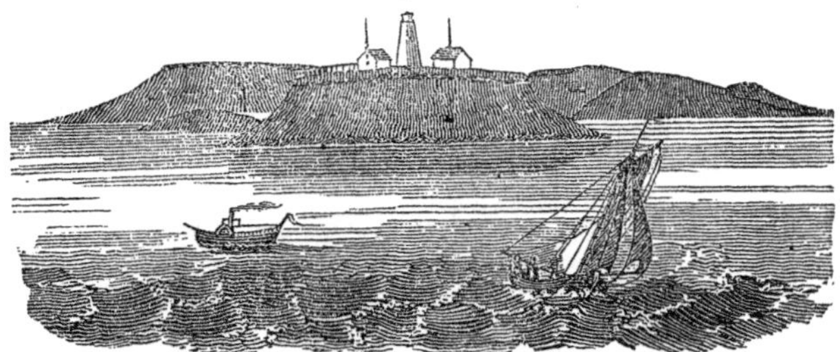

Denudation of Islands in Boston Harbor.

Beaches of Shingle and Sand.

Descr. The shingle, or perfectly water worn pebbles of a coast, and sand, are sometimes driven upon the shore by the waves, so as to form beaches; and sometimes even large bowlders are thus urged inland by powerful storms, so as to lie in a row on the shore. In some cases of this sort, after the beaches have been formed, the waves rather protect the coast than encroach upon it.

Dunes or Downs.

Descr. The sand which is driven upon the shore by the waves as above described, is often carried so far inland as to be beyond the reach of the returning wave; and thus an accumu lation takes place, which is the origin of most of those moving sand hills, known by the name of *dunes* or *downs.* When the sand becomes dry, the sea breezes drive it farther and farther inward; the land breezes not having equal power to force it back; and at length it becomes a formidable enemy, by overwhelming fertile fields, filling up rivers, and burying villages. Sometimes these dunes occur in the interior of a country.

Exam. Every one is familiar with the history of these dunes in Egypt. The westerly winds have brought in the sands from the Lybian desert, and all the west side of the Nile, with the exception of a few sheltered spots, has been converted into an arid waste. In Upper Egypt especially, the remains of ancient temples, palaces, cities, and villages are numerous among the drifting sands. In Europe, around the Bay of Biscay, a simi lar destructive process is going on. A great number of villages have been entirely destroyed; and no less than ten are now imminently threatened by sand hills, which advance at the rate of 60 or even 72 feet annually. On the coast of Cornwall in England, similar effects have taken place These dunes are also common on the coast of the United States, especially on Cape Cod in Massachusetts; where strenuous efforts have been made to arrest their progress, and to prevent the destruction of villages and harbors that are threatened.

Waves and Tides.

Rem. 1. It has generally been stated that waves do not affect the bottom of the ocean where the water is more than 20 feet deep. But the exact depth to which their disturbing influence extends, has not been accurately settled.

Rem. 2. It must be recollected in estimating the power of waves to remove rocks, that the weight of the latter in water is not much more than half their weight in air; and consequently that a much less force will remove them.

Descr. In large inland bodies of water, such as the Mediterranean, Black and Caspian Seas, and Lake Superior, tides are scarcely perceptible; never exceeding a few inches; and in the open ocean they are very small; not exceeding 2 or 3 feet: But in narrow bays, estuaries, and friths, favorably situated for accumulating the waters, the tides rise from 10 to 40 feet; and in one instance even 60 or 70 feet on the European coasts, and in the Bay of Fundy, in Nova Scotia, 70 feet. In such cases especially where wind and tide conspire, the effect is considerable upon limited portions of coast, both in wearing away and filling up. *De la Beche's Manual, p.* 85. *Lyell' Prin. Geol. Vol.* 2. *p.* 24.

Oceanic Currents.

Descr. Oceanic currents are produced chiefly by winds. The most extensive current of this kind is the Gulf Stream. This flows out of the Indian Ocean, around the Cape of Good

Hope, passes northward along the coast of Africa to the equator, thence across the Atlantic; being increased by the Trade winds: and impinging against South America, it is turned northward, and continues along the coast of the United States even to the banks of Newfoundland; from whence it turns east and southeast across the Atlantic, returning to the coast of Africa to supply the deficiency of waters there. It is estimated that this current covers a space 2000 miles in length, and 350 in breadth. Its velocity is very variable; but may be stated as from one to three and even four miles per hour; its mean rate being 1.5 mile. A current sets northward between America and Asia, through Behring's Straits, which passes around the northern extremity of America, and flows out into the Atlantic in two currents, one called the Greenland current, which passes along the American continent, at the rate sometimes of 3 or 4 miles per hour, until it meets and unites with the Gulf Stream, near the Banks of Newfoundland, where the velocity is two miles per hour: the other sets into the Atlantic between America and Europe. It is these two currents that convey icebergs as far south as the 40th degree of north latitude before they are melted. Among the Japanese Islands a current sets northeast, sometimes as strong as five miles per hour. Another sets around Cape Horn from the Pacific into the Atlantic Ocean. A constant current sets into the Mediterranean through the straits of Gibraltar, at less than half a mile per hour. It has been conjectured, but not proved, that an under current sets outwards through the same strait, at the bottom of the ocean. Mr. Lyell also suggests that the constant evaporation going on in that sea, may so concentrate the waters holding chloride of sodium in solution, that a deposit may now be forming at the bottom. But the deepest soundings yet made there, (5880 feet,) brought up only mud, sand, and shells. Numerous other currents of less extent exist in the ocean, which it is unnecessary to describe. They form, in fact, vast rivers in the ocean, whose velocity is usually greater than that of the larger streams upon the land. *De la Beche's Manual, p.* 91.

Descr. The ordinary velocity of the great oceanic currents is from one to three miles per hour: but when they are driven through narrow straits, especially with converging shores, and the tides conspire with the current, the velocity becomes much greater, rising to 8, 10, and even in one instance to 14 miles per hour. *Lyell's Prin. Geol. Vol.* 2. *p.* 29.

Descr. The depth to which currents extend has not been accurately determined. Some limited experiments seem to indicate that they may sometimes reach to the depth of nearly

500 feet. It ought to be remembered, however, that the friction of water against the bottom, greatly retards the lower portion of the current; so that the actual denuding and transporting power in these currents is far less than the velocity at the surface would indicate.

Descr. Alike uncertain are the data yet obtained, for determining what velocities of water at the bottom are requisite, for removing mud, sand, gravel, and bowlders. It has been stated, however, (and these are the best results yet obtained,) that 6 inches per second will raise fine sand on a horizontal surface, 8 inches, sand as coarse as linseed; 12 inches, fine gravel: 24 inches per second will roll along rounded gravel an inch in diameter: and 36 inches will move angular fragments of the size of an egg. The velocity necessary for the removal of large bowlders has not been measured. A velocity of 6 feet per second would be 4 miles per hour: of 8 feet per second, 5.4 miles per hour: of 12 feet per second, 8.2 miles per hour: of 24 inches per second, 16.4 miles per hour: of 36 feet per second, 24.6 miles per hour. Fine mud will remain suspended in water that has a very slight velocity, and often will not sink more than a foot in an hour; so that before it reached the depth of 500 feet, it might be transported by a current of 3 miles per hour, to the distance of 1500 miles. *De La Beche's Theoretical Geology, p. 56, and 64.*

Inf. 1. It appears that most rivers, in some part of their course, especially when swollen by rains, possess velocity of current sufficient to remove sand and pebbles; as do also some tidal currents around particular coasts: but large rivers and most oceanic currents can remove only the finest ingredients; and as to large bowlders, it would seem that only the most violent waves and mountain streams can tear them up and roll them along.

Inf. 2. Oceanic currents have the power greatly to modify the situation of the materials brought to the sea by rivers and tides, and to spread them over surfaces of great extent.

Exam. The waters of the Amazon, still retaining fine sediment, are found on the surface of the ocean 300 miles from the coast, where they are met by the Gulf stream, which runs there at the rate of 4 miles per hour. Thus are these waters carried northerly along the coast of Guiana, where an extensive deposit of mud has been formed, which extends an unknown distance into the ocean. In like manner, the muddy waters of the Oronoco and other rivers are swept northerly, and probably a deposit is going on along the whole coast of America as far north as the Gulf Stream extends. *Lyell's Prin. Geol. Vol. 2. p. 112.*

CHEMICAL DEPOSITES FROM WATER.

Calcareous Tufa, or Travertin.

Descr. In certain circumstances water holds in solution a quantity of carbonate of lime, which is readily deposited when those circumstances change. The deposit is called *travertin or calcareous tufa.*

Exam. At Clermont in France, a single thermal spring has deposited a mass of travertin 240 feet long, 16 feet high, and 12 feet wide. At San Vignone in Tuscany, a mass has been formed upon the side of a hill, half a mile long and of various thickness, even up to 200 feet. At San Filippo, in the same country, a spring has deposited a mass 30 feet thick in 20 years. And a mass is found there, 1.25 mile in length, one third of a mile wide, and in some places 250 feet thick. In the vicinity of Rome, some of the travertin can hardly be distinguished from statuary marble; and that which is constantly forming near Tabreez in Persia, is a most beautiful variety of semi-transparent marble, or alabaster. At Tivoli in Italy, the beds are sometimes from 400 to 900 feet thick, and the rock of a spheroidal structure. *Lyell's Prin. Geol. Vol.* 1. *p.* 397.

Marl.

Descr. The only kind of marl now in the course of formation, is that deposited at the bottom of ponds, lakes, and salt water, known by the name of *shell marl;* and which consists of carbonate of lime, clay, and peaty matter; as has been described in a preceding section. The marls in the tertiary strata are frequently indurated, and go by the name of rock marl. Much of the marl used in Virginia, and other southern States, is composed mostly of fossil marine shells; and this is a true *shell marl.* But that usually so called, contains only a small proportion of shells: the remainder being pulverulent carbonate of lime, except the clay and peaty matter, mixed with the carbonate. These beds of marl often cover hundreds of acres, and are several feet thick. In Ireland they contain bones of a large extinct species of elk, as well as shells of *Cypris, Lymnæa, Valvata, Cyclas, Planorbis, Ancyclus, &c.* The marls of this country contain shells of *Planorbis, Lymnæa, Cyclas,* and other small freshwater molluscs.

Rem. These alluvial deposits of marl have been generally supposed to be the result of the decomposition of the small shells which occur in them. But they seem to me only in part due to this cause. Carbonate of lime it is well known, is scarcely soluble in pure water. But if the water contain carbonic acid, and carbonate of lime be diffused in it, the acid will render it soluble. Yet the excess of acid is easily expelled, and then the salt will be deposited; as we know to be the case in many waters that are not thermal; as at the mouths of several of the streams that empty into the Mediterranean,

Nor will this deposit be necessarily crystalline ; for it may be pulverulent Now the waters in limestone regions frequently contain carbonic acid. They also often contain carbonate of lime, in a state of suspension, which has been worn from the rock. Hence the salt thus dissolved will be very likely to be deposited, when the solution containing it forms ponds, whose stagnant waters are liable to chemical changes sufficient for this purpose. Or if this be doubted, it is certainly very possible that the streams that empty into ponds, will carry thither minute particles of limestone, which have been worn from the rocks over which the waters have passed; and these will be deposited when the waters have become quiet. The largest part of these alluvial marls, that have come under my observation, appear to have been formed in one of these modes, and not by the disintegration of the shells. These are generally in a sound state, when the marl is first dug, whereas if the powdered part originated from them, we ought to find them in fragments of every size. *Henry's Chemistry, Vol.* 2. *p.* 612, *Eleventh Edition. Thomson's Inorganic Chemistry, Vol.* 2. *p.* 512.

Siliceous Sinter.

Descr. Thermal waters alone can contain silica in solution to any important amount. The most noted of these are the Geysers in Iceland, where a siliceous deposit, about a mile in diameter, and 12 feet thick, occurs ; and those of the Azores where elevations of siliceous matter are found 30 feet high. The stems and leaves of the frailest plants are converted into sinter or covered with it. Thermal springs, also, not in volcanic regions, as on the Washita river in this country, and in India, deposit a copious sediment of silica, iron, and lime.

Hydrate of Iron, or Bog Iron Ore.

Descr. It is probable, as will be shown in a subsequent section, that the greater part of the ferruginous deposits so widely diffused, originate from the fossil shields of animalcula. Yet in some instances we have direct evidence that they are produced by the decomposition of iron pyrites : for where such decomposition is going on, (as in the western part of Worcester County, in Massachusetts,) the rocks are coated over with the hydrate, and the surrounding soil deeply impregnated with it. Nor can there be any doubt but this iron would be often carried by water—although not directly soluble in it,—to the lowest places, and into ponds and rivers, so as to form deposits there.

Rem. The hydrate of manganese is almost as widely diffused through the rocks as the hydrate of iron ; but its quantity is so small that it exerts but a slight influence in the production of geological changes, and will therefore be passed without particular description. The same remarks will apply to sulphate of lime, carbonate of magnesia, chloride of calcium, &c. which occur in almost all natural waters, and sometimes form deposits of small extent.

Petroleum, Asphaltum, &c.

Descr. The great amount of bituminous matter with which certain springs are impregnated, renders them deserving of notice as existing causes of geological change, capable of explaining certain appearances in the older rocks; many of which are highly bituminous. In the Burman Empire, a group of springs or wells at one locality, yielded annually 400.000 hogsheads of petroleum. It is found also in Persia, Palestine, Italy, and the United States. In this country it has the name of Seneca Oil, from having been early observed on the surface of springs at Seneca in N. York. It is thrown up in considerable abundance, also, at the salt borings on the Kenhawa river in Ohio; where a few years ago a large quantity of it, floating on the surface of a small stream, took fire, and the river for a half a mile in extent appeared a sheet of flame. *Am. Journal of Science, Vol.* 24. *p.* 64. In Palestine the Dead Sea is called the lake Asphaltites, from the asphaltum which formerly abounded there. But the most remarkable locality of bituminous matter is the Pitch Lake in the island of Trinadad, in the West Indies. It is three miles in circumference, and of unknown thickness It is sufficiently hard to sustain men and quadrupeds; though at some seasons of the year it is soft. *Geological Transactions, Vol.* 1. *p.* 63.

Rem. Mineral pitch was a principal ingredient in the cement used in constructing the ancient walls of Babylon, and of the temple in Jerusalem. It has lately been employed in a similar manner, and it is said very successfully, to form a composition for paving the streets of cities.

Prin. The various bitumens are produced from vegetables by the processes by which these are converted into coal in the earth.

Inf. Hence the bitumens that rise to the surface of springs, or form inspissated masses on the earth's surface, or between the layers of rocks, are supposed to be produced from vegetable matters buried in the earth; and to be driven to the surface by internal heat; and the fact that such deposits usually occur in the vicinity of active or extinct volcanos, gives probability to this theory.

Phenomena of Springs.

Descr. Water is very unequally distributed among the different strata; some of them, as the argillaceous, being almost impervious to it; and others, as the arenaceous, admitting it to percolate through them with great facility. Hence when the former lie beneath the latter in a nearly horizontal position, the lower portions of the latter will become reservoirs of this fluid.

Inf. Hence if a valley of denudation cuts through these pervious and impervious strata, we may expect springs along their junction.

Illus. If B. B. Fig. 98, be the pervious, and C. C. the impervious stratum, and A. the valley of denudation, we may expect springs at E. E.

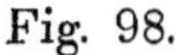

Fig. 98.

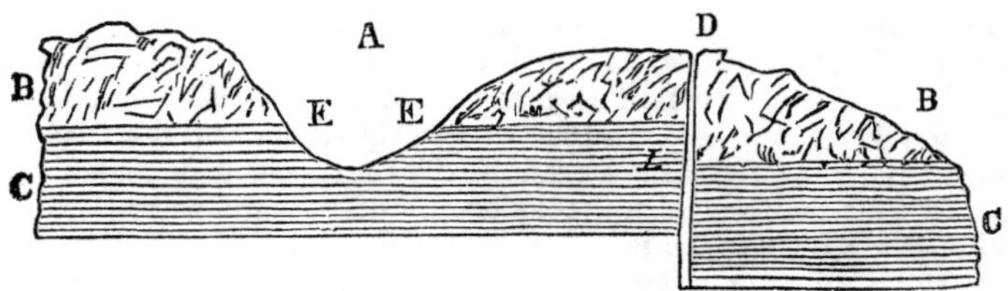

Descr. If a fault occur in these strata, as at D, whereby they are sunk on the right of D, and still dip towards L, the water will be accumulated at L, because it cannot pass into C, and a spring may be expected at L.

Rem. Sometimes the geologist can discover the line of a fault by the occurrence of springs, where nothing else indicates its existence at the surface.

Descr. In many parts of the world, if the strata be penetrated to a considerable depth by boring, water will rise, sometimes with great force, to the surface, and continue to flow uninterruptedly. Such examples are called *Artesian wells ;* from having been first discovered at Artois, the ancient Artesium.

Theory. The theory of these wells is simple. In Fig. 31, (p. 48.) suppose the formation marked as the upper coals, and also the millstone grit to be impervious to water : while the lower coal is pervious, or the water bearing stratum. Now if excavations be made at B, or E, till the coal strata are reached, it is obvious that water will be forced to the surface by hydrostatic pressure : because some part of the water bearing stratum is higher than the points B, and C.

Inf. 1. If any water bearing stratum, passing under a place where boring is attempted, rises higher at any point of its prolongation than the surface where the boring is made, the water will rise above that surface: and it will fall as much below that surface as is the level of the highest part of the pervious stratum.

Inf. 2. Hence borings of this sort may fail; first, because no water bearing stratum is reached; and secondly, because that stratum does not rise high enough above the place to bring the water to the surface.

Inf. 3. These explorations have proved that subterranean streams of water exist: some of which have a communication with water at the surface.

Exam. At St. Ouen in France, at the depth of 150 feet, the borer suddenly fell a foot, and a stream of water rushed up. At Tours the water brought up from the depth of 374 feet, fine sand, vegetable matter, and shells of species living in the vicinity, which must have been carried to that depth within a few months preceding. In Westphalia the water brought up several small fish, although no river existed at the surface within several leagues. The borings in the United States prove that cavities containing water exist even in granite.

Depths of the Borings. In England, Artesian wells have been carried to the depth of 620 feet with success. In France, they have been sunk 800 and even 1200 feet, and in one instance near Paris to 1800 feet before water was reached. In the United States, borings for salt water in the Western States, have been carried as deep as 800 or 900 feet. In the cities of New York, Baltimore, Albany, and in various parts of New Jersey, &c. borings for fresh water have been carried, and in most instances with success, to the depth of nearly 400 feet, though water has usually been obtained at a much less depth. An excavation in the city of New York, 100 feet deep and 16 feet diameter, yields 8000 gallons daily; and that in Bleeker Street, 442 feet deep, yields 44.000 gallons daily. *Am. Journal of Science, Vols. XII. p.* 136, *and XXIII. p.* 206.

Rem. 1. Until recently these borings have been generally performed by means of a continuous iron rod, sharpened like a drill at the lower end. But a far more convenient and economical method, which has long been in use in China, has lately been adopted: viz. to use a heavy cylinder of iron in the same manner, by means of a rope attached to its upper end; a borer with valves being connected with the lower end, for bringing up the comminuted materials. *Buckland's Bridgewater Treatise, Vol.* 1. *p.* 568.

Rem. 2. Thermal Springs will be considered under the eighth section.

Salt and other Mineral Springs.

Descr. All waters found naturally in the earth, contain more or less of saline matter: but unless its quantity is so great as to render them unfit for common domestic purposes, they are not called mineral waters.

Descr. The ingredients found in mineral waters, are the sulphates of ammonia, soda, lime, magnesia, alumina, iron, zinc, and copper: the nitrates of potassa, lime and magnesia: the chlorides of potassium, sodium, barium, calcium, magnesium, iron, and manganese; the muriate of ammonia; the carbonates of potassa, soda, ammonia, lime, magnesia, alumina, and iron; the silicate of iron; silica, strontia, lithia, iodine, bromine, and organic matter; the phosphoric, fluoric, muriatic, sulphurous, sulphuric, boracic, formic, acetic, carbonic, crenic, and apocrenic acids; also oxygen, nitrogen, hydrogen, sulphuretted hydrogen, and carbureted hydrogen. *Ure's Chemical Dictionary, Article, Water. See also Dr. Daubeny's admirable Report to the British Association, on Mineral and Thermal Waters*, 1837, *p.* 14.

Theory. Many of the above ingredients are taken up into a state of solution from the strata through which the water percolates: others are produced by the chemical changes going on in the earth, by the aid of water and internal heat; and others are evolved by the direct agency of volcanic heat.

Salt Springs.

Descr. The most important mineral springs in an economical point of view, are those which produce common salt. These are called salines, or rather such is the name of the region through which the springs issue. They occur in various parts of the world; and the water is extensively evaporated to obtain table salt. They contain also other salts; nearly the same in fact, as the ocean.

Exam. Some of these springs contain less, but usually they contain more salt, than the waters of the ocean. Some of the Cheshire springs in England yield 25 per cent.: whereas sea water rarely contains more than 4 per cent. In the United States they contain from 10 to 20 per cent. They are used in New York, Ohio, Virginia, Pennsylvania, Illinois, Michigan, Missouri, Arkansas, and Upper Canada. 450 gallons of the water at Boon's Lick in Missouri, yield a bushel of salt: 300 gallons at Conemaugh, Penn.: 280 at Shawneetown, Ill.: 120 at St. Catharine's, U. C.: 75 at Kenawha, Vir.: 80 at Grand River, Arkansas: 50 at Muskingum, Ohio: and 41 to 45 at Onondaga, N. Y.: 350 gallons of sea water yield a bushel at Nantucket. In 1829, according to a Report of the Secretary of the Treasury, 3.804.229 bushels of salt were made in the United States. Since that time the quantity has greatly increased. In 1841, no less than 3.340.769 bushels were made at the Onondaga Springs in New York alone; and 3.000.000 bushels at the Kenawha Springs in Virginia, in 1835. In all these places deep borings are necessary, sometimes even as deep as 1000 feet: and usually the brine becomes stronger the deeper the excavation. *Professor Beck's Geological Report to the Assembly of New York*, 1838. *See also Prof. W. B. Rogers's Report of the Geological Reconnoissance of the State of Virginia*, 1836. *Dr. Hildreth's first annual Geological Report to the Legislature of Ohio*, 1838: *Also his article on the Geology of Ohio, Am. Journal Science, Vol.* 29. *Also Mr. Foster's Geological Report on Ohio*, 1839: *and Dr. Houghton's Report on the Geology of Michigan*, 1838.

Origin of Salt Springs. In many parts of Europe salt springs are found rising directly from beds of rock salt; so that their origin is certain: but as yet no deposits of rock salt have been discovered in this country east of the Rocky Mountains except in Virginia: and Mr. Eaton has suggested (*Survey of the Erie Canal Rocks, p.* 110.) that the ingredients only for the formation of the salt exist in the saliferous rock, and are made to combine by chemical agencies, so that the water percolating through the strata would become impregnated. An English writer (*Annals of Philosophy*, 1829,) supposes that the salt is intimately disseminated through the saliferous rock, having been left there by the ocean that deposited the strata. Most American geologists, however, still maintain that our salt springs proceed from beds of rock salt, deposited so deep in the earth that they have not yet been discovered: and the fact that the brine increases in strength by descending, gives strong support to this theory, which is confirmed by the discovery of rock salt in Virginia. *Prof. Beck's Report*, 1838, *p.* 14. The springs in this country issue almost invariably from the Silurian rocks.

Gas Springs.

Descr. Carbonic acid and carbureted hydrogen are the most abundant gases given off by springs. They sometimes escape from the soil around the springs, over a considerable extent of surface, and produce geological changes of some importance. Carbonic acid, for example, has the power of dissolving calcareous rocks, and of rendering oxide of iron soluble in water. It contributes powerfully also, to the decomposition of those rocks that contain felspar. Carbureted hydrogen is sometimes produced so abundantly from springs, that it is employed, as at Fredonia in New York, in supplying a village with gas lights. At Kenawha, in Virginia, it is so abundant that it has been employed for boiling down the salt water that is driven up by it with great force. In almost all the States west of New England, this gas rises from springs in greater or less abundance, generally from salt springs.

Origin of these gases. Some of these gases, as carbonic acid, are given off most abundantly from springs in the vicinity of volcanos: and in such a case there can be no question but they are produced by decompositions from volcanic heat. When they proceed from thermal springs, there is a good deal of reason for believing that internal heat may have produced them. But where they rise from springs of the common temperature, they must generally be imputed to those chemical decompositions and recompositions that often occur in the earth without an elevated temperature. Although carbureted hydrogen may sometimes proceed from beds of coal, it may also proceed from other forms of carbonaceous matter; as from bitumen disseminated through the rocks.

Glacio-Aqueous Agency between the Tertiary and Historic Periods: formerly called Diluvial Action.

Rem. That remarkable deposit occurring between the tertiary and alluvial formations has generally been supposed to be the result of powerful currents of water, although the hypotheses founded upon this principle have been exceedingly unlike in their features. In the first edition of this work, I expressed the opinion that no theory hitherto proposed to explain the origin of this deposit, was satisfactory; and that the time had not yet come for a complete theory on the subject. But the subject has recently been so ably elucidated by Agassiz, Buckland, Lyell, and others, that it seems at least extremely probable, that all the phenomena of what has been called diluvial action, are the result of the joint agency of ice and water. This is what is meant by the term, *glacio-aqueous agency*. In the sequel it will be fully explained. But in order to judge of the correctness of this view, the facts respecting drift should be first presented. Its lithological characters have been already given. The remainder of the subject will embrace, 1. its dispersion: 2. its effects upon the earth's surface: 3. the theory of its origin.

1. *Dispersion of Drift.*

Descr. Drift is distinguished from tertiary deposits by three marks: 1. The tertiary strata were deposited in limited troughs and basins, whereas drift is found over almost every part of the northern regions of the globe, and at almost every altitude attained by mountains: and it has, therefore, been the result of some very general cause. 2. Such are the situation of the principal part of drift, and its general unstratified character, that it could not have been deposited in its present situation by water: and yet the sand and clay that constitute its upper portion, must have been deposited in quiet waters. 3. Drift is almost destitute of organic remains of animals and plants that lived during the time of its production, whereas the reverse is the case with the tertiary strata. The upper and regular layers of drift are almost entirely destitute of organic remains. 4. When the tertiary strata were deposited, the climate of northern countries was warmer than at present; but during the formation of drift, it was colder.

Descr. Drift is distinguished from alluvial deposits: 1. By its occurrence in situations where no agency at present in action could have produced it. 2. By requiring if not a different agency from any now in operation to produce drift, at least a greater intensity of action. 3. By the evidence of a great difference of climate between the two periods.

Descr. In the dispersion of drift we find the evidence of two distinct phases of action, which may however have been the result of the same general cause, operating in different circumstances. In the first case, the drift has been carried outward from the summits and axes of particular mountains, and spread over the neighbouring plains.

Exam. 1. The best example of this mode of dispersion exists in the Alps. The Bowlders there have usually been carried down the valleys, and they exist in the greatest abundance opposite the lower opening of those valleys. This shows that the valleys existed when the dispersion took place. Yet Elie de Beaumont seems to have shown that the Alps have experienced very great vertical movements since that period.

Exam. 2. Several similar instances have recently been pointed out by Dr. Buckland and Mr. Lyell as existing among the mountains of Scotland and the north of England, whose details cannot here be given. *Proceedings of the Geol. Soc. Lond. No.* 72, *Nov. and Dec.* 1840.

Exam. 3. Rozet describes the plain of Metidja, south of Algiers, as covered in its northern parts by bowlders derived from a long chain of hills running along its northern border: while its southern part is strewed over by bowlders from the Atlas chain, which stretches along its southern border. *Traite Elementaire de Geologie, p.* 259.

Descr. In the second case, the agency by which drift has been dispersed, has operated on a more extended scale, and driven it in a southerly direction over all the northern hemisphere, often to a great distance.

Proof. To begin with the American continent at the northeasterly point where observations to be depended upon have been made: we find that the bowlders spread over the southern part of Nova Scotia were derived, according to Sir Alexander Coke and Messrs. Jackson and Alger, from the ledges in the northern part of the province. *Trans. Amer. Acad. Vol.* 1. *New Series, p.* 302. Through the whole extent of Maine, the evidence is very striking of the southerly transport of the drift, the course being usually a few degrees east of south. And transported bowlders are even found towards the summit of Mount Katahdn, which is 5300 feet high. *Dr. Jackson's First and Second Reports on the Geology of Maine,* 1837 *and* 1838. *Also his Second Report on the Public Lands of Maine and Massachusetts, p.* 16.

Descr. In Massachusetts, the direction, taken by the drift, as shown by a multitude of examples, varied from north and south to northwest and southeast; the most usual course being a few degrees east of south. This course carried the current very obliquely across most of the precipitous ridges of mountains in the state; nevertheless, the bowlders held on in the general direction with remarkable uniformity. The largest blocks usually lie nearest to the bed from which they were derived, and they continue to decrease in size and quantity, in a southeasterly direction, for the distance of several miles; sometimes as many as 50 or 60; and not unfrequently even 100 miles, though usually the sea coast is reached short of that distance. But often bowlders from the continent are common upon the islands many miles distant from the coast; as on Nantucket, Martha's Vineyard, and Long Island. In the western part of Massachusetts the mountains are from 1000 to 3000 feet high: yet vast quantities of bowlders have been carried over these precipitous ridges, and both slopes are covered with them; the largest being upon the northern side. *Final report on the Geology of Massachusetts, p.* 379*: See also Plate* 53 *of that work; where the course of the drift is marked. Also American*

Biblical Repository, Vol. 10. *p.* 338. On Long Island the drift corresponds to the rocks on the continent: those of different kinds always lying south of the ledges from which they were derived. *Prof. Mather's first annual Report on the first Geological District of New York, p.* 88. 1837. In the eastern part of N. York, the course was southeasterly; as in the western part of Massachusetts. But towards the western parts of the State, its general course appears sometimes to have been west of south. *Mr. Hall's second annual Report on the Fourth Geological District of New York, p.* 308. In the southeasterly part of the state, bordering on Pennsylvania and New Jersey, its direction varied from south several degrees west, to southeast: and near the city of N. York the course was N. W. and S. E. *American Journal of Science, Vol.* 23. *p.* 243. *And Vol.* 16. *p.* 357. *Also Prof. Gale's Report for* 1839 *upon the Geology of the First District.* In the fossiliferous region of western New York, and in the states south of the western lakes, great numbers of bowlders of primitive rocks are strewed over the surface, significantly called *lost rocks.* These have been satisfactorily traced to the beds from which they were derived in the west part of Michigan and on the north side of the lakes in Upper Canada. *See the papers of the Messrs. Lapham in Vol.* 22, *and of Dr. Hildreth in Vol.* 29 *of American Journal of Science. Also the Geological Reports on the state of Ohio and Michigan.* Similar evidence of a southeasterly drift exists in Virginia. *Prof. W. Rogers's Report on the Geological Reconnoissance of the State of Virginia, p.* 16. According to Dr. Drake, primitive pebbles occur on the right bank of the Mississippi as far south as Natchez. *American Journal of Science, Vol.* 22, *p.* 209.

Descr. According to Mr. Catlin, (*American Jour. of Science, Vol.* 38, *p.* 143.) vast quantities of bowlders of primary rocks "are strewed over the great valley of the Missouri and Mississippi, from the Yellow stone almost to the Gulf of Mexico," which have been drifted thither from the northwest. At the Red Pipe Stone quarry on the Coteau des Prairies, which is several hundred miles west of Lake Superior, he describes five granite bowlders, from 15 to 25 feet in diameter, which he supposes must have been drifted several hundred miles from the north.

Descr. The distance to which bowlders have been driven southeasterly from their native beds in our country, has not been very satisfactorily determined. In New England they have been traced rarely more than 100 to 200 miles: But in the western states they are strewed over a greater distance. I am informed by the gentleman engaged in the geological surveys of those states, that primary bowlders are rarely found south of the river Ohio; but they are strewed over almost every part of Ohio and Michigan. Now the primary rocks from which they have been derived, are found on the north side of the great lakes. This would make their longest transit between 400 and 600 miles.

Rem. It may possibly be found that some examples of the dispersion of drift in our country have been down the slopes of particular mountains, or that the general course has been greatly deflected by the peculiar features of the surface. But this would not destroy the evidence, which is almost everywhere so abundant, of a very wide and powerful force from the north over this continent.

Descr. On the eastern continent the evidences of a southerly direction of the force seems to be decided: although from some of the highest mountains it was outward from the axis. In Great Britain the general course was a little east of south, modified, however, and sometimes very much changed, by the shape of the mountains; some of which, as the Penine chain, appear not to have been passed over by the bowlders, except at their lowest points. In the east part of England, the drift appears to have been derived from Scotland, and also from Norway. *De La Beche's Manual, p.* 189. *Phillips's Geology, p.* 208. *Also his Treatise on Geology; Vol.* 1. *p.* 274. On the continent of Europe, the Netherlands, Denmark, the plains of the north of Germany, of Poland, and Russia, are strewed over with bowlders and pebbles, which can be traced to the parent rocks in Sweden, Lapland, and Finland; in which countries they are yet more numerous upon the surface. In most cases these bowlders must have crossed the Baltic. In Sweden the current appears to have set S. S. W. The blocks decrease in size on going south, and finally at a great distance (more than 400 miles, *Greenough's Geology, p.* 138.) they disappear. *Tableau des Terrains par Al. Brongniart, p.* 77. *Traite Elementaire de Geologie par M. Rozet, Tome,* 1. *p.* 270. *De La Beche's Manual, p.* 189.

Descr. An interesting example of the dispersion of bowlders in a southerly direction in Northern Syria, is given by Mr. Beadle, American Missionary in that country. On the coast, 60 or 70 miles north of Beyroot, he "reached a volcanic region with a remarkable locality of greenstone. The pebbles from this locality are scattered the whole distance to Beyroot. At that place they are quite small, but gradually increase in size as you advance to the north, and terminate entirely in this locality." *Missionary Herald for May* 1841, *p.* 206. This is an important fact; because it proves the occurrence of glacio-aqueous action on the Asiatic continent much farther south (about 32° N. lat.) than had been before pointed out; unless it be upon the Himalayah Mountains.

Descr. According to Mr. Darwin, the equatorial regions of South America exhibit but few marks of glacio-aqueous action, or rather they are destitute of bowlders. But beyond 41° South latitude, they appear in Chili and Patagonia. Hence some geologists (Lyell and Darwin, see *Lyell's Elements*,) infer that the phenomena of drift are limited to the colder regions of the globe. But De La Beche describes drift as abundant in Jamaica in the West Indies; especially on the plain around Kingston; and says that it appears to have been brought from the north. *Geological Trans. Second Series, Vol. 7. p.* 182 A similar statement was made to me by the late Prof. Hovey, who resided two years in the West Indies. Prof. Struder states that in the hill country at the foot of the Himalayah Mountains in India, erratic bowlders occur. *American Journal of Science, Vol.* 36. *p.* 330. We have also seen above, that similar phenomena occur in Africa, near Mount Atlas, in N. latitude about 32°; Mr. Darwin, however, attempts to explain such cases by the operation of other causes, and very probably he is correct; though it is possible that high mountains, even within the tropics, may have been subject to glacio-aqueous agency, though no marks of it appear upon the surface generally. *Darwin's Journal of Researches, &c., p.* 289 *and* 615. *Also a paper by him on the Bowlders of S. America, and read April* 1841, *before Lond. Geol. Soc.* More recently Sir Robert Schombergh has described enormous, far transported bowlders in British Guiana. *Murchison's Geology of Russia, Vol.* 1. *p.* 551.

2. *Effects of Glacio-aqueous Action upon the Earth's Surface.*

Ancient Moraines.

Descr. The most abundant accumulations of drift are found in hilly regions, chiefly, however, in the valleys, and especially near gorges and defiles. This detritus in such places usually consists of rounded bowlders, pebbles, and sand, and even mud, piled up in ridges, straight, curved, and tortuous, and sometimes in regular tumuli. These correspond almost precisely to the moraines of the Alps, except that they are often on a much larger scale and more altered in shape by subsequent alluvial agencies.

Rem. In this work the term *moraine* includes those accumulations of detritus produced by the grating of icebergs along the bottom of an ocean, or any body of water, as well as those formed by common glaciers.

Illus. Prof. Agassiz, Dr. Buckland and Mr. Lyell, have recently pointed out numerous examples of ancient moraines and correspondent striæ upon the rocks, in Scotland and the North of England; and since they are familiar with the moraines of existing glaciers in the Alps, we can hardly suppose them mistaken respecting those in Great Britain. *Proceed. Geol. Soc. Lond. No.* 72.—1840—1841.

Illus. In New England these accumulations are very common, and sometimes they are so crowded together as to exhibit a picturesque appearance, being made up of tortuous and conical elevations with deep intervening cavities, as if scooped out by the hands of a Titan. The most remarkable examples that I have ever seen, are in the vicinity of Plymouth in Massachusetts, and near the extremity of Cape Cod in Truro, where they are sometimes 200 or 300 feet high. In Truro they are composed wholly of sand, and they give a singular aspect to the landscape. Fig. 99, represents a small portion of the surface near what is called the Harbour in Truro.

Fig. 99.

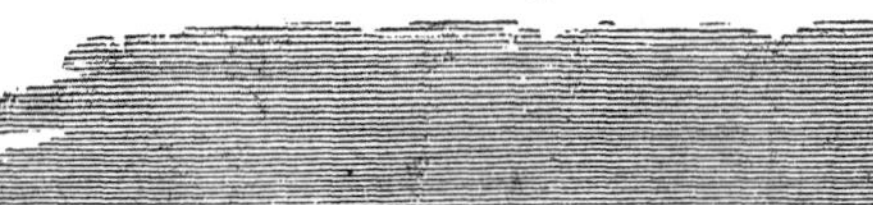

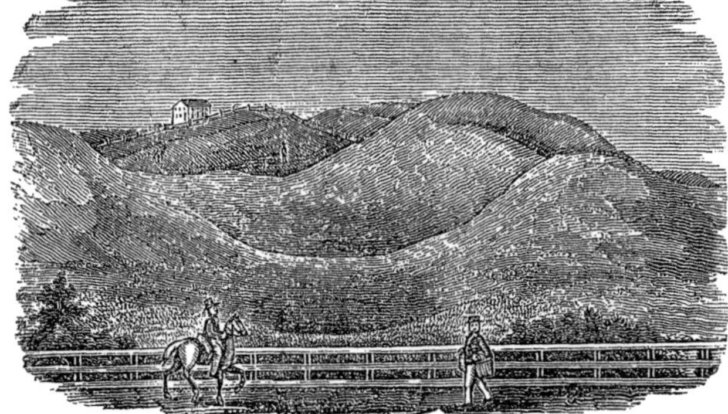

Moraines in Truro.

Descr. In Plymouth and Barnstable counties these moraines are distant from 50 to 100 miles from any mountains much more elevated than themselves.

Fig. 100, exhibits a similar group of moraines, though much less elevated, in the east part of Amherst, about two miles from the College. These are made up entirely of gravel and sand, with perhaps a few bowlders.

Fig. 100.

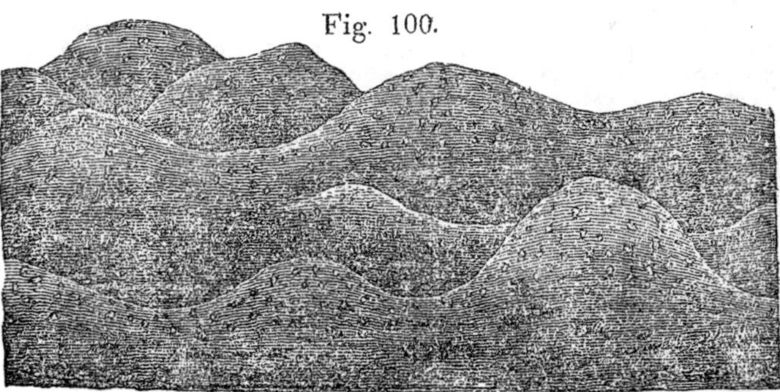

Moraines: Amherst.

Rem. 1. Before the Rev. Justin Perkins, now American Missionary at Ooroomiah in Persia, left this country, I showed him the moraines exhibited on the above figure, and requested him to notice whether any similar appearances exist in central Asia, especially in Armenia. In a letter received from him after he had passed through that country, he says, that before he reached Mount Ararat, and on the vast plain on its north side, "we passed many sections of diluvium much like the one we visited back of Amherst." Being now (March 1842,) on a visit to this country, he informs me, on the authority of Rev. Mr. Johnson, American Missionary at Trebrond, on the Black Sea, that such moraines are common in Cilicia in Asia Minor.

Rem. 2. In the above cases we may be sure that these moraines remain almost exactly as they were left by the ice: for it is impossible that water could have subsequently altered their form essentially: for the cavities between the elevations are not valleys, through which water may have flowed, but irregular depressions.

Descr. In the two following cases the round-topped tumuli of gravel may be only the remnants of a former lateral moraine, intersected by streams from the mountain. Fig. 101, shows a row of tumuli, some of them 100 feet high, a little south of the village of North Adams in Massachusetts, at the foot of Hoosac Mountain.

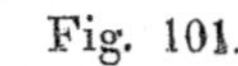

Fig. 101.

Moraine: N. Adams.

Illus. A similar group of moraines, but less conical, is shown on Fig. 102, as they appear at the east part of Monument Mountain, in Berkshire County, Massachusetts.

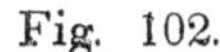

Fig. 102.

Moraines; Monument Mt.

Rem. 1. In the Biblical Researches of Robinson and Smith, tumuli similar to the above are described as existing near Jericho, on the west side of the river Jordan, not far from its mouth. I requested Dr. Robinson, on a recent visit to Amherst, to go to the spot in that town represented on Fig. 100: and he assures me that the materials are the same in the tumuli near Jericho as in the moraines in Amherst; but that the former are insulated and conical. *Researches, Vol.* 2. *p.* 284, 297, 298.

Rem. 2. I do not find any description of such moraines as are figured above in Agassiz's work on the Glaciers of the Alps. But Dr. Buckland says that they occur there, and there can be no doubt but what he describes as ancient moraines in Scotland, are identical with those accumulations of gravel in this country thus denominated above. Indeed, it is impossible to explain their existence by the action of currents of water alone; and ice is the only other known agent that could produce them.

Descr. Sometimes these moraines constitute ridges of considerable height, which extend a long distance with steep sides. The most remarkable examples which I have met with, occur about one mile west of the Theological Seminary in Andover, Mass. In company with Rev. Alonzo Gray of the Teacher's Seminary in that place, I recently (April, 1842,) traced one of these, called the Indian Ridge, nearly two miles. It is considerably tortuous, and in some places is interrupted for several rods. Mr. Gray informs me that he has found other ridges parallel to this. One exists, also, near the village of South Reding, ten miles from Boston. But an accurate description of these most interesting examples cannot be given in this place.

Descr. When moraines have been cut through and worn down by the action of streams of water, the detritus becomes more or less stratified by the sorting of the materials. Such accumulations are called the *detritus of moraines*, and they are not uncommon in this country.

Illus. Fig. 103, shows a cliff of this sort on the banks of Connecticut river near Cobb's ferry in Montague, in Massachusetts, about 4 rods long and 20 feet high. Not only have we here stratification, but that oblique lamination which is the result of deposition from water upon a steep slope. The materials are sand and gravel.

Fig. 103.

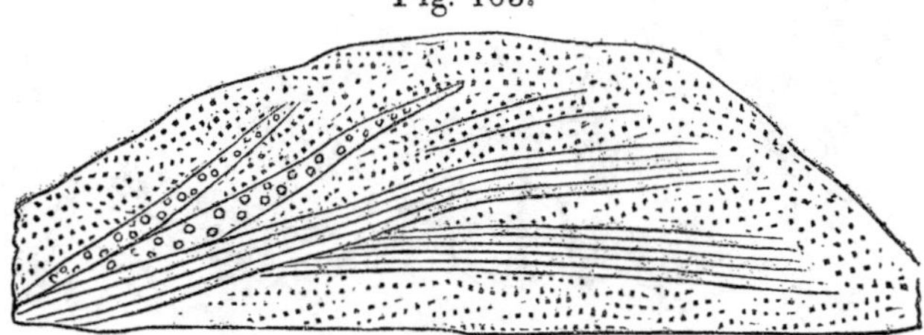

Fig. 104, shows a case developed in Palmer, Massachusetts, by the excavation of the western rail road. The bank is 20 rods long and 50 feet high; and although most of the cliff is gravel, sand, and coarse bowlders, yet in the midst are deposits of fine blue clay. Some part of the gravel is also stratified, with a dip of 25°. I suspect that in this case the hill was moved by a mass of ice after the clay was deposited.

Fig. 104.

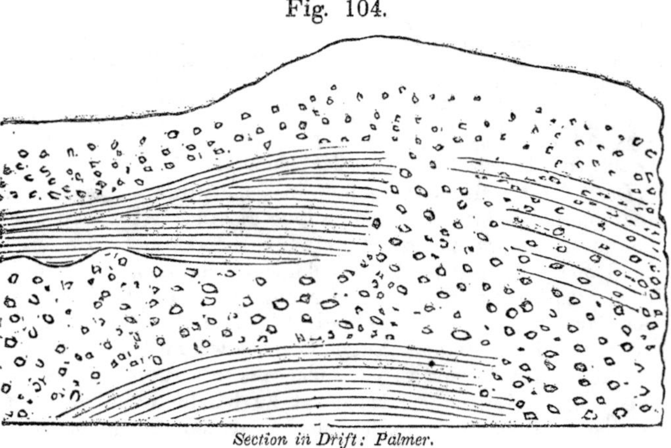

Section in Drift: Palmer.

Inf. There is reason to suspect that some of those mounds in our Western States that have been regarded as the work of man, are conical moraines, or the detritus of moraines. We have testimony that they are sometimes stratified, and that they are too large and numerous to be all artificial, although artificial mounds do undoubtedly exist there. *Illinois Magazine, Vol.* 2. *p.* 252. *Am. Jour. Sci., Vol.* 34. *p.* 88.

Descr. Frequently the surface is almost entirely covered for many square miles with large transported blocks of stone, which are but little rounded. These were most probably transported on icebergs.

Illus. One can hardly pass over any of the hilly parts of New England without meeting with more or less of these blocks. Perhaps the west part of Cape Ann in Massachusetts, exhibits them in as great abundance as any place. Fig. 105, will give some idea of a landscape near Squam in Gloucester.

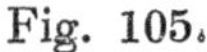

Fig. 105.

View in Gloucester, Mass.

Descr. The size of single bowlders is sometimes enormous. The block out of which was hewn a pedestal for the statue of Peter the Great, weighed 1500 tons. The Needle Mountain in Dauphiny, said to be a bowlder, is 1000 paces in circumference at the bottom, and 2000 at the top. Near Neufchatel is a block of granite 40 feet high, 50 feet long, and 20 broad, which weighs 3.800.000 pounds. The block called *Pierre a Martin*, contains 10.296 cubic feet. *Greenough's Geology, p.* 131. Prof. Forbes describes another bowlder in the Alps 100 feet long, and 40 to 50 feet high; also another 62 feet in diameter, containing 244.000 cubic feet. *Am. Jour. Sci., Vol.* 36. *p.* 184. The rock in Horeb, from which according to monkish tradition, Moses miraculously drew water, is a block of granite, 6 yards square, and contains 5823 cubic feet, lying in the plain near Mount Sinai, from which it was probably detached. This however has not probably been removed except by gravity. *Greenough,*

p. 127. In this country bowlders occur of equal dimensions. Thus, on Cape Ann and its vicinity, I have not unfrequently met with blocks of syenite not less than 30 feet in diameter; and in the southeast part of Bradford, I noticed one 30 feet square; which contains 27.000 cubic feet, and weighs not less than 2310 tons. In the west part of Sandwich on Cape Cod, I have seen many bowlders of granitic gneiss, 20 feet in diameter, which contain 8 000 cubic feet and weigh as much as 680 tons. Two graywacke bowlders of the same size lie a few rods distant from the meeting house in Norton, in Dr. Bates's garden. A granite bowlder of equal dimensions lies about half a mile southeast of the meeting house in Warwick; and one of similar dimensions lies on the western slope of Hoosac mountain in the northeast part of Adams, at least 1000 feet above the valley over which it must have been transported. One of granite lies at the foot of the cliffs at Gay Head on Martha's Vineyard, which is 90 feet in circumference and weighs 1447 tons. In Winchester, New Hampshire, I recently met with a block of granite 96 feet in circumference. It is near the road leading to Richmond. I noticed another in Antrim in that state 150 feet in horizontal circumference. Finally, at Fall River is a bowlder of conglomerate, which originally weighed 5400 tons or 10.800.000 pounds.

Descr. Glacio-aqueous action appears to have destroyed numerous species of animals that inhabited the northern regions of the globe at the time of its occurrence.

Proof. In accumulations of drift in the northern hemisphere, have been found the bones of several species of mastodon, hippopotamus, rhinoceros, bear, as well as the mammoth or elephant, megatherium, megalonyx, hyæna, deer, dinotherium, horse, ox, &c. animals of whose existence since that event we have no evidence. Not less than 100 species have already been found in drift, although not more than half are extinct.

Inf. A sudden fall of temperature took place in the northern hemisphere at the period of glacial action.

Proof. The animals whose remains are found in drift are mostly such as live in tropical climates; which shows that a higher temperature than now exists in these countries, prevailed at the commencement of glacial action. And that the change was sudden, appears from the occurrence of elephants and rhinoceros undecayed in the frozen mud of Siberia: for they must have been encased suddenly in ice to prevent their putrefaction.

Smoothed, Striated, and Furrowed Rocks in Place.

Descr. One of the most remarkable effects of glacial action, is the smoothing, rounding, scratching and furrowing of the surface of rocks in place.

Rem. 1. Although this is a very common phenomena, especially in the United States, yet until within a few years, scarcely any example had been pointed out by geologists: a proof of the little careful attention that had been given to the phenomena of drift. The following examples are only a selected few, out of the multitudes which have been observed in this country.

Rem. 2. Care must be taken by the observer, not to confound these furrows and striæ with those grooves on the surface of rocks produced in the direction of the cleavage planes, or the planes of stratification, by the unequal disentegration of the harder and softer parts; nor with the furrows between the veins of segregation, produced in the same manner; nor with ripple marks. In fine, it is best not to consider any case as of decided glacial origin, unless the grooves cross the planes of cleavage and stratification at a considerable angle. Such are all the cases mentioned below.

Exam. In the state of Maine is a good deal of slaty rock, often standing upon its edges, that admirably resists atmospheric agencies: and hence it presents a multitude of examples of well marked striæ. Around the city of Portland, they are very abundant and very distinct; having a direction N. 10° to 15° W. and S. 10° to 15° E. Farther east as at Hope and Appleton, they run nearly N.W. and S.E. and some of them are a foot in depth, and six inches wide. *First Report on the Geology of Maine, p.* 57. In other parts of the State, the direction is nearly north and south, or even inclining a few degrees to the N.E. and S.W. *Second Report on the Geology of Maine, p.* 91. In the eastern part of Massachusetts and Rhode Island, I might name fifty places where the striæ are obvious and distinct. In Essex County, Mass. they are very frequent on the hard syenite rocks: though often these are merely smoothed, and sometimes almost polished. They are visible on the gneiss at the top of the Wachusett mountain, the highest in the eastern part of Massachussetts; being 2000 feet above the ocean. The precipitous hills and the lower grounds of the valley of Connecticut river, are covered with them; and here as well as in nearly all the eastern part of the state, their direction is nearly north and south, usually however inclining a few degrees to the East of South, and West of North, and very rarely the other way. The high mountains west of Connecticut river, embracing the Hoosac and Taconic ranges, some points of which rise 3500 feet above the ocean, exhibit very numerous examples of the smoothing and furrowing effect of this supposed glacial agency. Graylock, the highest point in Massachusetts, is so covered with soil and trees, that the rocks are very rarely seen: but on that spur of the mountain called Bald Mountain, whose top is a few hundred feet below Graylock, the scratches are obvious, as they are also on the northeast side of the mountain. The high conical mountain in the town of Mount Washington, (called Mt. Everett,) which is 2600 feet above the ocean, has been worn over its whole surface, and the striæ are still visible in many places; although the rock has been so long exposed naked to atmospheric decomposing agencies. Similar markings are manifest on the top of Tom Ball, a high mountain in Alford, and they may be seen throughout the whole extent of the Taconic range, which on its west side is very precipitous. A large proportion of these grooves run N.W. and S.E. but some of them approach nearer to a coincidence with the meridian. The rock is mica and talcose slate, and in some instances is laid bare for a great distance, so as to show the smoothing and furrowing of its surface over a large extent. Indeed, one cannot stand upon these lofty and precipitous ridges, and witness this phenomenon, without being struck with the great power and extent of the agency that has thus left its traces upon some of the most elevated spots in N. England.—Between these mountains and Hudson river the graywacke is crossed by markings running N. W. and S. E. Near the city of New York, according to Prof. Gale, they run in the same direction. In the western part of New York, they are numerous, and they run sometimes S. S. W. and N. N. E. They are common also, upon the mountains of Pennsylvania; and also in Ohio, Michigan and Illinois, where their most usual course is from N. W. to S. E.

The hard gray limestone in the vicinity of Rochester and Buffalo in New York, and on the banks of the St. Lawrence, as well as in Ohio, is often smoothed and polished by this agency almost like marble. *Am. Jour. Science, Vol. XXXVII, p.* 240. *and Vol. XXXV, p.* 191. *Also Reports on the New York Geological Survey.*

Mount Monadnoc in New Hampshire, 3250 feet high, is little else than a naked mass of mica slate of a peculiar character, almost destitute of stratification. And from top to bottom it has been scarified on its northern and western sides. On its lower parts, especially on the southwest side, the striæ run about N. W. and S. E. by the magnetic needle; as they do in the country around the mountain; but when we approach its top, the course changes to N. 10° W. and S. 10° E. Other striæ are seen here on steeper slopes, both northern and southern, than I have found elsewhere. Deep furrows are also met with here, and some other peculiar phenomena connected with glacial action, which will be described farther on.

In a recent visit to the White Mountains of New Hampshire, I have discovered an example of striæ with other glacio-aqueous phenomena, at the height of nearly 5000 feet above the ocean, on the southern slope of Mount Pleasant, nearly at the highest point, indeed, where there is any rocks in place; all the peaks being broken into a multitude of fragments by frost. These striæ run N. 30° W. and S. 30° E.

Fig. 106, will convey some idea of a surface of rock which has been smoothed and striated in the manner that has been described. The lines running in the direction *a a*, show the divisions of the strata: those running in the direction *b b*, show the striæ, which are perfectly parallel to one another. This figure was copied from a rock of gneiss in Billerica in Massachusetts, on the turnpike from Boston to Lowell, near the sixteenth milestone from Boston. Thousands of similar examples may be seen in New England.

Fig. 106.

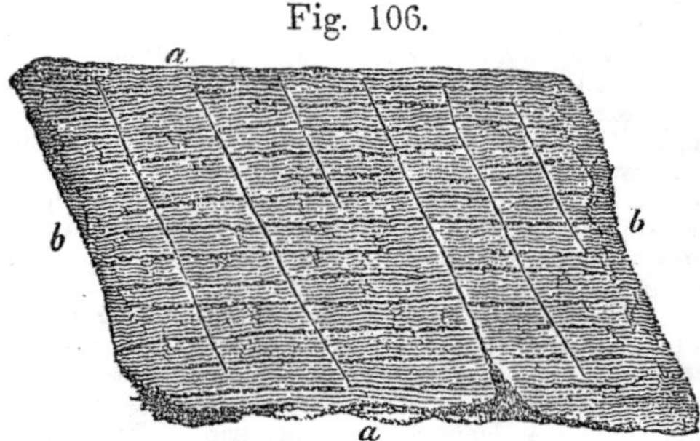

Descr. Sometimes we find two or more sets of striæ crossing one another at a small angle, the lines of each set preserving their parallelism very perfectly. Fig. 107, shows one of these cases copied from mica slate near the crest of Monadnoc, several hundred feet below the summit. The two sets intersect at an angle of 10°. An interesting example is also described by Prof. Locke, with a drawing as occurring in the S. W. part of Ohio, where the striæ was 5.26° E. and N. 26° W. *Second Report on the Geology of Ohio. p.* 230.

Descr. On the eastern continent these striæ and furrows appear to be less common than in this country; for notwithstanding the great ability and zeal displayed by European geologists, only a few cases of such markings have yet been recorded. In Scotland however, they were noticed long ago, by Sir James Hall, on greenstone and other rocks, having a direction N. W. and S. E. A similar case is mentioned in North Wales, and in the Brora coal region of Scotland, where they run N. N. W. and S. S. E. Dr. Buckland has recently pointed out several examples in that country, which he supposes resulted from the descent of ancient glaciers from the mountains. Their general course is N. N. E. and S. S. W. "Monsieur Sefstrom," says the distinguished Berzelius, "has found that the northeast parts of the mountains of Sweden, are, throughout rounded and worn from the base to the summit, so as to resemble at a distance, sacks of wool piled upon each other. The southwest sides of these mountains present almost fresh fractures of the rocks with their angles rounded little or none." Still further east in Northern Russia, the striæ have a direction N. W. and S. E. or N. N. W. and S. S. E. corresponding to the course taken by the bowlders; and in one instance these bowlders have been traced to their native bed 600 miles distant. *Murchison and Verneuil on the Geological Structure of the Northern and Central Regions of Russia in Europe, p.* 13. *Lond.* 1841.

Fig. 107.

Two Sets of Striæ: Monadnoc.

Descr. A careful examination of the mountains of New England shows that their northern and northwestern sides, like those in Sweden, are worn and rounded throughout. An interesting example is Monadnoc in New Hampshire; which is the more striking, because it is mostly naked rock. The surface of the mountain is very uneven; but the protuberances are nearly all rounded, and few are left angular, except on the south eastern side. The axis of the intervening hollows usually corresponds nearly to the direction of the striæ; so that the surface appears like the swell of the ocean after a storm. Seen in a certain direction these swells appear like domes. Fig. 108, will give some idea of a spot on the southwest part of this mountain about 5 rods square. So far as I can judge from the descriptions and drawings of Agassiz, this appearance corresponds precisely with that in the Alps denominated by Saussure, *roches moutonnees*, produced by glaciers.

Fig. 108.

Embossed Rocks, (Roches moutonnees), Monadnoc.

Fig. 109 shows similar embossed rocks near the top of the White Mountains in New Hampshire, near the peak called Mount Washington. It is the highest spot on those mountains where the rock is in place; being 5000 feet above the ocean. The peak rises 1200 feet higher; but the surface of the rock is entirely broken into fragments by frost. The loose fragments on the drawing are probably transported bowlders. (See note H.)

Fig. 109.

Embossed Rocks: south foot of Mount Washington, near the Lake of the Clouds: White Mountains.

Descr. The summit of Mount Holyoke in Massachusetts, which has been very much abraded by the agency under consideration, sometimes presents insulated hummocks of greenstone, resembling the "sacks of wool" described by Sefstroom, as shown in Fig. 110.

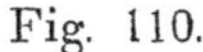

Hummock on Holyoke.

Descr. Sometimes instead of striæ, we find the summit of a mountain ploughed into deep furrows, which enlarge so as to form deep parallel valleys.

Exam. The most remarkable example of this kind that I know of, is the summit of Mount Holyoke, mentioned above. This is a narrow very precipitous ridge of greenstone, rising 700 or 800 feet above the valley of the Connecticut, and lying in the curvilinear direction shown on Fig. 111, where the line N S, represents the meridian, and corresponds to the direction taken there by the drift, which struck the mountain from the north. On that side the mountain is a nearly perpendicular wall of rock. Yet the summit is intersected by numerous grooves and valleys in the direction of the lines A, A, A, A, N S, from a few inches to several hundred feet deep. And not only do we see the marks of abrasion in the bottoms and on the sides of these valleys, but the fact that they preserve their parallelism so perfectly, although the mountain curves so much, shows that they were produced by some abrading agency rather than by the original structure or elevation of the mountain. For had they resulted from the latter causes, we might expect them to change their course to the lines B, B, B, as the mountain continued to curve more and more.

Fig. 111.

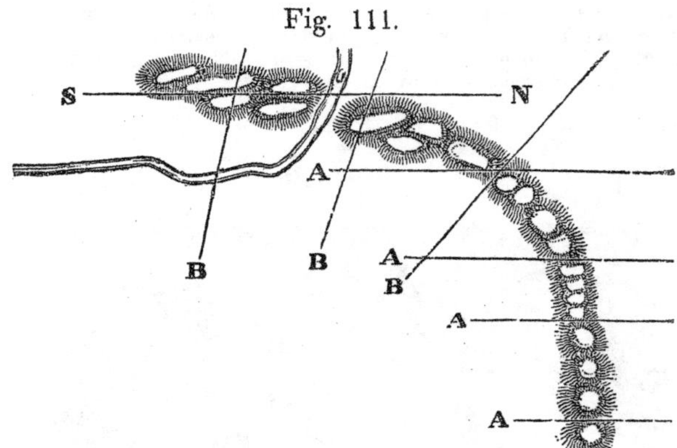

Inf. 1. These furrows and valleys must be imputed to the joint action of ice and water. If water alone were concerned, the valleys could not have so nearly preserved their parallelism. Indeed, unless the large valley around the mountain had been filled with ice, it is difficult to see how streams of water could have flowed over its summit so as to produce these valleys. Ice alone, moving over the top, might have begun the work, (and this would explain the parallelism of the valleys,) but could not have made so deep erosions without wearing down the intervening ridges. Moreover, these furrows and valleys correspond well to the joint action of glaciers and water in the Alps, and there they are called *lapiaz* or *lapiz.* See *Etudes Sur les Glaciers.*

Inf. 2. It appears that in all cases the striæ, furrows and valleys, that have been described upon the surface of rocks, correspond in direction to the course taken by the drift, and thus the two classes of phenomena are proved to have resulted from the same general cause.

Descr. The striæ are rarely met with on pure limestone, on account of its great liability to disintegration. Most of the coarse granites and conglomerates, as well as gneiss, are so much decomposed at the surface, as to have lost all traces of these markings. Greenstone, syenite, and porphyry, are frequently rounded and smoothed; but the markings are usually faint on account of their great hardness. Upon the whole, the upturned and smoothed ledges of talcose, micaceous, and argillaceous slates, retain these markings most distinctly. But where the soil has been removed, almost every rock presents them to view. And were the rocks of N. England to be entirely laid bare, I cannot doubt but a third of the surface would show marks of this scarification.

Transport of Drift from Lower to Higher Levels.

Prin. In some instances the drift has been transported from ower to higher levels.

Proof 1. On the northern slopes of mountains the striæ not unfrequently pass from the top to the bottom without essentially changing their parallelism. This is the case on Mount Monadnoc in New Hampshire, where the smoothed and striated surface is sometimes inclined northerly as much as 50° or 60°. In this case the angular rocks are rounded on the northern but not on the southern slope, except slightly; and this shows that the work was done by an ascending and not a descending body. Near the summit of that mountain, however, striæ are seen on quite steep southerly slopes. But this is an unusual case. On the west side of Taconic and Hoosac mountains in Massachusetts, the striæ may be seen sometimes, commencing several hundred feet below the top: and all the drift in that region has been carried southeasterly; which proves the force producing the abrasion to have been upwards. *Final Report on the Geology of Massachusetts, Vol.* 2, *p.* 393. A similar case occurs in Russia. *Murchison* and *Verneuil on the Geol. of Russia, p.* 12. 2. Bowlders are found transferred from a lower to a higher level. For example: the quartz rock of Berkshire county has been strewed all over the top of Hoosac mountain, which must be at least several hundred feet above any ledge of quartz, from which the bowlders could have been derived: and the Silurian rocks of New York have been carried over Taconic mountain and lodged upon the Highlands in New York, as well as upon Hoosac mountain.

Lakes produced by Moraines.

Descr. The great accumulation of drift which took place near the gorges of valleys, produced extensive lakes, which were filled with water upon the melting of the ice, and in whose bottoms thick deposits of clay and sand were frequently made above the body of the drift, while the waters were gradually wearing away passages through the detritus.

Proof. In all valleys of considerable size in the northern parts of the United States, such as those of the Hudson, the Connecticut, the Penobscot, &c., we find such deposit of clay above the drift; and above the clay more or less of sand. These deposits must have been made in quiet waters, either during the glacial or the alluvial period; and the fact that scarcely no organic remains are found in them except a few shells of a highly arctic character, makes it extremely probable that it took place during the glacial period. Similar deposits occur in the valley of the Rhine, which are there called *loess*, and are from 100 to 200 feet thick. *Lyell's Prin. Geology, Vol.* 2. *p.* 289. In the county of Suffolk, England, a deposit of similar clay exists, according to Rev. W. B. Clark, sometimes 400 feet thick. *Proceed. London Geolog. Society, No.* 50, *p.* 532.

Inf. We see why there is such an insensible passage from glacio-aqueous into alluvial deposits. For the drainage of the lakes produced by the moraines, has been going gradually on from the time when the waters first made a breach through the drift to the present. It is difficult, therefore, to say precisely where the glacial period terminated and the alluvial commenced, unless we fix it at the time when the present races of organic beings were created.

Inf. 2. We see the origin of the numerous small ponds that exist in regions abounding in moraines, as in Plymouth and Barnstable counties in Massachusetts. They occupy the depressions left between successive moraines.

Ledges of Rocks fractured by Glacial Action.

Descr. In several instances I have found the perpendicular strata of slaty rocks broken and partially knocked over, near the summit of hills, to the depth of 10 or 15 feet, so as to produce horizontal fissures of several inches wide; and I cannot doubt but they are the result of glacial agency. Fig. 112, will give an idea of one of these cases, in a quarry of clay slate, in Guilford, Vt. In this instance the force which thus crushed the top or western slope of the hill, must have been directed towards the west. Another case is in Middlefield, in Massachusetts, in hornblende slate, and the spot is covered by 20 feet of drift. *Final Report on the Geol. Mass. p.* 396, *Vol.* 2. It was brought to light by an excavation for the western railroad. The force to produce such an effect must have been prodigious.

Fig. 112.

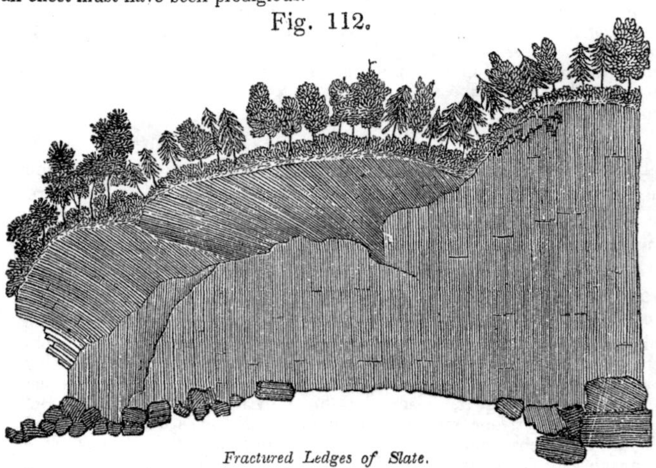

Fractured Ledges of Slate.

Descr. A third case occurs in Lowell, and is described in the Transactions of the Association of American Geologists, Vol. 1. p. 209. Prof. C. B. Adams has, also, discovered similar cases in the slate of Dummerston, Vermont. *Second Annual Report on the Geology of Vermont, p.* 131. Analogous cases have been given by Mr. Charles Darwin, as occurring in Scotland, which he refers to the action of icebergs, lifted up and down by the tides; and we might add, by the waves. This theory seems very plausible, and is certainly the best that has been proposed. *New Phil. Jour., Oct.* 1842, *p.* 358.

Trains of Blocks not Rounded.

Descr. Another very remarkable series of facts, evidently the result of the drift agency, has lately been brought to light in New England. It consists of trains of large fragments of rock, not at all rounded, yet streaming off in the direction in which the drift agency operated, in a straight line from the eminence from which they were broken. The first case of this kind was first noticed by Dr. S. Reid, in Berkshire County, Massachusetts, and described by the author of this work, in *Amer. Jour. Sci., Vol.* 49. *p.* 258. The mountain from which the blocks of hard talcose slate have been torn off, lies in Canaan, New York; and from thence they lie in trains, running for a few miles S. 56° E., and then changing to S. 34° E. and extending yet farther, making in the whole distance, not less than 15 or 20 miles: at least one of them extends that distance, passing obliquely over mountain ridges some 600 or 800 feet high. Its width is not more than 30 or 40 rods. The blocks are of all sizes, from two or three feet in diameter to those containing 16,000 cubic feet, and weighing nearly 1400 tons; and in some places they almost cover the surface. The trains lie upon the surface of the common drift and are not mixed with it. An analogous case has lately been described by Prof. C. B. Adams, in Huntington, Vermont. *Second An. Rep. on Geol. of Vermont, p.* 128.

Inf. 1. These blocks must have been scattered as one of the latest results of the drift agency. 2. Water alone could not have done it. 3. It was probably the result of water and ice. 4. But the precise manner in which this agency operated, appears to me not yet understood. The Professors Rogers, however, have ingeniously endeavored to explain it. *Boston Journal of Natural History for June,* 1846.

Vertical and Horizontal Limits of Glacio-aqueous Agency.

Descr. It can hardly be doubted that this agency was mainly confined to the colder regions of the globe. In the United States drift rarely appears south of the Ohio river in north latitude from 38° to 40°, though some bowlders are seen in Kentucky, and the gravel of drift is seen much farther south in the valley of the Mississippi. Farther east the southern limit of drift runs along the northern part of Pennsylvania, though it appears along the Susquehanna and Delaware rivers farther south. Perhaps 40° north latitude is its average limit, and according to Mr. Darwin, drift is first met with at about the same degree of south latitude in S. America. *Prof. H. D. Rogers' Address at Washington, before Geol. Assoc., May,* 1844, *p.* 45. *Lyell's Travels in N. America, Vol.* 2. *p.* 60.

Descr. In some parts of Europe the drift agency did not extend to the tops of high mountains, nor was that upper limit horizontal. In the Alps this upper limit varies from 3000 to 8000 feet, and its inclination is never quite three degrees. In this country this upper limit has not been ascertained; for the White Mountains are the highest point east of the Rocky Mountains where we should look for a solution of the problem, and the highest rocks in place there (5000 feet) have been smoothed and rounded by this agency. In the Alps, which formed a centre of dispersion for drift, the inclination of the upper limit of this action is outward from their centre. But as none of our high mountains formed such centres, we cannot predict how high glacio-aqueous agency may nave reached. *Moxon's Geologist for Dec.* 1842, *p.* 356.

Descr. From the German Ocean and Hamburgh on the west, to the White Sea on the east, a region 2000 miles long and from 400 to 800 miles wide, is more or less covered with detritus; all of which has been derived from Scandinavia. *Horner's Address before Lond. Geol. Soc., Feb.* 1846, *p.* 60.

Descr. It now appears that Scandinavia formed a centre of dispersion for drift, and that the bowlders were thrown off eccentrically, towards the north in Lapland, towards the east and southeast in Russia, towards the south in

Germany, and perhaps towards the southwest, into Great Britain. The greatest distance any of them have travelled, is from 700 to 800 miles. In no case have they reached the Uralian mountains, which are entirely free from them; as is also, it is said, the whole of Siberia. *Murchison's Geology of Russia, Vol.* 1. Prof. Forchhammer has endeavored to show, from the phenomena of drift in Denmark, that the period of this agency reached as far back as the commencement of the tertiary deposits, and even of the newer cretaceous beds. *Geological Journal for* 1845, *p.* 263.

Theories of Drift.

Prin. Any theory of drift to be satisfactory, must explain the four following varieties of the phenomena: 1. Erratic Bowlders: 2. Common Drift: 3. Striæ, Furrows and Valleys: 4. Moraines. Some theories might satisfactorily explain particular phases of the phenomena, but fail in their application to the entire group.

Descr. In no part of geology have opinions been so unsettled and changing, as in respect to the origin of drift and its attendant phenomena. Until recently, powerful currents of water have been regarded as the sole agent concerned. To account for these currents, it was a favorite theory with many, that a comet once impinged against the earth. *Geologie Populaire, par N. Boubee, p.* 46. *Paris,* 1833. But since it has been ascertained that comets have "no more solidity or coherence than a cloud of dust, or a wreath of smoke, through which the stars are visible, with no perceptible diminution of their brightness," it is no longer necessary to give a formal refutation of this hypothesis. *Whewell's Bridgewater Treatise, p.* 152, 153, *first American Edition.* Another theory, very widely adopted, imputed these effects to the deluge of Noah. But it is now almost universally abandoned by geologists, because the remains of man have not been found in drift; because most of the animals found in it belong to extinct species; because the period occupied by the Noachian deluge was much too short; and because the phenomena cannot be explained by water alone. In short, every theory which imputes these effects to currents of water alone, has been abandoned by geologists; and they now almost universally refer them to the conjoint action of water and ice. Three varieties of opinion as to the origin and mode of operation of these agents, exist among eminent geologists. The two first of these theories may be called the *Iceberg Theories:* and the last, the *Glacier Theory.*

First Theory.

Descr. This theory imputes most of the phenomena of drift to icebergs carried southerly by the currents of the ocean, while the continents, where drift occurs, were yet beneath the ocean. As they were gradually raised from the deep, the mountains, which would form islands, would send down glaciers to their shores, and thus masses of ice would be broken off to be floated away, loaded with detritus. In many places large bodies of water would remain after the ocean had retired, in which deposits of clay and sand would take place. *Lyell's Elements Geol., Vol.* 1. *Chap. X. and XI.*

Proof. 1. In high northern and southern latitudes, the process which this theory assumes is daily going on. Icebergs frequently transport towards the equator blocks of great size, which are dropped upon the bottom of the ocean. Scoresby saw upon several icebergs in 70° N. latitude, masses of earth and rock weighing from 50.000 to 100.000 tons: and a deposit of drift is now actually accumulating in the southern hemisphere, in latitudes no higher than Northern Italy, Switzerland, and England. *Lyell's Elements Geol. Vol. 1. p.* 232 and 257. 2. There is evidence daily accumulating of the existence of a much lower temperature in northern latitudes when the drift was depositing than is now found in, the same latitudes, and, therefore, glaciers might have existed in much lower latitudes than at present, and icebergs might have been carried nearer to the equator than they now are, before melting.

Obj. 1. The uniformity of direction taken by drift over all the northern part of Europe and America, and the rare occurrence of centres of dispersion are explained with difficulty by this theory. There has yet been discovered no mountain in North America from which detritus has been scattered on different sides; and the force was southeasterly over the top of the White Mountains, the highest point, with one exception, east of the Mississippi. Nor have any such centres of dispersion been pointed out in the North of Europe. But if this theory be true, it is difficult to conceive why such mountains as the White Mountains, as they were raised from the ocean, should not have proved centres of dispersion. 2. Several facts, such as the transference of bowlders from lower to higher levels, and the smoothing and furrowing of the northern slopes of mountains even when quite steep, agree better with the idea of the ocean's rising over the land than the land's rising from the ocean. 3. It is certain, that the relative levels of the earth's surface have not essentially changed since the deposition of drift: that is, the same mountains and valleys which existed at that period exist now; and, therefore, if our continents have experienced any vertical movements since, they must have been raised as a whole, by such an extremely slow movement as some regions are now experiencing. 4. But this would make the period of the formation of drift immensely remote: coeval indeed with the time when the highest granite mountains of northern countries were beneath the ocean. Whereas we know that the drift period occurred after the deposition of the most recent tertiary rocks. Now at that period we know that a large part of our continents,—all that was covered by rocks older than the tertiary, were above the waters. If we suppose these mountains to have sunk down again at the tertiary period, then they must have been suddenly raised in order to have reached their present height so soon. But as just mentioned, the facts do not admit of such a sudden rising.

Second Theory.

Descr. This theory supposes the phenomena of drift to have resulted from the rise of large areas beneath the Arctic and Antarctic oceans, whereby their waters have been driven southward, over a considerable part of Europe and America, bearing along masses of ice loaded with detritus. And further, that there may have been a succession of vertical movements, which produced successive waves; so that the waters may have repeatedly fallen and risen again, and while at their ebb, they may have been frozen to the surface, so that as they subsequently rose, vast masses of ice may have been driven along, loaded with detritus, which may have been forced up declivities considerably steep, and thus the surface have been powerfully and rapidly abraded, and the rocks scoured and furrowed. *De La Beche's*

Geological Manual, p. 172. *Also his Theoretical Geology, p.* 319. *Biblical Repository, Vol.* 11, *p.* 23.

Proof. 1. It is established, that in early times repeated upheavings of the bottom of the ocean have taken place; and at the present day, also, similar elevations occur, (ex. gr. Hotham or Graham Island in the Mediterranean, in 1831: Sabrina, near the Azores, in 1811: and among the Aleutian Islands in 1814, an island, said to be 3000 feet high:—another in 1806, which is permanent: and another in 1795. *Lyell's Prin. Geol., Vol.* 2. *p.* 253, 266, 318.) And in some instances, we are able to see the effects in the tremendous waves that follow; as during the earthquake at Lisbon in 1765. The force of such waves, as they reached the shore, and successively rose higher and higher, were a large area of the northern ocean to be suddenly upraised, can scarcely be estimated; loaded as they would be with ice; and it seems almost the only conceivable agency by which large bowlders could be forced up hills of considerable steepness, and deep groves be formed on the northern slopes of hills in Sweden and America. 2. In the latter country certainly, if not in the former, it seems scarcely possible to doubt that the surface is essentially the same now as when this agency was exerted upon it; except perhaps those few local minor and slow elevations and depressions of which probably every country furnishes some examples: and hence the source of this action must be sought out of the country. 3. We have evidence that volcanic agency, whereby large areas may have been uplifted, has been very active in high northern and southern latitudes. 4. This theory explains the absence of animal and vegetable life, and the presence of vast masses of ice in comparatively low latitudes during the glacio-aqueous period.

Obj. 1. It is difficult to conceive how these vertical movements around the poles, should have produced floods of such vast extent and magnitude, as to inundate with water and ice, certainly all the northern, and probably also all the southern parts of the globe, as far at least as the 40th degree of latitude. 2. The action by which drift was accumulated, bowlders transported, and rocks furrowed, must have been long continued; whereas any vertical movement around the poles, sudden enough to produce such deluges, must have been comparatively transient in its effects. 3. If the waters thus thrown over existing continents southward did not return to the polar regions, what has become of them? If they did return, as the elevated portions subsided, why have they not left traces of their northerly currents on the land?

Rem. This theory, somewhat modified, has been sustained with great ability by Professors H. D. and W. B. Rogers. *See the An. Add. of the former before the Assoc. of Amer. Geologists and Naturalists*, 1844.

The Glacier Theory.

Descr. The history of glaciers, which has been given in the first part of this section, forms the groundwork of the glacier theory. It supposes that at the close of the tertiary period, a sudden reduction took place in the temperature of the surface of the globe, whereby all organic life was destroyed; and in high latitudes at least, glaciers were formed on mountains of moderate altitude; indeed, that vast sheets of ice were spread over almost the entire surface, extending south as far as the phenomena of drift have been observed. The northern regions, especially around the poles, are supposed to have formed one vast Mer de Glace, which sent out its enormous glaciers in a southerly direction by the force of expansion; and the advance and retreat of these glaciers, accumulated the moraines and produced the striæ and embossed appearance (*roches moutonnees*) upon the rocks. When the temperature was raised, the melting of the immense sheet of ice produced vast currents of water, which would lift up and bear along huge icebergs loaded with detritus, and thus scatter bowlders over wide surfaces. The blocking up of the gorges by moraines, would form lakes and ponds, in which clay and sand, such as now lie above the drift, might have been deposited, and afterwards the barriers of these lakes, consisting of loose matter, may have been cut through, and the waters gradually drained off, and assumed their present levels. In some parts of the world the elevation of mountains, as the Alps for instance, during the same period, might have increased the effects that have been described.

Proof. 1. The perfectly preserved elephants and rhinoceros of Siberia in frozen mud, show that the change of climate there must have been very sudden from quite warm to intense cold. 2. The general absence of organic remains in the clay and sand lying above the drift, makes it probable that during their deposition the climate was too cold to favor the existence of animals and plants, while the highly arctic character of the few species of shells that have been found in these deposits in New York, Canada, Scotland, Sweden, and Russia, confirms this conclusion. 3. The history of the effects of glaciers is the history of the phenomena of drift in miniature. In the first place, the moraines of glaciers correspond to the accumulations of drift that are so common in northern regions. The latter are, indeed, somewhat modified, partly by subsequent aqueous agency, and partly by a somewhat differ ent mode of production; so that the distinct varieties of moraines accompanying glaciers are not always to be distinguished. Secondly, the smoothing, rounding, and polishing of the rocks, are the same beneath the glaciers as over the whole northern hemisphere. Thirdly, the parallel striæ upon their surfaces are perfectly explained by the passage of ice over them in unbroken sheets, with angular fragments fixed into their lower surface. Fourthly, the parallel furrows and valleys produced by the agency under consideration, upon the crests and sides of steep mountains are very analagous to those beneath the glaciers, the result of the joint action of ice and water. Fifthly, this same joint action may have transported bowlders to great distances and lodged them upon precipitous ridges and on sandy plains. Finally, these effects are inexplicable by currents alone. 4. This theory furnishes an adequate agency for smoothing and furrowing the slopes of mountains, and for the transportation of drift from lower to higher levels by an ascending force; facts more difficult to explain than almost any other phenomena connected with drift. This might have been done both by the expansive force of ice, pushing one extremity of the sheet up the hill, and by water, lifting up icebergs with detritus from the bottom of the valleys, and as it rose, carrying them to higher levels. 5. It shows how deposits of clay and sand might have been formed above the coarse detritus in lakes produced by the moraines and melting of the ice, and how their barriers afterwards might have been removed. 6. It gives a reason why those clays and sands are so destitute of organic remains, viz, a cold climate. 7. It provides an agent sufficiently powerful to break down the tops of ledges of rocks, as appears to have been done at least in a few instances in New England, by an enormous force operating obliquely downwards in connection with the formation of drift. The expansion and great weight of a huge sheet of ice might exert a force upon obstacles almost irresistible.

Obj. This theory furnishes a very inadequate cause for the southerly direction taken by drift over so large a part of the northern hemisphere. It supposes that as the earth was cooling, most of the water that was evaporated in tropical regions would go to be condensed in the polar regions, and thus accumulate there a vast quantity of ice, so long as the glacial period continued. Admitting such a process, it might produce a moderate southerly movement in the ice as it expanded. But the distance southerly to which the drift has been transported, even that principal mass of it which must have been pushed along by the ice, is very great to refer to such a cause. And besides, the southerly direction of the force seems to have continued from the beginning to the end of the work; so that the bottom of the valleys as well as the tops of the mountains have been equally affected by it. This fact seems to indicate some other cause besides mere expansion to give an austral direction to the force. It is difficult, also, to conceive how currents, formed by the mere melting of ice, could have transported bowlders from 400 to 600 miles to the south. 2. The largest and most remarkable moraines (ex. gr. those of Plymouth and Barnstable counties in Massachusetts,) are situated at a great distance from any mountains. This shows at least that the idea of their production by common glaciers, descending from higher to lower ground, must be given up: though this idea is not, perhaps, essential to the theory. If we can only suppose the sheet of ice fixed immoveably on its northern side, then, as it expanded, it would enlarge on its southern side. 3. It is difficult to imagine a cause for a sudden change on the earth's surface from an almost tropical to an ultra-arctic climate, without supposing a total change in the present arrangements of the solar system: without supposing, for instance, the earth left without the light and heat of the sun. If facts prove such a change, however, it must be admitted even if we can imagine no cause.

Rem. 1. This theory was first suggested by Venetz, a Swiss engineer; then advocated by Charpentier; and more recently brought out in its full proportions by Agassiz, in his *Etudes sur les Glaciers*.

Rem. 2. It gives nearly as erroneous an idea of this theory to call it simply the *Glacial Theory*, as it would to call it the *Diluvial theory*. For it is agreed by all that immense currents of water, and even extensive deluges, must have been the result of the melting down of vast accumulations of ice. These must have been sufficient to carry blocks and gravel from 400 to 600 miles southerly; and to form extensive lakes in the depressions of the surface. And if the return of heat was as sudden as Agassiz supposes its departure to have been, vast deluges must have been the consequence. Hence it would seem more appropriate to call this theory the *Glacio-aqueous theory*.

Rem. 3. A curious example to illustrate the effect of the sudden melting of large masses of snow and ice, has just been communicated to me by Rev. Justin Perkins, American Missionary at Ooroomiah, in Persia, not far from Mount Ararat, in a letter of Nov. 6, 1840. In giving an account of two very powerful earthquakes, experienced on and around that mountain in the summer of 1840, he says, "the vast accumulation of snow which had been increasing on and about the tops of the mountains for centuries, was broken into pieces, and parts of it shaken down on the sides of the mountains in such immense quantities, that (it being midsummer and the snow descending down as far as a warm climate and suddenly melting,) torrents of water came rolling down the remainder of the mountain, and flooded the plains for some distance around its base."

Inf. 1 Very probably the Iceberg Theories and the Glacier

Theory are not irreconcileable with one another: or rather, the ultimate and true theory of drift may probably be compounded of these three theories.

Proof 1. It is impossible to explain the phenomena of drift in North America and Northern Europe, without admitting the transit over the entire surface of deep and long-continued currents of water, loaded with islands of ice. Such currents and such icebergs are supposed in all these theories to have existed. 2. Mr. Maclaren has rendered it probable, that if most of the globe was covered with an ocean, there would exist between the tropics a westerly current, and without the tropics, two easterly currents. If, now, ice had been accumulated around the poles to a great thickness, (10,000 feet, says Maclaren,) as Agassiz supposes, its fusion, upon the return of heat, would produce currents towards the equator; and these, meeting with the easterly currents, would form a southeasterly current in the northern hemisphere, and a northeasterly current in the southern hemisphere. This would explain the southeasterly direction taken by the drift in the northern hemisphere.—*Geology of Fife and the Lothians, p.* 225. *Also the Scotsman for Dec.* 1, 1841. 3. Essentially the same effects would result, if we suppose deep and long continued inundations of the land to have taken place from the melting of polar ice; so that it is not necessary to suppose the glacio-aqueous period to have existed before the rise of our present continent from the ocean. 4. Let us make a further supposition. Suppose the earth to have ceased to revolve on its axis during that period, and to have been removed from solar influence. The consequence would have been, that the water would accumulate around the poles to the depth of 13 miles, and all of it be converted into ice; while the equatorial regions would be left bare. Suppose next, that the earth was brought into its present orbit and commenced its diurnal revolution. The melting ice would flow rapidly towards the equator, inundating the land, and the currents would take the directions above named. 5. Murchison and Verneuil regard it as most probable, that the striæ upon the rocks of Northern Europe, were produced by "icefloes and detritus, grating upon the bottom of the sea," and set in motion by currents produced by the elevation of northern continental masses." *On the Geological Structure of Russia in Europe, by Murchison and Verneuil, p.* 13. *London*, 1841. 6. The Alps are an example of the rise of a mountain to a considerable height during the period in which drift was accumulating; and their elevation produced important modifications of glacio-aqueous agency in that part of Europe, where they exist. In fact, it seems probable, that the conditions of the several theories that have been described, may all have existed in some part of the globe during the glacial period; and, therefore, it may be found that all those theories are in harmony with one another.

Inf. 2. The proximate cause of the phenomena of drift seems at length to be determined; viz. *the joint action of ice and water:* so that the dynamics of this most difficult subject appear to be at last nearly settled.

Proof. In all the above named theories, that are now adopted by geologists, viz. the three last, ice and water are the agents supposed to have been employed; and the chief difference between them seems to be concerning their relative agency.

Rem. This is certainly a very great and happy advance in a subject which only a few years ago presented an aspect truly chaotic.

Inf. 3. Not improbably all advance which geologists will ever be able to make beyond the true dynamics of this subject, will be only vague conjecture: that is, while they may under stand perfectly by what agencies the phenomena of drift have been produced, and even be able to calculate the intensity of those agencies, they may never ascertain their origin.

Rem. The case is the same in many other parts of geology; and it is often because writers are not satisfied with ascertaining the proximate causes of phenomena, and will speculate upon antecedent causes, that doubt and perplexity are thrown over a plain subject, and the whole science appears more unsettled to the superficial reader than is actually the case

---

SECTION VII.

OPERATION OF ORGANIC AGENCIES IN PRODUCING GEOLOGICAL CHANGES.

Rem. Many facts naturally belonging to this Section, have been necessarily anticipated in the preceding Sections; and will therefore, need only to be referred to in this place.

Agency of Man.

Prin. The human race produce geological changes in several modes: 1. By the destruction of vast numbers of animals and plants to make room for themselves. 2. By aiding in the wide distribution of many animals and plants, that accompany man in his migrations. 3. By destroying the equilibrium between conflicting species of animals and plants: and thus enabling some species to predominate at the expense of others. 4. By altering the climate of large countries by means of cultivation. 5. By resisting the encroachments of rivers and the ocean. 6. By helping to degrade the higher parts of the earth's surface. 7. By contributing peculiar fossil relics to the alluvial depositions now going on, on the land and in the sea: such as the skeletons of his own frame, the various productions of his art, numerous gold and silver coins, jewelry, cannon balls, &c. that sink to the bottom of the ocean in shipwrecks, or become otherwise entombed.

Exam. The only example of the entire extinction of the larger animals coeval with man, and probably through his agency, are the following. 1. The *great Irish elk*, which was 10 ft. high to the top of the horns, which are from 10 to 14 ft. between their tips. 2. The *dodo*, a bird larger than the turkey, which existed in Mauritius and the adjacent islands when they were colonized by the Dutch 200 years ago: but it is no longer to be found; and even all the stuffed specimens that were brought to Europe are lost: so that a head and a foot of one individual in the Ashmolean museum at Oxford, and the leg of another in the British museum, are all that remains of it, except some fossil bones lately found in the Isle of France. 3. The *apteryx australis* of New Zealand appears to be on the point of extinction, if not actually extinct: as only one specimen of it exists in Europe; and no others can be obtained, though the missionaries say that its skin is still worn by the natives. *Wonders of Geology, Vol.* 1. *p.* 105. 4. The six species of Dinornis formerly inhabiting New Zealand, of which a brief description and sketch has been given in Sect. V. *American Journal of Science, Vol.* 48. *p.* 194.

In particular countries it is a more common occurrence for species to become extinct; as the beaver, wolf, and bear, in England. In this country the animals of the forest are disappearing or moving westward as the forests are clearing up. Since the discovery of the island of South Georgia, in 1771, one million two hundred thousand seal skins have been annually taken from thence; and nearly as many more from the Island of Desolation. The animal is becoming extinct at these islands. How many of the smaller animals may have become extinct, through the agency of man, it is impossible to ascertain. It has been maintained with great confidence, that the climate of Europe is very much warmer than in the days of ancient Rome: and this has been imputed to the clearing away of the forests. *Rees' Cyclopedia, Article Climate.* But M. Arago has lately rendered it probable, that no such change has taken place. *Lyell's Prin. Geol., Vol.* 3. *p.* 256. In North America, however, the extremes of heat and cold have probably been modified by the clearing away of the forests.

In Italy, the Po, Adige, and other rivers, are prevented from overflowing the adjacent country by embankments. These have been carried so high, and the bed of the river has so much filled up in some places, that the surface of the Po is more elevated at Ferrara than the roofs of the houses. *Lyell's Prin. Geol., Vol.* 1. *p.* 348.

Inf. Some writers maintain that as species of animals and plants disappear from the earth, new species are created to take their place at once, that the proper equilibrium of organic nature may be preserved. But as no certain example of the creation of a new species has yet been discovered within historic times, this opinion can be regarded only as an hypothesis. And the majority of authors suppose that in general no new creation takes place, until nearly the entire race inhabiting a country at any one period have been destroyed, either by a sudden catastrophe, or in the slow manner that has been described.

Obs. This question, with the whole subject of the permanence, distribution, and mutual influence of species of animals and plants, is fully and ably discussed in *Lyell's Principles of Geology, Book* 3. The general principles regulating their distribution have been given in Section V.

Agency of Other Animals.

Prin. Very many other animals exert an influence on geological changes analogous to that of man, though less in degree, except the following.

Polyparia or Polyps.

Rem. The lithological character of the stony habitation erected by these minute animals, has been described in Section III. Some account of the polyps has also been given in Section VI. The history of *coral reefs* remains to be given.

Descr. Coral reefs are ridges of calcareous rock, whose basis is coral, (chiefly of the genera, porites, astrea, madrepora meandrina and caryophillia,) and whose interstices and surface are covered by broken fragments of the same, with broken shells and echini, and sand, all cemented together by calcareous matter. They are built up by the polyparia, apparently on the tops of submarine ridges and sometimes perhaps, though not generally, on the margins of ancient volcanic craters, beneath the ocean, not generally from a depth greater than 25 or 30 feet, yet sometimes 120 or 130 feet. The polyparia continue to build until the ridge gets to the surface of the sea at low water; after which, the sea washes upon it fragments of coral, drift wood, &c. and a soil gradually accumulates, which is at length occupied by animals with man at their head. The reefs are sometimes arranged in a circular manner, with a lagoon in the centre, where, in water a few fathoms deep, grow an abundance of delicate species of corals, and other marine animals, whose beautiful forms and colors rival the richest flower garden. Volcanic agency often lifts the reef far above the waters, and sometimes covers one reef with lava, which in its turn is covered with another formation of coral. The growth of coral structures is so extremely slow, that centuries are required to produce any important progress. The rate of increase has not been determined.

Descr. The diameter of the circular reefs has been found to vary from less than one to 30 miles. On the outside, the reef is usually very precipitous, and the water often of unfathomable depth. Fig. 113. is a view of one of these circular islands in the south seas, called Whitsunday Isle; so far reclaimed from the waters as to be covered with cocoa nut trees and with some human dwellings.

Fig. 113.

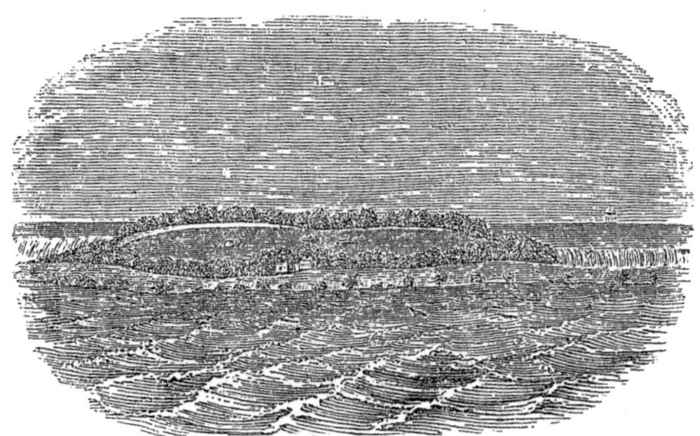

Whitsunday: a Coral Island.

Descr. These islets occur abundantly in the Pacific Ocean, between the thirtieth parallels of latitude. They abound also, in the Indian Ocean, in the Arabian and Persian Gulfs, in the West Indies, &c. Usually they are scattered in a linear manner over a great extent. Thus, on the eastern coast of New Holland, is a reef 350 miles long. Disappointment Islands and Duff's Group are connected by 500 miles of coral reefs, over which the natives can travel from one island to another. Between New Holland and New Guinea, is a line of reefs 700 miles long, interrupted in no place by channels more than 30 miles wide. A chain of coral islets 480 geographical miles long, has long been known by the name of the Maldivas. Some groups in the Pacific, as the Dangerous Archipelago, are from 1100 to 1200 miles long, and from 300 to 400 miles broad. *Lyell's Prin. Geol. Vol.* 3. *p.* 367. Especially See *Couthoy on Coral Formations in the Pacific, in the Bost. Journ. Nat. History, Vol.* 4. *No.* 1.

Infusoria.

Rem. It is certainly one of the most astonishing discoveries of modern science, that the animalcula of infusions, of which 500.000.000 may live and sport in a drop of water, should originate extensive formations of rocks and soils by their skeletons. But the mystery is explained when we learn how astonishing is their power of multiplication. The facts on this subject, however, have been so fully detailed in Section V. that nothing more can be profitably added in this place.

Agency of Plants.

Descr. Animal and vegetable substances, when buried in the earth, or the waters, sometimes undergo an almost entire decomposition: at other times, this is very partial; and sometimes the change is so slow that for years scarcely no apparent progress is made. Different substances will be the result of these different degrees of decomposition.

Descr. Berzelius embraces all the organic matter of soils in the generic term *humus.* Dr. Dana uses the term *geine* as synonomous with humus, when he speaks agriculturally. Berzelius supposes humus to contain humic acid, humin, crenic and apocrenic acid, and traces of glairin. Dr. Dana regards these as different forms of geine, which, when in a soluble state, may be taken up by the roots of plants: but in an insoluble state, can afford no nourishment. When he uses the term geine, however, in a strict chemical sense, he means by it the same as the humic acid of Berzelius; or a compound of 16 atoms of oxygen, 2 atoms of hydrogen, and 16 atoms of carbon. This unites with bases and forms geates, or humates. In some places, as on the western prairies, these organic matters of soils increase so as to form a layer several feet thick: but in general they are so much used in the nourishment of plants, that they rarely become more than a few inches thick.

Descr. The prevalent opinion among vegetable physiologists has been, that plants are nourished partly by imbibing carbonic acid through their leaves and their roots, and partly by taking up soluble geine from the soil. Recently, Liebig, Raspail and other distinguished chemists have advanced the opinion, that they are nourished solely by the absorption of carbonic acid, and that geine acts no other part in sustaining them but to furnish that gas. *Liebig's Organic Chemistry of Agriculture and Physiology, with notes, by Professor Webster, Cambridge,* 1841. *Also Dana's Muck Manual, Lowell,* 1842.

Peat.

Descr. Peat usually consists of soluble and insoluble geine, with a mixture of undecomposed vegetable matter, and some earths. Most of it results from the decomposition of certain mosses; especially of the genus *sphagnum;* which decay at their lower extremity, while the top continues to flourish with vigor. Trees and whatever other organic matter happen to get into these peat bogs, soon become enveloped and assist to swell the amount. In some instances the beds have acquired a thickness of more than 40 feet.

Descr. In tropical climates, except on high lands, the decomposition of vegetable matter is so rapid that it is resolved into its ultimate elements before peat can be produced. Hence peat is limited chiefly to the colder parts of the globe. In Ireland, the peat bogs are said to occupy one tenth of the surface and one of them, on the Shannon, is 50 miles long and two or three broad. In Massachusetts, exclusive of the four western counties, the amount of peat has been estimated at not less than 120 millions of cords; and probably this falls far short of the actual amount.

Descr. By the long continued action of water and other agents, the geine of peat is changed into bitumen and carbon, which constitute lignite and bituminous coal. In a few instances the process of bituminization has been found considerably advanced in the beds of peat. *Macculloch's System of Geology, Vol.* 2. *p.* 352. *Dr. C. T. Jackson's Second Report on the Geology of Maine, p.* 80. *American Journal of Science, Vol.* 35. *p.* 345.

Descr. Peat bogs are remarkable for their antiseptic power, or the power of preserving animal substances from putrefaction; some remarkable cases of which are on record. *Lyell's Prin. Geol. Vol.* 3. *p.* 271. Whether this be owing to the existence of acetic acid in the peat, or to the conversion of the animal muscle into adipocere, seems to be not well ascertained.

Descr. Peat bogs sometimes burst their barriers in consequence of heavy rains, and produce extensive inundations of black mud.

Descr. The increase of peat varies so much under different circumstances, that it is of no use to attempt to ascertain its rate of growth. On the continent of Europe, it is stated to have gained seven feet in 30 years. *Macculloch's System of Geology. Vol.* 2. *p.* 344.

Descr. Where peat is formed in, or transported into estuaries, it is sometimes covered with a deposit of mud; over this another layer of peat forms, and in this way several alternations may occur.

Descr. In some peat bogs, large trees have been found standing where they originally grew, yet immersed to the depth of 20 feet: as in the Isle of Man. *Lyell's Prin. Geol. Vol.* 3. *p.* 269.

Descr. The following analysis of three specimens of marsh peat from Massachusetts, will give an idea of the composition of this substance.

	From Sunderland.	From Westboro'.	From Hadley.
Soluble Geine,	26.00	48.80	34.00
Insoluble, do.	59.60	43.60	60.00
Sulphate of Lime,	4.48	1.88	1.36
Phosphate of do.	0.72	0.12	0.24
Silicates,	9.20	5.60	4.40
	100.00	100.00	100.00

Drift Wood.

Descr. Large rivers, which pass through vast forests, carry down immense quantities of timber. When these rivers overflow their banks, this timber is in part deposited upon the low grounds. But much of it also collects in the eddies along the shores, or is carried into the ocean. After a time it becomes *water-logged;* that is, saturated with water; and sinks to the bottom. Thus a deposit of entangled wood is often formed over large areas. This is subsequently covered by mud; and then another layer of wood is brought over the mud: so that in the course of ages, several alternations of wood and soil are accumulated. The wood becomes slowly changed into what Dr. Macculloch terms *forest peat;* that is, peat which retains its woody fibre.

Exam. 1. The Mississippi furnishes the most remarkable example known of these accumulations. In consequence of some obstruction in the arm of the river called the Atchafalaya, supposed to have been formerly the bed of the Red river, a raft had accumulated in 35 years, which in 1816 was 10 miles long, 220 yards wide, and 8 feet thick. Although floating, it is covered with living plants, and of course with soil. Similar rafts occur on the Red river: and one on the Washita, concealed the surface for 17 leagues. At the mouth of the Mississippi, also, numerous alternations of drift wood and mud exist, extending over hundreds of square leagues. *Lyell's Prin. Geol. Vol.* 2. *p.* 8. *American Journal of Science, Vol.* 3. *p.* 17.

Exam. 2. Similar deposits of wood and mud are found in the river Mackenzie, which empties into the North Sea; and in the lakes through which it passes. At the mouth of the river, which is almost beyond the region of vegetation, are extensive deposits brought from the more southern regions, through which the river passes. *Lyell's Prin. Geol. Vol.* 3. *p.* 310.

Exam. 3. A part of the drift wood which is brought down the Mississippi and other rivers, along the coast of America, is carried northward by the Gulf Stream, and thrown upon the coasts of Greenland. The same thing happens in the bays of Spitzbergen, and on the coasts of Siberia. *Lyell's Prin. Geol. Vol.* 3. *p.* 313.

Inf. In the history of common peat and drift wood, we see the origin of the beds of coal which exist in the older strata: for it needs only that the layers of peat (in which term I include submerged drift wood,) should be bituminized, and the intervening layers of sand and mud be consolidated, in order to produce a genuine coal formation. Common marsh peat alone, can have originated but a small part of the beds of coal. *Phillips's Geol. p.* 116.

Consolidation of Loose Materials

Rem. Having in this and the preceding sections described a variety of natural processes by which just such materials as form the fossiliferous rocks are produced, it remains to enquire whether any agents are now in operation to effect their consolidation.

Prin. A considerable degree of solidity is sometimes produced by mere desiccation.

Exam. 1. When clay is exposed for a long time to the sun, it becomes as hard as some rocks:—ex. gr. the marly clay dug from the bottom of Lake Superior. *Lyell's Prin. Geol. Vol.* 1. *p.* 317. 2. Some rocks, when dug from a considerable depth in the earth, in so soft a state as to be readily cut with a knife, become very hard on exposure to the atmosphere.

Prin. Carbonate of lime, conveyed in a state of solution among the loose particles of gravel, sand, clay, or mud, and there precipitated, becomes a very efficient agent of consolidation.

Exam. 1. On the shores of the Bermuda and West India Islands, extensive accumulations of broken shells, corals, and sand, are formed upon the shores by the waves; and these are subsequently consolidated, frequently into very hard rock, by the infiltration of the water which contains carbonate of lime in solution. The famous Guadalope rock, in which human skeletons, along with pottery, stone arrow heads, and wooden ornaments, are found, is of the same kind. 2. The Mediterranean delta of the Rhone, is ascertained to be in a good measure solid rock, produced by the numerous springs that empty into it, that contain carbonate of lime in solution. The same is true of the deposits at the mouths of other rivers in the south part of Italy: but more especially on the east coast of the Mediterranean; where the ancient Sidon, formerly on the coast, is now two miles inland. *Lyell's Prin. Geol. Vol.* 1. *p.* 433. 3. In Pownal, Vermont, three miles north of Williams College, large masses of drift are cemented by carbonate of lime. 4. I have specimens of a calcareous breccia from West Stockbridge, in Massachusetts, which was formed by the chips thrown off in hewing marble, cemented together by the stream that passed over them, so as to be nearly as solid as the original limestone. This was accomplished in 17 years.

Prin. Another agent of consolidation is the red or per oxide of iron; or rather the carbonate of iron; since the per oxide is not soluble in water, without carbonic acid.

Exam. 1. On the northern coast of Cornwall, Eng., large masses of drifted sand have been cemented by iron into rocks, solid enough sometimes to be employed for building stones. 2. A similar case occurs on the coast of Karamania, and other parts of Asia Minor. *De La Beche's Manual, p.* 78. In the United States it is common to find the sand and gravel of the drift and tertiary strata, more or less consolidated by the hydrated per oxide of iron.

Prin. Silica dissolved in water, appears to have been in former times, an important agent in consolidating rocks: but at the present day it seems to be limited chiefly to deposits from thermal waters; since it is only water in this condition that will dissolve silica in much quantity.

Exam. The deposits around the Geysers in Iceland, the Azores Islands, &c.

Prin. Heat is an important agent in the consolidation of rocks: the most so when it produces complete fusion: yet this is not necessary to the production of a good degree of solidification.

Exam. The instances are so common in the arts, (as in burning bricks pottery, porcelain, &c.) where heat solidifies; and also in the vicinity of volcanos, where loose materials have become very hard by the proximity of lava, that particular instances need not be pointed out.

Prin. In many of the cases that have been described, great pressure assists in the work of consolidation. Indeed, it is sometimes sufficient of itself to bring the particles within the sphere of cohesive attraction.

Exam. This principle is too well understood to require particular instances to be pointed out.

General Inference from this and the Preceding Section.

Inf. From the facts detailed in this and the preceding section, it appears that all the stratified fossiliferous rocks of any importance, may have resulted from causes now in operation.

Proof and Exam. 1. Beds of clay need only to be consolidated to become clay slate, graywacke slate, or shale. 2. The same is true of fine mud. 3. Sand consolidated by carbonate of lime, will produce calcareous sandstone: by iron, ferruginous sandstone. 4. Drift, in like manner, will form conglomerates of every age, according to variations in the agents of consolidation. 5. Marls need only to be consolidated to form argillaceous limestones; and if sand be mixed with marl, the limestone will be siliceous. 6. Coral reefs and deposits of travertin, subjected to strong heat under pressure, will produce those secondary limestones that are more or less crystalline: but more of this under the next section. 7. We have already seen how beds of lignite and coal may be produced from peat and drift wood. 8. The formation of such extensive beds of rock salt and gypsum, as occur in the secondary and tertiary rocks, is more difficult to explain by any cause now in operation. And yet in respect to the former, it is said that the lake of Indersk, 20 leagues in circumference, on the steppes of Siberia, has a crust of salt on its bottom more than six inches thick, hard as stone, and perfectly white. The lake of Penon Blanco in Mexico, yearly dries up, and leaves a deposit of salt sufficient to supply the country. *Ure's Geology, p.* 373. I have also described a somewhat similar case at the lake of Ooroomiah in Persia. (*See Section III.*) And I have just been informed (April, 1842) by Mar Yohanna, bishop of Ooroomiah, who is now on a visit to this country, and who resides near the north end of that lake, that a small pond, covering about an acre, exists near his residence, which has a permanent deposit of salt several inches thick on its bottom. It is separated from the lake by a low narrow ridge of sand. According to Dr. Daubeny, (*Report on Mineral Waters, p.* 7,) thick beds of rock salt exist at the bottom of lake Elton, and of several other lakes adjoining the Caspian Sea. Probably, however, volcanic action was concerned in the deposition of rock salt. The origin of gypsum is still more difficult to explain by agencies now at work, since we know of but very few springs that deposit it, and these (ex. gr. at Baden near Vienna,) in small quantity.

Rem. 1. It does not follow from the preceding inference that the causes of geological change now in action, have not operated during the deposition of the fossiliferons rocks with greater energy than at present; but only that they have been identical in nature, during the past and present periods.

Rem. 2. We can better judge whether existing agents have produced the older stratified rocks, and the unstratified class, when we have examined the dynamics of igneous agencies.

SECTION VIII.

OPERATION OF IGNEOUS AGENCIES IN PRODUCING GEOLOGICAL CHANGES.

Def. Volcanic action in its widest sense, is the influence exerted by the heated interior of the earth upon its crust. Igneous agency has a still more extensive signification; embracing all the action exerted by heat upon the globe, whether the source be internal or external. The history of the former will prepare us better to appreciate the influence of the latter.

Prin. Volcanic agency has been at work from the earliest periods of the world's history; producing all the forms and phenomena of the unstratified rocks, from granite to the most recent lava. Modern volcanos will first come under consideration.

Def. These are of two kinds, *extinct* and *active.* The former have not been in operation within the historic period: the latter are constantly or intermittingly in action.

Def. A *volcano* is an opening in the earth from whence matter has been ejected by heat, in the form of lava, scoria, or ashes. Usually the opening called the *crater*, is an inverted cone; and around it, there rises a mountain in the form of a cone, with its apex truncated, produced by the elevation of the earth's crust and the ejection of lava. The volcanic cones vary in height from 600 feet (Stromboli,) to 17.730 feet. (Cotopaxi.) *Humboldt on the Superposition of Rocks, p.* 408.

Def. When nothing but aqueous and corrosive vapors have been emitted from a volcanic elevation for centuries, such elevation is called a *solfatara, or fumerole.*

Def. When volcanos exist beneath the sea, they are called *submarine;* when upon the land, *subaërial.*

Descr. As a general fact, volcanic vents are not insulated mountains, but are arranged in extensive lines, or zones; often reaching half around the globe.

Exam. 1. Perhaps the most remarkable line of vents is the long chain of islands commencing with Alaska on the coast of Russian America, which passes over the Aleutian Isles, Kamschatka, the Kurilian, Japanese, Phillippine, and Moluccan Isles, and then turning, includes Sumbawa, Java, and Sumatra, and terminates at Barren Island in the Bay of Bengal. 2. Another almost equally extensive line, commences at the southern extremity of S. America, and following the chain of the Andes, passes along the Cordilleras of Mexico, thence into California, and thence northward as far at least as Columbia river; which it crosses between the Pacific Ocean and the Rocky Mountains. *Parker's Tour beyond the Rocky Mountains.* 3. A volcanic region, 10 degrees of latitude in breadth, and 1000 miles long, extending from the Azore Islands to the Caspian Sea, abounds in volcanos though very much scattered. The region around the Mediterranean, is perhaps better known for volcanic agency than any other on the globe; because no eruption occurs there unnoticed.

Def. Volcanos not arranged in lines or zones, are called central volcanos, and are more or less insulated.

Exam. Iceland, the Sandwich Islands, Society Islands, Island of Bourbon, Jorullo in Mexico, and a region in Central Asia, of 2500 square geographical miles, from 800 to 1200 miles from the ocean. *De la Beche's Theoretical Geology, p.* 130.

Descr. The number of active volcanos and solfataras on the globe, is estimated at a little over 300: (303) *Considerations Generales sur les Volcans, &c. par M. J. Girardin, p.* 28. *Paris,* 1831, and the number of eruptions about 20 in a year, or 2000 in a century: though on both these points there is room for considerable uncertainty.

The following table will show how the active volcanos and solfataras are distributed on the globe.

	On Continents.	On Islands.	Total.
Europe,	4	20	24
Africa,	2	9	11
Asia,	17	29	46
America,	86	28	114
Oceanica,		108	108
Total.	109	194	303

Descr. 194 of these volcanos, or about two thirds, are situated upon the islands of the sea: and of the remaining third, the greater part are situated upon the borders of the sea, or a little distance from the coast. *Girardin's Considerations, &c. p.* 25.

Inf. Hence it is inferred that water acts an important part in volcanic phenomena: indeed, it seems generally admitted that the immediate cause of an eruption is the expansive force of steam and gases. It ought not to be forgotten, however that some volcanos are far inland: as Jorullo in Mexico, and the volcanos in central Asia.

Intermittent Volcanos.

Descr. Only a few volcanos, are constantly active: in most cases their operation is paroxysmal; and is succeeded by longer or shorter intervals of repose. This interval varies from a few months to seventeen centuries. In the Island of Ischia, the latter period has been known to intervene between two eruptions.

Inf. Hence some of the volcanos of America, generally regarded as extinct, (as Chimborazo, and Carguairazo in Quito, Tacoza in Peru, and Nevado de Toluca in Mexico,) may yet break forth and show themselves to belong to the class of active volcanos.

Phenomena of an Eruption.

Descr. A volcanic eruption is commonly preceded by earth quakes in the vicinity; stillness of the air, with a sense of oppression; noises in the mountain; and the drying up of fountains. The eruption commences with a sudden explosion, followed by vast clouds of smoke and vapor, with flashes of lightning, and showers of ashes and stones; and finally by red hot lava; which flows over the rim of the crater and spreads over the surrounding country.

Descr. Probably the most remarkable eruption of modern times took place in 1815, in the island of Sumbawa, one of the Molucca group. It commenced on the 5th of April, and did not entirely cease till July. The explosions were heard in Sumatra, 970 geographical miles distant in one direction, and at Ternate in the opposite direction, 720 miles distant. So heavy was the fall of ashes at the distance of 40 miles, that houses were crushed and destroyed beneath them. Towards Celebes, they were carried to the distance of 217 miles; and towards Java, 300 miles, so as to occasion a darkness greater than that of the darkest night. On the 12th of April, the floating cinders to the westward of Sumatra, were two feet thick; and ships were forced through them with difficulty. Large tracts of country were covered by the lava: and out of 12.000 inhabitants on the island, only 26 survived.

Descr. During the great eruption of the volcano of Cosiguina in Guatimala, on the shores of the Pacific, in 1835, ashes fell upon the island of Jamaica, 800 miles eastward: and upon the deck of a vessel 1200 miles westward. *American Journal of Science, Vol.* 33. *p.* 53.

Descr. The situation of Vesuvius and Etna has made their history better known than that of most volcanos. Eighty one eruptions of the latter are on record, since the days of Thucydides; and thirty seven of the former, since the first century of the Christian era. That which occurred in Vesuvius A.D. 79, is best known, from the fact that it buried three cities. Herculaneum. Pompeii, and Stabiae; which were flourishing at its base. Not much melted lava appears to have been thrown out at the eruption, which consisted chiefly of lapilli, sand, and stones. Hence it is, that almost every thing enveloped in those cities;—streets, houses, inscriptions, papyri, (manuscripts,) grain, fruit, bread, condiments, medicines, &c. &c. are in a most perfect state of preservation. They are, indeed, perfect examples of fossil cities! *Lyell's Prin. Geol. Vol.* 2. *p.* 189. *Dr. James Johnson's Philosophy of Travelling, p.* 232.

Rem. The vast quantity of aqueous vapor that escapes during a volcanic eruption, is often condensed and descends in torrents of rain, which falling upon the ashes, which the volcano has cast out, converts them into mud; and it was probably mostly mud that enveloped Pompeii, though Herculaneum appears to have been covered with melted matter. Mud ready formed, however, not unfrequently is emitted from volcanos.

Descr. In the year 1759, in the elevated plain of Malpais in Mexico, which is from 2000 to 3000 feet above the ocean, and at the distance of 125 miles from the sea, a volcanic eruption took place, producing six volcanic cones; now varying in height from 200 to 1600 feet. Around these cones, and covering several square miles, are a multitude of small cones, from 6 to 2 feet high, called *hornitos*, which continually give off hot aqueous vapor and sulphuric acid.

Rem. There is still a diversity of opinion as to the manner in which volcanic cones are formed. Von Buch, the distinguished Prussian geologist, maintains that a large part of the cone is produced by the upheaving of the strata, and that the crater, which in such cases he calls a *crater of elevation*, results from the fracture at the summit. Upon this elevated mass lava accumulates. Other geologists suppose the cone to be entirely formed of lava.

Descr. Sometimes during a violent eruption, the whole mountain, or cone, is either blown to pieces, or falls into the gulf beneath, and its place is afterwards occupied as a lake.

Exam. 1. In 1772, the Papandayang, a large volcano in the island of Java, after a short and severe eruption, fell in and disappeared over an extent 15 miles long and 6 broad; burying 40 villages, and 2957 inhabitants. 2 In 1638, the Pic, a volcano in the island of Timor, so high as to be visible 300 miles, disappeared, and its place is now occupied by a lake. 3. Many lakes in the south of Italy are supposed to have been thus formed; and this is probably the origin of the Dead Sea in Palestine, of lake Ooroomiah in Persia, &c. *Bakewell's Geology, p.* 265. 4. A volcano occupying the same spot, as the present Vesuvius, is supposed thus to have been destroyed in 79, and its remains to constitute the circular ridge, called Somna, which is several miles in diameter.

Dynamics of Volcanic Agency.

Prin. We can form an estimate of the power exerted by volcanic agency, from three circumstances: first, the amount of lava protruded: secondly, from the distance to which masses of rock have been projected: and thirdly by calculating the force requisite to raise lava to the tops of existing craters from their base.

Descr. Vesuvius, more than 3000 feet high, has launched scoria 4000 feet above the summit. Cotopaxi, nearly 18.000 feet high, has projected matter 6000 feet above its summit; and once it threw a stone of 109 cubic yards in volume, to the distance of nine miles.

Descr. Taking the specific gravity of lava at 2.8, the following table will show the force requisite to cause it to flow over the tops of the several volcanos, whose names are given, with their height above the sea. The initial velocity which such a force would produce, is also given in the last column.

Name.	Height in feet.	Force exerted upon the Lava.	Initial velocity per second.
Stromboli, highest peak,)	2168	176 Atmospheres.	371 feet.
Vesuvius,	3874	314	496
Jorullo, Mexico,	2942	319	502
Hecla, Iceland,	5106	413	570
Etna,	10892	882	832
Teneriffe,	12464	1009	896
Mouna Kea, Sandwich Isles,	14700	1191	966
Popocatapetl, Mexico,	17712	1435	1062
Mount Elias,	18079	1465	1072
Cotopaxi, Quito,	18869	1492	1104

Rem. There can be but little doubt but the chimney of a volcano extends generally as much below the level of the sea, as it does above; and often probably fifty times as deep: so that the actual force pressing upon the lava in its reservoir, may be far greater than the second column of the preceding table represents: and the initial velocity much greater than in the third column.

Descr. The amount of melted matter ejected from Vesuvius in the eruption of 1737, was estimated at 11.839.168 cubic yards: and in that in 1794, at 22.435.520 cubic yards. But these quantities are small compared with those which Etna has sometimes disgorged. In 1660, the amount of lava was 20 times greater than the whole mass of the mountain; and in 1669, when 77.000 persons were destroyed, the lava covered 84 square miles. Yet the greatest eruption of modern times was from Skaptar Jokul in Iceland, in 1783. Two streams of lava flowed in opposite directions; one of them 50 miles long and 12 broad; and the other 40 miles long and 7 broad: both having an average thickness of 100 feet: which was sometimes increased to 500 or 600 feet. Twenty villages and 9000 inhabitants were destroyed. *Lyell's Prin. Geol. Vol.* 2. *p.* 254.

New Islands formed by Volcanic Agency.

Descr. History abounds with examples of new islands rising out of the sea by volcanic action. Such were Delos Rhodes, and the Cyclades, situated in the Grecian Archipelago, and described by Pliny the naturalist, and other ancient writers. In more modern times, small islands have risen in the Azore group: such as Sabrina in 1811, which was 300 feet high, and a mile in circumference: but after some months it disappeared: another in 1720, was six miles in circumference. In 1707, the island called Isola Nuova, was thrown up near Santorini, and continues to this day. Just before the great eruption of Skaptar Jokul in Iceland in 1783, a new island appeared off the coast; which, however, subsequently disappeared. In 1796, a new island rose to the height of 350 feet; having two miles of circumference, in the Aleutian group, east of Kamtschatka, which is permanent. In 1806 another permanent island rose in the same vicinity, four geographical miles in circumference. In the same archipelago, in 1814, another peak arose, which was 3000 feet high; and which remained standing a year afterwards. In those cases where the cone does not sink back beneath the sea, it is probably composed of the more solid lavas, such as trachyte, or basalt. So late as 1831, a new island appeared near Sicily in the Mediterranean, rising to the height of 220 feet, and after exhibiting volcanic phenomena for sometime, it disappeared. *Girardin Considerations, &c. sur les Volcans, p.* 31—*Poulett Scrope's Considerations on Volcanos, p.* 172, *London*, 1825—*Lyell's Prin. Geol. Vol.* 1. *p.* 288. *Bakewell's Geology, p.* 261.

Rem. 1. In some instances the islands thus raised, are composed mainly of the rocks which form the bottom of the sea; and which has been upheaved; as the island of New Kamenoi, near St. Erini, which rose in 1707, and which was composed partly of limestone and covered with living shells.

Rem. 2. These islands are not always raised to their full height by a single paroxysm of the volcanic force; but by a succession of efforts for months and even years.

Descr. Very many large islands appear to be wholly, or almost entirely, the result of volcanic action; and to be composed chiefly of lava and rocks upheaved by volcanic action; such as sandstone and limestone: ex. gr. the Sandwich Islands: of which Hawaii, the largest, contains 4000 square miles of surface and rises 18.000 feet above the ocean: Teneriffe, 13.000 feet high: Iceland, Sicily, Bourbon, St. Helena, Tristan d'Acunha, the Madeira, Faroe and Azore Islands; a great part of Java, Sumatra, Celebes, Japan, &c.

Character of Molten Lava.

Descr. Lava in general is not very thoroughly melted; so that when it moves in a current over the country, its sides form walls of considerable height, and a crust soon forms over its surface, which serves still more to prevent its spreading out laterally.

Rem. Hence a lava current may be deflected from its course by breaking away its crust on one side: and in this way it has sometimes been turned away from towns that were threatened by it. In one instance, the inhabitants of Catania attacked a lava current and turned it towards Paterno; whose inhabitants took up arms and arrested the operation. *Lyell's Prin. Geol. Vol.* 2. *p.* 213.

Descr. The crust that forms upon lava soon becomes a good non-conductor of heat; and hence the mass requires a long time to cool: ex. gr. the case of Jorullo in Mexico, 1600 feet high, which was ejected almost 100 years ago, but is not yet cool. The lava thrown out of Etna in 1819, was in motion, at the rate of a yard in a day, nine months after the eruption: *Scrope on Volcanos, p.* 101. and it is stated that lava from the same mountain, at a previous eruption, was in motion after the lapse of 10 years. *De La Beche's Theoretical Geology, p.* 135.

Rem. This explains a curious fact. In 1828, a mass of ice was found on Etna, lying beneath a current of lava. Probably before this flowed over it, the ice might have been covered by a shower of volcanic ashes; which are a good non-conductor of heat; and might prevent the immediate melting of it; while the superimposed lava has preserved it from the period of its eruption to the present.

Descr. When lava is thrown out upon the dry land, with only the pressure of the atmosphere upon it, it is apt to become vesicular and scoriaceous: but when cooled slowly and under great pressure, it becomes compact and mây be even crystalline.

Volcanos constantly Active.

Descr. A few volcanic vents have been constantly active since they were first discovered. They always contain lava in a state of ebullition; and vapors and gases are constantly scaping.

Exam. 1. Stromboli, one of the Lipari Islands, has been observed longer probably than any volcano of this class; and for at least 2000 years it has been unremittingly active. The lava here never flows over the top of the crater: though it is sometimes discharged through a fissure into the sea, killing the fish, which are thrown upon the shore ready cooked. It is said to be more active in stormy than in fair weather; likewise more so in winter than in summer: a fact explained by the different degrees of pressure exerted by the air upon the lava at different times. When the air is light, the internal force predominates; but when heavy, it restrains the energy of the volcano

Exam. 2. In lake Nicaragua is a volcano which is constantly burning Villarica in Chili, so high as to be seen 150 miles, is never quiet. Th same is said to be the case with Popocatepetl, in Mexico, which is nearly 18.000 feet high. Ever since the conquest of Mexico it has been pouring forth smoke. *Girardin's Considerations, &c. p.* 183.

Exam. 3. But the most remarkable volcano on the globe is that of Kirauea or Kilauea, in the Sandwich Islands, on Hawaii; for the first accurate account of which we are indebted to American missionaries. *American Jour. of Science, Vol.* 11, *p.* 1, *and* 362. Rev. Messrs. Stewart and Ellis, the first an American, and the latter an English missionary, have both given us most graphic and thrilling descriptions of it. It appears to be situated upon a plain 8000, or 10.000 feet above the ocean; and at the foot of Mouna Roa. In approaching the crater, it is necessary to descend two steep terraces, each from 100 to 200 feet high, and extending entirely around the volcano. The outer one is 20 and the inner one 15 miles in circumference; and they obviously form the margin of vast craters, formerly existing. Arrived at the margin of the present crater, the observer has before him a crescent shaped gulf, 1500 feet deep; at whose bottom, which is from 5 to 7 miles in circumference, the top being from 8 to 10, is a vast lake of lava, in some parts molten, in others covered with a crust; while in numerous places (some have noticed as many as 50 at once.) are small cones, with smoke and lava issuing out of them from time to time. Sometimes, and especially at night, such masses of lava are forced up, that a lake of liquid fire, not less than two miles in circumference, is seen dashing up its angry billows, and forming one of the grandest and most thrilling objects that the imagination can conceive. Fig. 114, is a view of this volcano taken by Mr. Ellis.

Rem. A powerful eruption of this volcano took place in May and June, 1840. For several years the great gulf had been gradually filling up until it was not more than 900 feet deep. At length the lava found a subterranean passage, and flowed 8 miles under ground, when it reached the surface, and then advanced 32 miles further; and for three weeks continued to pour into the sea a stream of red hot lava with frightful hissings and detonations. *Missionary Herald for July,* 1841. *Account by Rev. Mr. Coan, American Missionary.*

Fig. 114.

Volcano of Kirauea: Sandwich Islands.

Seat of Volcanic Power.

Prin. Volcanic power must be deeply seated beneath the earth's crust.

Proof. 1. The melted lava is forced out from beneath the oldest rocks; as gneiss and granite: for masses of these rocks are frequently broken off and thrown out. 2. Lines or trains of volcanos indicate some connection between the vents; and the great length of these lines, several thousand miles in some instances, can be explained only by supposing that the fissure or cavity by which the connection is made, must extend to a great depth. 3. When in 1783, a submarine volcano, on the coast of Iceland, ceased to eject matter, immediately another broke out 200 miles distant, in the interior of the island. 4. Were not the power deep seated, volcanos would become exhausted; as they sometimes throw out more matter at a single eruption, than the whole mountain melted down could supply.

Earthquakes.

Descr. Earthquakes almost always precede a volcanic eruption; and cease when the lava gets vent.

Inf. 1. Hence the proximate cause of earthquakes is obvious: viz. the expansive efforts of volcanic matter, confined beneath the earth's surface.

Inf. 2. Hence too the ultimate cause of volcanos and earthquakes, is the same: whatever that cause may be.

Descr. Earthquakes are said to be preceded by great irregularities of the seasons; by redness of the sun, haziness of the air, violent winds, succeeded by dead calms, and the like but it is doubtful, whether these precursors be not more imaginary than real. That electric matter, or inflammable gas, or fire, should issue from the soil with mephitic vapors; that noises, like the trundling of carriages, and the discharge of artillery, should be heard beneath the ground; that something like sea sickness should be experienced; and that animals should show greater alarm than men, are all easily believed; because they are effects naturally resulting from the known phenomena.

Descr. During the paroxysm of the earthquake, heavy rumbling noises are heard; the ground trembles and rocks; fissures open on the surface, and again close; swallowing up whatever may have fallen into them; fountains are dried up; rivers are turned out of their courses; portions of the surface are elevated and portions depressed; and the sea is agitated and thrown into vast billows.

Exam. The cases that might be mentioned, of cities and towns, wholly or in part submerged by the ocean, in consequence of earthquakes, are very numerous. In the year 876, Mount Acraces is said to have fallen into the sea: in 541, Pompeiopolis was half swallowed up: in 1692, a part of Port Royal in the West Indies was sunk: in 1755, a part of Lisbon: in 1812, a part of Caraccas. About the same time numerous earthquakes agitated the valley of the Mississippi, for an extent of 300 miles, from the mouth of the Ohio, to that of the St. Francis; whereby numerous tracts were sunk down and others raised, lakes and islands were formed, and the bed of the Mississippi was exceedingly altered. In 1819, the bed of the Indus, at its mouth, was sunk 18 feet, and the village and port of Sindree submerged. At the same time a tract of the delta of the Indus, 50 miles long and 16 broad, was elevated about 10 feet. In Caraccas, in 1790, a forest was sunk over a space of 800 yards in diameter, to the depth of 80 or 100 yards. In 1783, a large part of Calabria was terribly convulsed by earthquakes, over an area of 500 square miles. The shocks lasted for four years: in 1783, there were 949, and in 1784, 151. A vast number of fissures of every form were made in the earth, and of course a great many local elevations and subsidences; which, however, do not appear to have exceeded a few feet. In some sandy plains, singular circular hollows a few feet in diameter, and in the form of an inverted cone, were produced by the water, which was forced up through the soil. Some of these are exhibited on Fig. 115.

Rev. Mr. Parker has described a remarkable subsidence, 20 miles in length, and a mile in width, just above the falls in Columbia river. Through this whole distance the trees are standing in the bottom of the stream, at an average depth of twenty feet; only that part of them above high water mark being broken off. He could discover no evidence that this tract had separated from the bank of the river. But the whole region appears to be one of extinct volcanos, and the river passes through hills and walls of basalt, and most probably this is a case of subsidence from an earthquake. The banks are too high and rocky to admit of the explanation that a lake has been formed by the river cutting through its *levee*, and overflowing the adjacent low ground, whereby, along the Mississippi, a lake with trees standing in it, is sometimes produced. *Parker's Exploring Tour beyond the Rocky Mountains, p.* 132.

Fig. 115.

Holes formed by an Earthquake.

The most extensive elevation of land on record by means of earthquakes, took place on the western coast of South America in 1822. The shock was felt 1200 miles along the coast: and for more than 100 miles the coast was elevated from 3 to 4 feet; and it is conjectured that an area of 100,000 square miles was thus raised up. This case, originally noticed by Mrs. Graham, and subsequently by Dr. Meyen and Mr. Freyer, has excited a great deal of discussion among European geologists; nor can it yet be regarded as absolutely settled; for Mr. Cumming, an able naturalist, who resided at Valparaiso at the time of the earthquake, (whose greatest power was exhibited there,) says that the spring tides rose to the same height upon a wall near his house after the event as before. *Lyell's Prin. Geol. Vol.* 2. *p.* 302.

Rem. 1. In estimating the permanent effects of earthquakes, it ought to be recollected that the changes of level which they produce, often balance one another, after the lapse of a few months or years.

Rem. 2. The number of earthquakes is probably about the same as that of volcanic eruptions, viz. about 20 annually.

Vertical movements of Land without Earthquakes.

Descr. It seems to be pretty well established, that various parts of our present continents are subject to vertical movements, either of elevation or depression, or of both, in alternation; and that too in districts not known to be subject to the action of earthquakes, or of volcanic agency in any form.

Exam. 1. The most certain example of elevation of an extensive tract of country in comparatively recent times, is that of the northern shores of the Baltic, investigated with great ability by Von Buch and Lyell. Some parts of the coast appear to have experienced no vertical movement. But from Gothenburgh to Torneo, and from thence to North Cape, a distance of more than 1000 geographical miles, the country appears to have been raised up from 100 to 200 feet above the sea. The breadth of the region thus elevated is not known, and the rate at which the land rises (in some places towards 4 feet in a century) is different in different places. The evidence that such a movement is taking place, is principally derived from the shells of mollusca now living in the Baltic, being found at the elevations above named; and some of the barnacles attached to the rocks. They have been discovered inland in one instance 70 miles. *Lyell's Prin. Geol. Vol.* 1. *p.* 437.

2. Several examples are quoted in England of elevated sea beaches with the remains of shells; such as now occur in the ocean. Also in Italy; at the Cape of Good Hope; and in the West Indies. But I may be permitted to doubt whether some of these cases are not rather to be explained by depositions during the glacio-aqueous period; since clay and sand of great thickness were then deposited. Yet Prof. Phillips says, that in most of the cases under consideration, the sea shells occur beneath the drift. If so, must it not be quite difficult to distinguish the formation from the most recent tertiary? 3. On the margin of Lubec Bay in the state of Maine, Dr. C. T. Jackson describes a deposit of recent shells, in clay and mud, with the remains of balani attached to the trap rock, 26 feet above the present high water mark. *First Report on the Geology of Maine, p.* 18. 4. A remarkable case of subsidence seems pretty well established as having occurred on the coast of Greenland for a distance of 600 miles north and south. *Lyell's Prin. Geol. Vol.* 1. *p.* 420, 474. 5. The history of submarine forests furnishes another example of subsidence. But this phenomenon deserves a fuller description.

Submarine Forests.

Descr. On the shores of Great Britain, France, and the United States, usually a few feet beneath low water mark, there occur trees, stumps, and peat, seeming to be ancient swamps, which have subsided beneath the waters, sometimes to the depth of 10 feet. In many cases the stumps appear to stand in the spots where they originally grew; yet it requires great care to ascertain this fact. *For localities, see De La Beche's Manual, p.* 151. *Lyell's Prin. Geol. Vol.* 2. *p.* 46. 80. *Vol.* 3. *p.* 315, *and Vol.* 2. *p.* 140. *Final Report on the Geology of Massachusetts, p.* 307.

Origin of Submarine Forests. It is probable that this phenomenon results from several causes. 1. When the barrier between a peat swamp and the sea is broken through, so that the water may be drained off, a subsidence of several feet may take place in the soft spongy matter of the swamp, sufficient to bring it under water. 2. From a case which I have described on Hogg Island, in Casco Bay. *Boston Journal of Natural History, Vol.* 1. *p.* 338. I have inferred that some submarine forests may have been produced by the gradual removal of the contents of a peat swamp, by the retiring tide, after the barrier between it and the ocean has been removed so as to form a slight slope into the sea. At the spot referred to, the process may be seen partly completed. 3. But probably most submarine forests were produced by earthquakes, or other causes of subsidence, which we find to have operated on the earth's surface: and the explanation of which will be better understood after the statement of more facts relating to igneous agency.

Thermal Springs.

Descr. Hot springs are very common in the vicinity of volcanos: such as the well known geysers in Iceland. Some of these are intermittent, probably in consequence of the agency of steam within subterranean cavities. The great geyser consists of a basin 56 by 46 feet in diameter; at the bottom of which is a well 10 feet in diameter and 78 feet deep. Usually the basin is filled with water in a state of ebullition: but occasionally an eruption takes place, by which the water is thrown up from 100 to 200 feet, until it is all expelled from the well, and there follows a column of steam with amazing force and a deafening explosion, by which the eruption is terminated. These waters hold silica in solution; as do those of the Azore Islands; and extensive deposits are the result. The coating over of vegetables by this siliceous matter, has given rise to the common opinion that certain rivers and lakes possess the power of rapid petrifaction.

Descr. Thermal springs are not confined to the vicinity of volcanos. They occur in every part of the globe; and rise out of almost every kind of rock. They frequently contain enough of mineral substances to constitute them mineral waters. But one of their most striking properties is the evolution of gas; such as carbonic acid, nitrogen, oxygen, sulphureted hydrogen, &c. in a free state.

Theory of Thermal Springs. When these springs occur in volcanic districts, their origin is very obvious. The water which percolates into the crevices of the strata, becomes heated by the volcanic furnace below, and impregnated with salts and gases by the sublimation of matter from the same focus. Dr. Daubeny, who has devoted great attention to this subject, has also endeavored to show that the thermal springs not in volcanic districts, in a large majority of cases rise either from the vicinity of some uplifted chain of mountains, or from clefts and fissures caused by the disruption of the strata; and therefore, in all such cases are probably the result of deep seated volcanic agency, which may have been long in a quiescent state. If this view of the subject be not absolutely proved, it is at least extremely probable. *Report on Mineral and Thermal Waters, by Prof. Daubeny*, 1836, *p.* 62.

Extinct Volcanos.

Descr. Many writers maintain that there is a marked difference between the matters ejected from active and extinct volcanos. It is said that the more modern lavas have a harsher feel, are more cellular, and more vitreous in their appearance, and also less felspathic than the ancient. *Girardin's Considerations sur les Volcans, p.* 13. *De La Beche's Manual p.* 126. But it is doubtful whether any character will satisfactorily distinguish them, except the period of their eruption.

Descr. The extinct volcanos are of very different ages. Some of them were active during the tertiary period, some during the drift period; and some since that time. The lava, especially in the most ancient, was not always ejected from conical elevations, so as to form regular craters, but along extended fissures. In some instances, as in a mountain called the Puy de Chopine in Auvergne, which stands in an ancient crater, and rises 2000 feet above an elevated granitic plain, itself about 2800 feet above the sea, there is a mixture of trachyte and unaltered granite.

Exam. I. The extinct volcanos of Auvergne, and the south of France, have long excited deep interest; and have been fully illustrated by Scrope, Bakewell, and others. Near Clermont, the landscape has as decidedly a volcanic aspect as in any part of the world; of which Fig. 116 will convey some idea.

Fig. 116.

Extinct Volcanos: Auvergne.

2. Extinct volcanos exist also in Spain, in Portugal, in Germany, along the Rhine, in Hungary, Styria, Transylvania, Asia Minor, Syria, and Palestine. To the east of Smyrna in Asia Minor, is a region called the *Burnt District*, (Katakekaumena of the Greeks,) because it shows such striking marks of extinct volcanos. In the valley of the Jordan, especially around lake Tiberias, extending as far northwest as Safed, volcanic rocks abound,

with warm springs and occasional earthquakes. *Robinson's and Smith's Biblical Researches in Palestine, &c. Vol. 3. p.* 312.

Descr. That the region occupied by the Dead Sea has been some time or other the seat of volcanic action, can hardly be doubted. For a lake so charged with saline matter as this, and producing bitumen, and having moreover cliffs of rock salt upon its margin, with sulphur occasionally, and being probably depressed more than 500 feet below the level of the Mediterranean, must be referred to such an agency. *Robinson and Smith, Vol.* 2. *p.* 221. And it has been usual to suppose that the five ancient cities, Sodom, Gomorrah, Admah, Zeboim, and Zoar, which undoubtedly occupied what is now the southern part of the Dead Sea, were destroyed by a volcanic eruption. There is a passage in the Book of Job, (Chap. 22, v. 15 to 20.) which, if it refer to that catastrophe, lends much probability to this supposition. Dr. Henderson thus translates the passage.

"Hast thou observed the ancient tract
That was trodden by wicked mortals,
Who were arrested on a sudden;
Whose foundation is a molten flood?
Who said to God, depart from us,
What can Shaddai do to us?
Though he had filled their houses with wealth.
(Far from me be the counsel of the wicked!)
The righteous beheld and rejoiced,
The innocent laughed them to scorn:
Surely their substance was carried away,
And their riches devoured by fire."

Rem. It may still be doubted whether an eruption has taken place in the vicinity of the Dead Sea so recently as the time of the overthrow of the cities of the plain. Indeed, my own convictions are, after reading the accounts given us by modern travellers and missionaries, especially by Robinson and Smith, that this region is to be ranked among the extinct, and not among the active volcanos. Dr. Robinson states that the rocks which constitute the mountains around the Dead Sea, are limestone, and says that "he is not aware that the dark basaltic stones so frequent around the lake of Tiberias have even been discovered in its vicinity," except a large specimen of augitic vesicular lava, found by Rev. Mr. Hebard, near the mouth of the Jordan. *Researches, &c. Vol.* 2. *p.* 221. I possess, however, a specimen of dark porous lava, presented to me by the Rev. Mr. Homes with this label: "from a mound once surrounded by water on the shores of the Dead Sea." I understood from that gentleman, who, himself, picked up the specimen, that the mound, or hummock, was composed of the same. It is impossible, however, to determine from the appearance of a single specimen, whether it is the product of an active or an extinct volcano, or whether it may not belong to a still earlier period. It does prove, however, that igneous rocks have been protruded near the Dead Sea at a comparatively recent epoch. The specimen found by Mr. Hebard, of which I possess also a fragment, is more decidedly vesicular than I have ever seen any greenstone or basalt.

Descr. "Professors Michaelis and Busching suggest, that Sodom and Gomorrah were built upon a mine of bitumen, that lightning kindled the combustible mass, and that the cities sunk in the subterranean conflagration." *Encyc. Relig. Knowledge, Art. Dead Sea.* Messrs. Robinson and Smith have recently suggested this hypothesis anew, and more fully illustrated it, by reference to the pitch lake of Trinidad, and the remarkable protrusion of an unmelted basaltic dyke in 1820, in the island of Bonda. *Am. Bib. Repos. Jan.* 1840, *p.* 24, *also Biblical Researches in Palestine, &c. Vol.*

2. *p.* 669. *See Objections to this hypothesis by Dr. Lee. Am. Bib. Repos. April,* 1840, *p.* 324. The principal difficulties in the way of this hypothesis are, first, to see how the bitumen, buried beneath a considerable thickness of soil, could have burnt rapidly enough suddenly to destroy the cities and their inhabitants; and secondly, to conceive of a bed of bitumen so thick, as by its combustion to sink the surface from the present high water mark to the bottom of the sea. Dr. Robinson describes the high water mark as seen by him "a great distance," south of the margin of the Sea at that time. The surface, therefore, must have suffered a great depression. Would it not somewhat relieve these difficulties to suppose volcanic action combined with the combustion of the bitumen. No geologist will doubt the correctness of Von Buch's opinion, that a fault extends from the Red Sea through the valley of Arabah and the Jordan to mount Lebanon; and along that fissure we might expect volcanic agency to be active. But it might have produced very striking effects without the ejection of lava. Earthquakes sometimes cause the surface to sink down many feet, and flames have been seen to issue through the fissures which they produce. Thus might the *slime pits* (literally *wells of asphaltum,*) have been set on fire, immense volumes of steam, smoke, and suffocating vapors have been set at liberty, perhaps, too, the remarkable ridge of rock salt called usdum, have been protruded, and finally, by the subsidence of the surface after the destruction of the cities, might the waters of the lake have flowed over the spot. In a similar manner was the city of Euphemia, in Calabria, destroyed in 1638. "After some time," says Kircher, who was near the spot, "the violent paroxysm (of the earthquake) ceasing, I stood up, and turning my eyes to look for Euphemia, saw only a frightful black cloud. We waited till it had passed away, when nothing but a dismal and putrid lake was to be seen, where once the city stood." Dr. Robinson has, indeed, made it almost certain that the Dead Sea existed before the catastrophe of Sodom, and that the Jordan, before that period, did not flow through El Ghor and Arabah into the Red Sea; but a sinking down of the surface a few feet at that time, is not inconsistent with these facts. But I will not enlarge further on a subject so difficult. Nor should I have said so much, had I not read the very accurate descriptions of that country by Messrs. Robinson and Smith.

Desc. According to Prof. Parrot, Mount Ararat in Asia, is an extinct volcano. A specimen sent me by Rev. Justin Perkins from that mountain is decidedly vesicular lava.

Descr. A large proportion of the lofty peaks of the Andes and the mountains of Mexico belong to the class of extinct volcanos: and it is very probable, from the statements of Rev. Mr. Parker and others, that a vast region between the Rocky Mountains and the Pacific Ocean is of the same character. For although he describes the prevailing rock as basalt, and only incidentally alludes to volcanic cones and craters, yet in personal conversation he assures me that regular craters are not unfrequent; and having shown him specimens of trachytes from the continent of Europe, he at once identified them with rocks found associated with the basalt of that region.

Descr. The size of ancient volcanic cones and craters was often very large.

Exam. In the middle and southern parts of France, extinct volcanos cover several thousand square miles. Between Naples and Cumea, in the space of 200 square miles, according to Brieslak, are 60 craters; some of them larger than Vesuvius. The city of Cumea has stood three thousand years in a crater of one of these volcanos. Vesuvius stands in the midst of a vast crater, whose remains are still visible, called Somna. The volcanic peak of Teneriffe stands in the centre of a plain, covering 108 square miles, which is surrounded by perpendicular precipices and mountains, which were probably the border of the ancient crater. According to Humboldt, all the mountainous parts of Quito, embracing an area of 6300 square miles, may be considered as an immense volcano, which now gets vent sometimes through one, and sometimes through another of its elevated peaks; but which must have been more active in former times, to have produced the results now witnessed. We have seen that the great volcano of Kirauea, on the Sandwich Islands, is surrounded by two circular walls, one 15 and the other 20 miles in circumference; which must have marked the limits of the crater in early times. Two other ancient craters exist in the island of Maui, one 24 and the other 27 miles in circuit. *Couthoy on Coral Formations, p.* 9

Inf. From such facts it has been inferred by many geologists, (ex. gr. Poulett Scrope, Bakewell, Phillips, Brongniart, Girardin, &c.) that volcanic agency in early times was more powerful than at present; and that it is gradually diminishing. Mr. Lyell, however, considers this view as entirely erroneous; and quotes the eruption from Skaptar Jokul, in 1783, as equalling any that is known to have occurred in ancient times. *Lyell's Prin. Geol. Vol.* 2. *p.* 254.

The Older Unstratified Rocks.

Rem. So rapid has been the change of opinion respecting the origin of the unstratified rocks, that from an almost universal belief in their deposition from water, geologists are now nearly or quite unanimous in ascribing them to igneous agency. A brief summary of the arguments that sustain this latter opinion, will be, therefore, all that is now necessary to present.

Prin. The different unstratified rocks appear to be the result of volcanic agency, exerted at different periods under different circumstances.

Proof 1. *Identity of lithological characters between recent lavas and several varieties of unstratified rocks.* The amygdaloids of the trap rocks often exactly resemble those vesicular lavas which are cooled in the open air: while the compact trap rocks can scarcely be distinguished from the compact lavas of submarine production. Some varieties of trachyte very much resemble granite; and the two rocks often pass insensibly into each other: so that it is difficult to say whether trachyte be melted granite, or a portion of the materials out of which granite was originally produced, cooled in a different manner.

Proof 2. *The insensible gradation of the different unstratified rocks into one another.* In the same continuous mass we find a gradual passage from trap into all the other unstratified rocks; so that the same general cause that produced one variety, must have produced the whole. It is very rare, however, that coarse granite, destitute of hornblende, graduates into trap rocks of the same age. In general, they appear to have been formed at different epochs. *Macculloch's System of Geology, Vol. 1. p.* 157.

Rem. It must not be inferred from this statement, that all the unstratified rocks have resulted from the same melted mass, cooled under different circumstances: for the difference in their chemical composition is too great to admit of a such conclusion. *See Section* 4. *p.* 78.

Proof 3. *The mode of occurrence of the unstratified in relation to the stratified rocks.* We have seen, (section IV.) that the former exist as protruding, intruding, and overlying masses, and occupying veins in the latter. Now these are the precise modes in which recent lava occurs when connected with stratified rocks: whereas no example can be produced in which rocks have been made to take these forms by aqueous agency. Indeed, it is difficult to conceive how this would be possible.

Proof 4. *The columnar structure of the trap rocks.* This structure is not uncommon in lavas. The experiments of Mr. Watt also, upon 700 pounds of melted basalt, which on cooling assumed the columnar form, as detailed in section IV, confirms this view: whereas no example of such a structure from aqueous agency has ever been found.

Proof 5. *The crystalline structure of some of the unstratified rocks.* When several substances are contained in an aqueous menstruum, it is difficult to make them crystallize except in succession; whereas in granite the different ingredients appear to have crystallized simultaneously. And if the materials of granite, or of glass, be melted and slowly cooled, especially under pressure, most if not all the ingredients will assume more or less of a crystalline form at the same time.

Proof 6. *The mechanical effects produced by the unstratified upon the stratified rocks.* In the vicinity of veins and irregular masses of the unstratified rocks, the stratified ones are bent and twisted in every conceivable manner, and sometimes broken entirely. Not unfrequently also, fragments of the stratified rocks are broken off and entirely imbedded in the veins of the unstratified. In almost every case an upward or a lateral force appears to have been exerted; showing that the veins were filled from beneath. For examples, *See Macculloch's Geology of Glen Tilt in the geological transactions. Also his Geology of the Western Islands. Also Final Report on the Geology of Mass. pp.* 414, 416, 418, 464, *&c.* Fig. 117, shows a vein of greenstone, or indurated wacke, passing through a ledge of sandstone near New Haven, on the road to Middletown, the sandstone being bent upwards.

Fig. 117.

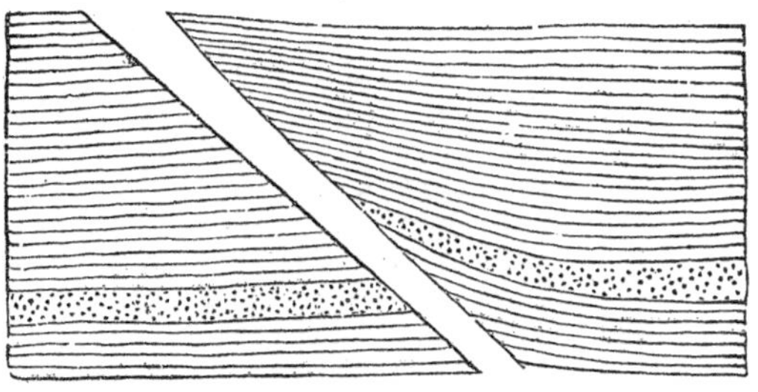

Proof 7. *The chemical effects produced upon the stratified rocks by the contact of the unstratified.* These effects are precisely the same as those produced by dykes of recent lava. Thus, compact fossiliferous limestone and chalk, where dykes of trap rocks, porphyry, and granite pass through them, are changed into crystalline limestone: shale and sandstone are indurated and converted into siliceous slate or jasper; as at Nahant, Mass. Newport, Rhode Island, and Rocky Hill, near Hartford, Ct.: micaceous sandstone and other slaty rocks are changed into mica slate, or hornblende slate. Now these are effects that could not result from any other agency with which we are acquainted except heat. And in respect to chalk, an experiment of Sir James Hall is decisive. He confined some of it in a strong iron tube, and subjected it to a strong heat; which, liberating the carbonic acid, produced a powerful pressure, and the result was crystallized carbonate of lime. *Bakewell's Geology, p.* 146.

Obj. The older unstratified rocks present no example of a volcanic cone or crater.

Ans. 1. They appear in general to have been produced by a force acting along extended fissures, and not directed to particular foci. 2. If these cones and craters once existed, the powerful denuding agencies that have operated on the globe may have destroyed them.

Prin. The greater degree of crystallization in the older unstratified rocks, may be explained, by supposing a more perfect fusion of the materials than in recent lavas, and greater slowness in cooling, under perhaps the more powerful pressure of a deep ocean.

Temperature of the Globe.

Prin. The principal circumstances that determine the temperature of the globe and its atmosphere, are the following: 1

Influence of the sun. 2. Nature of the surface. 3. Height above the ocean. 4. Temperature of the celestial spaces around the earth. 5. Temperature of the interior of the earth, independent of external agencies.

1. *Solar heat.* The solar rays exert no influence as a general fact, at a greater depth than about 100 feet. (Baron Fourier mentions 130 feet as the maximum depth: Poisson fixes it at 76 feet. *Am. Jour. Science, Vol.* 32. *p.* 5, *and Vol.* 34. *p.* 59.) A thermometer placed at that depth, remains stationary all the year. The diurnal effect does not extend more than 3 or 4 feet. In receding from the tropics, the amount of solar heat diminishes. During six months it continues to increase, and to diminish the remaining six months. The decrease of the mean temperature from the equator towards the poles, is nearly in the proportion of the cosines of latitude. Prof. Forbes has recently made some observations near Edinburgh, from which it appears that the oscillations of annual temperature would cease at the depth of 49 feet in trap tufa, 62 feet in incoherent sand, and 91 feet in compact sandstone. *American Journal of Science, Vol.* 38. *p.* 109.

Rem. Solar heat is the fundamental element on which depends the surface temperature of the globe and the character of the climate.

Prin. The amount of solar heat is actually though very slightly diminishing in consequence of a change in the eccentricity of the earth's orbit. The possible amount of this diminution is not known, because the limits of the eccentricity of that orbit are not known. But there is no probability that the annual temperature ever has changed or ever will change from this cause, more than 3° or 4°. *Am. Jour. Sci. Vol.* 36. *p.* 332. Hence this cause is insufficient to account for the extra-tropical heat of the present cold regions of the earth in early times.

2. *Nature of the surface.* The radiating and absorbing power of land is quite different from that of water. Ice and snow are still more different; and the nature of the soil affects sensibly its power to imbibe or give off heat. Hence low islands have a higher temperature than larger continents in the same latitude; and the ocean possesses a greater uniformity of climate than the land.

Rem. On these facts Mr. Lyell has founded an hypothesis for explaining the high temperature of the surface of the globe in northern latitudes in early times. He supposes that but little land then existed in the northern parts of the globe, and that this produced so great an elevation of temperature, above what it is at present, that tropical plants and animals might then have inhabited regions now subjected to almost perpetual winter. That the quantity of dry land in the northern hemisphere during the deposition of the older fossiliferous rocks was much less than at present, is very probable; and that this would render the climate warmer and more uniform is made certain by comparing the climate of Great Britain with that of the United States. But that from this cause the climate of Canada, of the North West Coast of America, between 60° and 70° of North Latitude, and even of Greenland and Melville Island, where the thermometer now descends to 58° below zero, was so mild and uniform as to produce tropical ferns, lepidodendra, &c. is a position which will need strong proof; especially when we recollect that for several months annually they must have been most of the time in darkness. Those, however, who wish to see this hypothesis ably defended, may consult *Lyell's Prin. Geol. Vol.* 1. *p.* 160, *&c.*

3. *Height above the ocean.* The temperature of the air diminishes one degree (Fahr.) for 300 feet of altitude: two degrees for 595 feet: three degrees for 872 feet: four degrees for 1124 feet: five degrees for 1347 feet: and six degrees for 1539 feet. Hence at the equator perpetual frost exists at the height of 15.000 feet, diminishing to 13.000 feet at either tropic. Between latitudes 40° and 59°, it varies from 9000 to 4000 feet. In almost every part of the frigid zone this line descends to the surface. These results, however, are greatly modified by several circumstances: so that in fact, the line of perpetual congelation is not a regular curve, but rather an irregular line descending and ascending. *American Journal of Science, Vol.* 33. *p.* 52. *Introduction a la Geographie Mathematique et Physique, par S. F. Lacroix, p.* 289.

4. *Temperature of the celestial spaces around the earth.*—This cannot be much less than the temperature around the poles of the earth; where the solar heat has scarcely no influence. Now the lowest temperature hitherto observed on the globe, (at Melville Island,) is 58° below zero: and this has been assumed as the temperature of the planetary spaces. Hence it follows that there must be a constant radiation of heat from the earth into space.

5. *Temperature of the interior of the Earth, Independent of External Agencies.*

Prin. In descending into the earth, beneath the point where it is affected by the solar heat, we find that the temperature regularly and rapidly increases.

Proof 1. *The temperature of springs which issue from the rocks in mines, as shown in the following table.*

Temperature of Springs in Mines.

COUNTRIES.	MINES.	Depth in Feet.	Temperature.	Mean temperature at Surface.	Depth for one degree Fahrenheit.
Saxony,	Lead and Silver Mine of Junghohe Birk.	256	48°9	46.9	102.4
	do of Beschertgluck,	712	54.5	46.4	87.
	do do	840	56.8		80.7
	do Himmelfahrt,	735	57.9		63.9
	do Kuprinz,	634	80.1		18.8
Brittany,	do Poullauen,	128	53.4	52.7	182.
	do do	246	53.4		351.
	do do	459	58.3		82.
	do Huelgoet	197	54.	51.8	89.5
	do do	262	59.		36.4
	do do	394	59.		54.7
	do do	755	67.5		48.4
Cornwall,	Dolcoath Mine,	1440	82.	50.	45.
Mexico.	Guanaxato, Silver Mine.	1713	98.2	68.8	45.8

Proof 2. *Temperature of the rock in mines;* as shown in the following table.

COUNTRIES.	MINES.	Depth in Feet.	Temp. Observed.	Mean temp. of surface.	Depth for one degree
1. *In loose matter near the face of the rock.*					
Cornwall,	United Copper Mines,	1142	87°.4	50°	30.5
		1201	88.		31.1
Carmeaux France.	Coal Pit of Ravin,	597	62.8	52	55.3
	do of Castellan,	630	67.1	52	40.8
Littry, do.	do of St. Charles,	325	61.		36.1
Decise, do.	do of St. Jacobi,	351	64.		29.2
do. do.	do do	561	71.7		28.5
2. *In the rock near its surface.*					
Saxony,	Mine of Beschertgluck,	591	52.2		101.
	do do	813	59.	46.4	67.
	do do	236	47.7		174.7
	do do	552	55.		63.7
	do do	880	59.		69.8
	do do	1246	65.7		64.4
3. *Three feet three inches within the rock.*					
Cornwall,	Dolcoath Mine. Register kept 18 Months.	1381	75.6	50.	54.
Saxony,	Lead and Silver Mine of Kurpinz,	413	59.6		31.3
	do	686	62.5		42.6
	do	1063	67.7		49.9

In a colliery at Wigan, in Lancashire, England, at 150 feet deep, the temperature was constantly 53°: at 450 feet, it was 56°75; at 750 feet it was 63°. This would give an increase of one degree for every 48 feet. *Am. Jour. of Science, Vol.* 34. *p.* 36.

A single experiment in the deepest coal mine in Great Britain, near Sunderland, gave the following results: depth of the place of observation, 1584 feet: below the level of the sea, 1500 feet. Mean annual temperature at the surface, 47°6: temperature on the day of observation, (Nov. 15, 1834.) 49°: do. of the air at the bottom of the pit, 64°: close to the coal, 68°: do. of water collected at bottom, 67°: do. of salt water issuing from a hole made the same day, 70°1: do. of gas rising through the water, 72°6, do. of the front of the coal, 68°: do. of the same, left in a bore hole for a week, 71°2. Hence the heat increases at the rate of about a degree for every 60 feet.

Proof 3. *Temperature of Artesian Wells*, as shown in the following table.

LOCALITIES.	Depth in Feet.	Mean temp. at Surface	Temp. of Wells.	Depth for 1 degree in Feet.
Paris: Fountain de la Garde St. Ouen	216	51°1	55°2	52.7
Do. near the Barriere de Grenelle.	1800	51.1	83.	50.
Dept. du Garde et des Pas Calais				
Fountain Artesienne de Marguette.	184	50.5	54.5	46.
Do. d'Aire.	207		55.9	38.3
Do. de St. Venant.	328		57.2	49.
Sheerness, England, mouth of the Medway,	361	50.9	59.9	40.1
Tours,	459	52.7	63.5	42.5
A well at La Rochelle,	369	53.4	64.6	33.
Near Berlin, Prussia, at	675	49.1	67.66	36.3
Do. the same well, at.	516		63.95	34.7
Do. do. at.	392		62.82	28.5
Near New Brunswick, N. Jersey, at the depth.	250		52.	72.
do. at.	394		54.	
South Hadley, Mass.	180	46.7	52.	34.

Rem. 1. Near Vienna, in Austria, are from 40 to 50 Artesian Wells, whose temperature varies from 52° to 58°; the mean temperature at the surface being 50°54. At Heilbronn in Wurtemburg, five wells sunk from 60 to 112 feet, have a temperature of 55°.

Rem. 2. Artesian wells have lately been applied with success in Wurtemburg, to prevent frost from stopping machinery, which was moved by running water; and also for warming a paper manufactory. Who knows but this application may prove of immense benefit to some regions of the globe? *Buckland's Bridgewater Treatise, Vol.* 1. *p.* 567.

Proof 4. *Thermal springs.* Vast numbers of these occur in regions far removed from any modern volcanic action; generally upon lofty mountain ranges; as upon the Alps, the Pyrenees, Caucasus, the Ozark mountains in this country, where are nearly 70, &c. Their temperature varies from about summer heat nearly up to that of boiling water. Nor can their origin be explained without supposing a deep seated source of heat in the earth. This argument is not indeed, as direct and conclusive as those previously mentioned, but it confirms the others.

Proof 5. *The existence of numerous deep seated volcanos.* This argument is of the same kind as the last, and does not need any farther illustration here.

Proof 6. *Not one exception to this increase of internal temperature has ever occurred, where the experiment has been made in deep excavations.*

Inf. 1. The increase of temperature from the surface of the earth downwards, does not appear to be at the same rate in all countries. The mean of all the observations recorded in the preceding tables, which have been made in England, gives 44 feet for a change of one degree. In some mines in France, the increase is much slower, and in a few it is faster. The mean is reckoned at about 45 feet for each degree. In Mexico, according to the only observation given above, it is 45.8 feet. In Saxony it is considerably greater, not far from 65 feet to a degree. The few observations in this country given in the preceding table, indicates an increase of 54 feet to a degree.

Inf. 2. The average increase for all the countries where observations have been made, is stated by Kupffer, to be 36.81 feet for each degree. *Edinburgh Journal of Science, April,* 1832.

Inf. 3. At this rate, and assuming the temperature of the surface to be 50°, a heat sufficient to boil water would be reached at the depth of 5962 feet, or a little more than a mile: a heat of 7000°, sufficient to melt all known rocks, would be reached at 48 miles; and at the centre of the earth, it would amount to 577.000°. *Cordier's Essay on the Temperature of the Interior of the Earth. Amherst,* 1828, *p.* 73. *Moffatt's Scientific Class Book, by Prof. Johnson, Philadelphia,* 1836, *Vol.* 2. *p.* 311.

Rem. It has been thought by many, and probably with reason, that the rate of increase in the subterranean heat, as deduced by Kupffer, is too rapid. From careful observation upon the Artesian Wells of Scotland, Dr. Patterson finds the mean increase to be one degree for every 47 feet: and from a more extended comparison given in Jameson's Journal, (April to July, 1839,) the mean increase is one degree for 65 feet: per haps the rate of 45 feet to a degree, fixed upon by the British Association, ought to be considered the best hitherto obtained.

Inf. 4. From the preceding facts, and other collateral evidence, it has been inferred that all the interior of the earth, except a crust from 50 to 100 miles thick, is at present in a state of fusion: that originally the whole globe was melted, and that its present crust has been formed by the cooling of the surface by radiation.

Illus. Fig. 118 is intended to represent the proportion of melted and unmelted matter in the earth, agreeably to the above inference; and on the supposition that the solid crust is 100 miles thick. This is shown by the broad line that forms the circumference. According to the mean increase of subterranean heat stated above, this crust should be only half as thick.

Fig. 118.

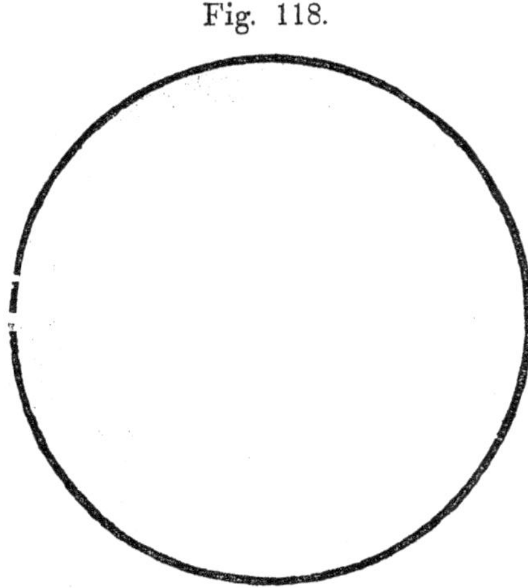

Proof 1. Until some fact can be adduced showing that the heat of the earth ceases to increase beyond a certain depth, nothing but hypothesis can be adduced to prove that it does not go on increasing, until at least the rocks are all melted: for when they are brought into a fluid state, it is not difficult to see how the temperature may become more equalized through the mass, in consequence of the motion of the fluid matter; so that the temperature of the whole may not be greatly above that of fused rock. Now if the hypothesis of internal fluidity have other arguments (which follow below) in its favor, while no facts of importance sustain its opposite, the former should be adopted.

Proof 2. It appears from the experiments and profound mathematical reasoning of Baron Fourier, that even admitting all the internal parts of the earth to be in a fused state, except a crust of 30 or 40 miles in thickness, the effect of that internal heat might be insensible at the surface, on account of the extreme slowness with which heat passes through the oxidized crust. He has shown that the excess of temperature at the surface of the earth, in consequence of this internal heat, is not more than 1–17th of a degree, (Fahr.) nor can it ever be reduced more than that amount by this cause. This amount of heat would not melt a coat of ice 10 feet thick, in less than 100 years: or about one inch per annum. The temperature of the surface has not diminished on this account, during the last 2000 years, more than the 167th part of a degree: and it would take 200.000 years for the present rate of increase in the temperature as we descend into the earth, to increase the temperature at the surface one degree: that is, supposing the internal heat to be 500 times greater than that of boiling water. From all which it follows, that if internal heat exist, it has long since ceased to have any effect practically upon the climate of the globe. *Annals de Chimie et de Physique, No.* 27. *American Journal of Science, Vol.* 32. *p.* 1. *Phillips's Treatise on Geology, Vol.* 2. *p.* 275.

Rem. These results of Fourier require the application of very profound mathematical investigations. And it may not be amiss to mention that the late lamented Dr. Bowditch informed me, that he had followed Fourier through all his intricate analyses of this subject; and that the reasoning was entirely conclusive: nor did he consider his results at all invalidated by the papers of Prof. Parrot, which he had also read. Those of M. Poisson, in opposition to Fourier, have appeared since Dr. Bowditch's death.

Proof 3. *The existence of* 300 *active volcanos, and many extinct ones, whose origin is deep seated, and which are connected over extensive areas.* If these were confined to one part of the globe, or if after one eruption the volcano were to remain forever quiet, we might regard the cause as local and the effect of particular chemical changes at those places, aided perhaps by electro-magnetic agencies. But if the internal parts of the earth are in a melted state, that is, in the state of lava; and if this mass be slowly cooling, occasional eruptions of the matter ought to be expected to take place by existing volcanos. Assuming the thickness of the earth's crust to be 60 miles, the contraction of this envelope one 13.000th of an inch, would force out matter enough to form one of the greatest volcanic eruptions on record. More probably, however, the percolations of water to the heated nucleus, or other cause of disturbance, more frequently produces an eruption than simple contraction.

Other Hypotheses of Volcanic Action.

Hypothesis of the metalloids. This hypothesis, originally proposed, though subsequently abandoned, by Sir Humphrey Davy, supposes the internal parts of the earth, whether hot or cold, fluid or solid, to be composed in part of the metallic bases of the alkalies and earths, which combine energetically with oxygen whenever they are brought into contact with water, with the evolution of light and heat. To these metalloids water occasionally percolates in large quantities through fissures in the strata, and its sudden decomposition produces an eruption. Dr. Daubeny, the most strenuous advocate of this theory at the present time, has brought forward a great number of considerations which render it quite probable that this cause may often be concerned in producing volcanic phenomena, even if we do not admit that it is the sole cause. *Daubeny on Volcanos.*

Rem. 1. It is interesting to notice how the hypothesis of central heat or the Mechanical Theory of Cordier, as it is often called, and this Chemical Theory of Daubeny, apply almost equally well to the explanation of volcanic phenomena. Both agree as to the necessity of water being brought in contact with a heated mass in the earth. Both explain equally well the formation of vapor, the extrication of gases, and the sublimation of sulphur, salts, &c. Both show why volcanos are usually in the vicinity of water; "why their action is intermittent, and why the volcanic power appears to have decreased in energy. The constantly active volcanos, especially such an one as Kirauea, are more difficult to explain by the chemical theory. It must also be considered a strong objection to this hypothesis, that silicum, the most abundant of all the metalloids in the earth, "is incombustible in air and in oxygen gas; and may be exposed to the flame of the blowpipe without fusing, or undergoing any other change." *Turner's Chemistry, p.* 323, and that aluminium, the most abundant metal in the earth next to silicum, "is not oxidized by water at common temperatures: though on heating the water to near its boiling point, oxidation of the metal commences:—the oxidation, however, is very slight." *Turner, p.* 316.

Rem. 2. There is not necessarily any discrepancy between these two theories: for admitting even igneous fluidity in the earth, the nucleus may nevertheless be metals uncombined. Hence some distinguished advocates for the doctrine of central heat, have also adopted partially or wholly the other theory. *Ex. gr. De la Beche, in his Theoretical Geology; and Professor Phillips, in his Treatise on Geology in the Encyclopædia Brittanica.*

Modified Chemical Theory. Some geologists have called in the aid of electricity to assist in the decompositions and recompositions that result from volcanic agency. By this means the temperature of the uncombined metals is raised, so as to cause them to become more readily oxidized. This is the view advanced by Mr. Lyell. *Principles of Geology, Vol.* 2. *p.* 449. This hypothesis includes of course, as one of its elements, the earlier hypothesis of Lemery and others, who imputed volcanic phenomena to the combustion of coal, bitumen, &c. and the decomposition of the sulphate of the metals. *For an account of numerous modifications of opinion respecting the cause of volcanic agency, see Girardin Sur les Volcans, p.* 84.

Proof 4. *The Spheroidal Figure of the Earth.* Its form is precisely that which it would assume, if while in a fluid state, it began to revolve on its axis with its present velocity; and hence the probability is strong that this was the origin of its oblateness. But if originally fluid, it must have been igneous fluidity: for since the solid matter of the globe is at present 50.000 times heavier than the water, the idea of aqueous fluidity is entirely out of the question.

Other suppositions. 1. Some maintain that the earth was created in its present oblate form. This is indeed possible; because God could have given it any form he pleased. But there is no proof that such was the fact: while on the other hand, we may always assume that whenever we can see natural causes for natural phenomena, they were produced by those causes: unless we can see some reason for special Divine interference. 2. Sir John Herschel has suggested the possibility of accounting for the flattening at the poles, by causes now in action; though Sir John does not maintain that such was actually the mode in which it took place. He supposes the earth to have been created a uniform sphere, covered by an uniform ocean: and to have commenced a rotation on its axis, as at present. The water of course would rush towards the equator, leaving the polar regions dry, and very much elevated. But as this great equatorial ocean wore down its shores, the land would gradually be carried towards the equator, and spread over the bottom of the sea, and ultimately be elevated so as to form the present continents. It is hardly necessary to say, that the present distribution of land and water, and the form of continents, do not accord with such a mode of formation: and so improbable is the idea that two vast continents, around the poles, with a height of nearly 12 miles, have been thus worn down and carried thousands of miles towards the equator, that though theoretically possible, it must be regarded as practically impossible. *Prof. Phillips's Treatise on Geology, p.* 8.

Proof 5. *The tropical and ultra tropical character of organic remains found in high latitudes.* If the globe has passed through the process of refrigeration, as the hypothesis of original igneous fluidity implies, there must have been a time, before reaching its present statical condition, when the surface had the high temperature denoted by these remains: and that period must have been very remote; since no essential change of temperature from internal causes has taken place for thousands of years. A climate, also, chiefly dependent on subterranean agency, would be more uniform over the whole globe, than one dependent on solar influence; and such appears to have been the climate of those remote ages. Hence we may reasonably impute that temperature to internal heat; if some other more probable cause cannot be found.

Other Suppositions. 1. It has already been stated, that Mr Lyell has proposed an hypothesis, dependent upon the relative height of land in high latitudes at different periods, to explain the tropical character of organic remains, without the aid of secular refrigeration. But that hypothesis has been already sufficiently explained. 2. Another hypothesis has been advanced with much confidence by certain writers, not however practical geologists, to the same effect. It supposes these organic remains to have been drifted after death from the torrid zone. But their great distance in general from the torrid zone, the perfect preservation, in many cases, of their most delicate parts, with other evidences of quiet inhumation near the spot where they lived, such as the preservation in several cases of the softer parts of the animals, render such a supposition wholly untenable.

Proof 6. *The fact that nearly all the crust of the globe has been in a melted state.* As to the unstratified rocks, there will scarcely be a dissenting voice among geologists, to the opinion that they are of igneous origin, and have been melted. As to the detrital, or fossiliferous rocks, also, it will be admitted by all, that they were originally made up of fragments derived from the primary stratified or unstratified rocks; and that consequently, so far as derived from the latter, they have been melted. And in regard to the primary stratified rocks, also, although there are two different theories as to the mode in which they have been produced, yet both admit either of the entire fusion of these rocks, or of their having been so highly heated as to be able to assume a crystalline arrangement. Hence if the entire crust of the globe has been fused, it is a fair presumption that it was the result of the fusion of the whole globe

Proof 7. *This theory furnishes us with the only known adequate cause for the elevation of mountain chains and continents*

Other supposed causes of elevation.

1. *Earthquakes.* Examples have been given in another place (p. 236.) of small and limited elevations of land, produced by earthquakes. And it has been maintained that an indefinite repetition of such events might elevate the highest mountains, if they took place on no larger scale than at present. But it seems to be satisfactorily proved, that some elevations at least, such as those producing the enormous dislocations in the north of England, have occurred to an extent of several thousand feet, by a single paroxysmal effort; whereas the mightiest effects of

a modern earthquake have produced elevations only a few feet, and in most cases the uplifted surface has again subsided. Again, there is little probability that a succession of earthquakes should take place along the same extended line through so many ages, as would be necessary to raise some existing mountain chains. Earthquakes may explain some slight vertical movements of limited districts; but the cause seems altogether inadequate to the effect, when applied to the elevation of continents.

2. *Expansion of the rocks by heat.* Col. Totten, who is now at the head of the Topographical Bureau in this country, has made some accurate experiments on the expansion of rocks by heat. A block of granite, five feet long, by a change of temperature of 96° F. expanded 0.027792 inch: crystalline marble, 0.03264 inch: sandstone, 0.054914 inch. By these data it appears, that were the temperature of a portion of the earth's crust 10 miles thick, to be raised 600°, it would cause the surface to rise 200 feet. This would be a greater thickness than could be produced by the accumulation of detritus at the bottom of any ocean, whose temperature might be raised on the hypothesis of Prof. Babbage. Yet a still greater thickness might be heated, provided any new and extensive foci of heat should be produced deep beneath the surface of the globe. Still this accession of heat would finally be dissipated by radiation; and then the surface would again subside. This cause, therefore, though it may perhaps explain such vertical movements of particular regions as are taking place in Scandinavia, Greenland, Italy, England, &c. seems inadequate to account for the permanent elevation of large continents. If they had been raised in this manner, and the same remark applies to some extent to earthquakes, we should hardly expect to find several distinct systems of elevation on the same continent, nor so many examples of vertical strata. *Ninth Bridgewater Treatise, by Prof. Babbage, p.* 193, *Am. Ed.*

3. *Unequal contraction and expansion of land and water by cold and heat.* Assuming the mean depth of the ocean to be 10 miles, and that it had cooled from boiling heat to 40° F. its volume would contract about 0.042; while the contraction of the land would be only 0.00417. This would produce a sinking of the ocean of 697 feet. *Phillips's Geology, p.* 277. An increase of temperature would produce an opposite effect: viz. the partial submersion of the land; though it would be less than the desiccation, because of the greater area over which the water would flow. Admitting these changes of temperature to have taken place, and the theory of central heat supposes

the former, that is, the refrigeration, they could not account for the desiccation of the globe, because the tilted condition of the strata shows that the land has been raised up: whereas this theory implies a mere draining of the waters.

4. *A change in the position of the poles of the globe.* This hypothesis,—not long since so much in vogue,—would explain how continents once beneath the ocean are now above it, if we admit the form of the earth before the change, to have been the same as at present: viz. an oblate spheroid. But it would not explain the tilted condition of the strata, nor is it sustained by any analogous phenomena which astronomy describes.

Elevation by Central Heat.

First mode. It is possible to conceive that volcanic power, acting as at present, but with vastly greater intensity, might have lifted up continents: for their elevation, in part at least, appears to have been the result of local forces acting beneath the earth's crust.

Second mode. A more probable hypothesis, suggested by Beaumont, imputes the present ridged and furrowed condition of the earth's surface to *a collapse of its consolidated crust upon its contracted interior nucleus.* This may be illustrated by Fig. 119.

Fig. 119.

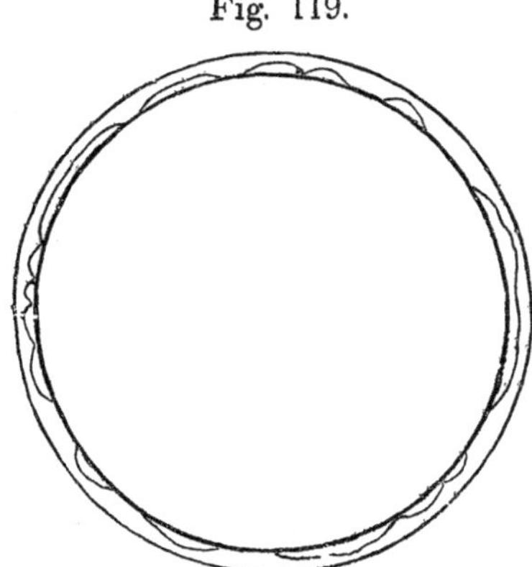

The outer circle represents the crust of the earth, after it had become consolidated above the liquid mass within. This heated nucleus would go on contracting as it cooled, while the crust would remain nearly of the same size. At length, when it became necessary for the crust to accommodate itself to the nucleus, contracted say to the inner circle, it could do this only by falling down in some places and rising in others; as is represented by the irregular line between the two circles. Thus would the surface of the earth become plicated by the sinking down of some parts by their gravity, and the elevation of correspondent ridges by the lateral pressure. The principal ridges thus produced, must coincide very nearly with a great circle: and as the earth's crust made successive efforts to accommodate itself to the constantly contracting nucleus, ridges would be produced in different directions, crossing one another; and thus the various systems of elevation known to exist on the globe, be formed at various epochs.

Obj. Such a shortening of the earth's diameter as this hypothesis supposes, would increase the rapidity of its rotary motion, and shorten the length of the day; whereas astronomy shows that for 2000 years no such change has taken place.

Ans. That period is too short fairly to test the point; since it requires a long time for the tension upon the crust of the globe to become so great as to produce a fracture; and this may not have occurred since that time. If there be any flexibility however, in the earth's crust, gravity must produce some depression of it in some places, and elevation in others, before the tension is great enough to produce a fracture. And possibly this may be the origin of some cases of slight subsidence or elevation on record.

8. *Other Worlds appear to be passing through the Changes which this Theory supposes the Earth to have Undergone.*

Proof. 1. Comets and some nebulæ appear to be in a gaseous state, from excess of heat; yet gradually condensing. 2. The sun is apparently in a state of igneous fusion. 3. Still farther advanced in the process of refrigeration are the planets. Around the asteroids is an unusual nebulosity: Jupiter is perhaps covered by water; and Saturn by a fluid lighter than water. 4. But the state of the moon, as the nearest heavenly body, has been most accurately ascertained; and it exhibits the most astonishing examples of volcanic action, though it is not certain that any volcanos upon it are now active. But craters, cones, and circular walls, or mountains, exist, of extraordinary dimensions. Some of the cones are nearly 25.000 feet high; and some of the craters, 25.000 feet deep, below the general surface; and the latter are of various diameters, even up to 150 miles. The inside of some of these craters presents all the wild and jagged appearance of similar rocks on the globe. Of the mountains and cavities of the moon, about 1100 have been measured with great accuracy, and we have a more accurate map of the surface of the moon turned towards us, than of our own planet. Yet as no water exists in the moon, and perhaps no atmosphere, it must be destitute of such inhabitants as people this world. Below is a sketch of one of the most remarkable volcanic regions, called Heinsius, seen a little obliquely. *See an admirable paper on the Volcanos of the Moon, by J. D. Dana, in Am. Jour. Sci., New Series, Vol. 2.* Most of the facts on this subject will be repeated a few pages forward, in another connection.

Volcanos in the Moon.

Origin of the Primary Stratified Rocks.

Rem. The way has not previously been prepared for a full understanding of the hypotheses above alluded to, concerning the origin of the primary stratified rocks; because both of these depend more or less upon internal heat.

First hypothesis. According to this hypothesis the stratified primary rocks are merely the mechanical or fossiliferous rocks altered by heat. As these accumulated at the bottom of the ocean, being much poorer conductors of heat than water, they would confine the internal heat that was attempting to escape by radiation, until it became so great as to bring the matter into a crystalline state: but not great enough to produce entire fusion, so as to destroy the marks of stratification.

Arguments in favor of this Hypothesis.

1. Numerous facts show that the molecular constitution of solid bodies may undergo great changes, without much change of the general form; and even without any great elevation of temperature. Thus the heat of the sun alone, will change prismatic crystals of zinc into octahedrons: and the same takes place with sulphate of nickel. *Connection of the Physical Sciences by Mrs. Somerville, p.* 130. Indeed, Dr. Macculloch says he has completely proved by experiments, that "every metal can completely change its crystalline arrangements while solid, and many of them at very low temperatures." *System of Geology, Vol.* 1. *p.* 190. Analogous changes have taken place in sandstone beneath trap rocks: in trap rocks after they have become solid: and in solid glass. Hence the presumption is in

favor of these internal changes in rocks of mechanical origin from internal heat.

2. The heat requisite for the conversion of mechanical into crystalline rocks, without destroying the stratified structure may have been derived either from an internal heated nucleus in the earth, when the crust was thinner than at present, as it was during the period in which the primary strata were deposited, or from local nuclei of heat, propagated upwards through detritus, according to the theory of Prof. Babbage.

3. Geology furnishes numerous examples in which the mechanical or fossiliferous rocks have been converted by heat into primary crystallized rocks in limited spots by the agency of heat. When dykes of granite, porphyry, trap rocks, or recent lava, pass through fragmentary deposits, for a certain distance on the sides of the dyke these conversions have taken place. Chalk and earthy limestones are in this manner in Ireland, converted into crystallized marble: and the same effect was produced upon chalk by heating it powerfully in a sealed gun barrel. Experimental proof has also been furnished by the chemist, that quartz rock is merely sandstone altered by heat; as is shown also at Salisbury Craig, Teesdale, and Shropshire, in Great Britain, where sandstone and basalt come in contact. In Shetland, argillaceous slate, when in contact with granite, is changed into hornblende slate. Clay slate is obviously nothing but clay that has been subjected to strong heat and pressure.

4. The primary stratified rocks still retain marks of a mechanical origin. The general appearance of gneiss and mica slate is that of fragments of crystals, more or less worn and rounded, and then re-cemented by heat. But real conglomerates occur which yet have all the characters of the primary stratified rocks, except perhaps gneiss. Thus, I have in my cabinet from Bellingham in Massachusetts, a perfect and highly crystalline mica slate, which contains as perfectly rounded pebbles of quartz as any secondary conglomerate; also a talco micaceous slate of the same character, from Rhode Island, which abounds in crystals of magnetic oxide of iron. These rocks are connected with transition slates on the one side, and with primary slates on the other. I have also, perfectly distinct conglomerates of quartz rock, made up of rounded fragments of quartz, cemented by comminuted materials of the same kind. The strata of this rock in Berkshire County, in Massachusetts, are associated with gneiss and mica slate; all of which at the spot, (in Washington,) stand upon their edges.

Obj. 1. There is little probability that detritus is conveyed to the bottom of the ocean in quantities sufficient to cause such an accumulation of internal heat, as would convert mechanical into crystalline rocks:—a degree of heat nearly equal to that which would melt them. True, the heat would accumulate in these deposits to a certain degree: but not beyond what exists in the solid crust of the earth generally; and this would require us to descend nearly 50 miles, before a temperature would be reached sufficient for the purpose. Unless, therefore, this theory supposes a much higher temperature on the globe when this change took place, than at present, (and most of its advocates deny this,) the requisite heat could not have been obtained, especially as in many cases the primary rocks extend to the surface, and do not appear to have ever been covered with newer ones; so that there must have been heat enough to produce this transformation immediately beneath the waters of the ocean.

2. The difference in chemical composition between the primary and the newer rocks, is opposed to the idea that the former are only modifications of the latter. For we find that some of the ingredients, lime and carbon for instance, are far more abundant in the newer than in the older rocks. This difference points of course to a different origin.

3. If all the stratified primary rocks are metamorphic, we ought to find in them occasionally, especially in the limestones, traces of organic remains. For examples are not uncommon, in which the traces of such remains are found, (of which a description has been given in Section V,) in calcareous rocks which have become perfect crystalline limestones, as in the encrinal limestone: and in other rocks which are converted into vesicular trap by the agency of heat. It is incredible, therefore, that if the remains of animals and plants once existed in these rocks, as numerous as they now exist in the secondary rocks, they should all have vanished; since it is certain, that the heat which produced the metamorphosis, was not great enough to obliterate the stratification.

Second hypothesis. This hypothesis supposes the primary stratified rocks to have been formed, partly in a mechanical, and partly in a chemical mode, by aqueous and igneous agency, when the temperature of the crust of the globe was very high and before organic beings could live upon it.

Arguments in favor of this Hypothesis.

1. It shows why, amid so much evidence of chemical agency in the formation of the primary rocks, there is still so much proof of the operation of mechanical agencies For in that

state of the globe, when its crust had cooled only so far as to allow water to exist upon it in a fluid state, volcanic agency must have been far more active than at present: and consequently the agitated waters must have worn away the granite at their bottom extensively. But as the heated waters would contain a great deal of silica, and other ingredients which would readily fall down as chemical deposits, the abraded materials would be consolidated before they had become entirely rounded into pebbles; so that the compound might, upon the whole, be regarded as of chemical origin; and yet not be destitute, as gneiss and mica slate are not, of the marks of attrition. Indeed, it would be strange, if in some instances the attrition did not proceed so far as to produce the materials for a perfect conglomerate; as the facts mentioned under the last hypothesis show was sometimes the fact.

2. It shows us why silicates predominated in the earlier periods of the globe; and why limestone and carbon were more abundant at later periods. Thermal waters, it has been shown in another place, often contain an abundance of silica in solution; but cold water never does. Again, by heating water to the boiling point, the carbonic acid is all driven off: and without this acid, carbonate of lime could not be held in solution to much extent: and farther, hot water will dissolve much less quicklime than cold; the proportion being as 778 to 1270. Hence the heated seas of those early times would contain and deposit more of silica, but less of lime, or carbonate of lime, than under existing circumstances. Another cause why less of carbonate of lime is found in the older rocks, is, that animals did not then exist to eliminate lime from its other combinations, and convert it into the carbonate.

3. It explains the absence of organic remains in the primary stratified rocks. It shows that the temperature was too high, and the surface too unstable, to allow of the existence of animals and plants. And if they had existed in as great abundance as at present,—an assumption which is made by the preceding hypothesis,—it is incredible that some traces of them should not remain: for if the fusion of these rocks was not so entire as to obliterate all marks of mechanical agency, if, in fact, perfectly rounded pebbles still occur in them, there is no reason why the harder parts of animals should not also remain: We have examples where the traces of organic remains exist in rocks, that have been almost entirely fused:—at least so much melted, as in the case of a vegetable stem in trap, in the valley of the Connecticut, that it is converted into decided vesicular amygdaloid; and yet its vegetable character can scarcely be

doubted. (*See a fine Specimen in Amherst College, and another in the State Collection of Massachusetts, at Boston.*) We may hence infer, with no little confidence, that organic life did not exist on the globe when the primary rocks were in a course of deposition, and this hypothesis explains the reason.

4. It explains too the reason why carbon is much less abundant in the older than in the newer rocks. Organic beings are undoubtedly the source of most of the carbon in the rocks :— and of course it would be found in small quantities where neither animals nor plants existed.

Rem. Dr. Macculloch does indeed state that he found organic remains (*orthoccrata*,) in quartz rock, connected with gneiss, in Sunderland; but other distinguished geologists (Sedgwick and Murchison,) have failed in finding any at that spot. He thinks, also, that fragments of shells occur in hornblende slate in Glen Tilt. *System of Geology, Vol.* 1. *p.* 418. Von Dechen also mentions fossiliferous graywacke, interstratified with gneiss and mica slate in Bohemia. But facts of so anomalous a character need still further confirmation. *Phillips's Geology, p.* 78.

5. It explains the imperceptible gradation of gneiss into granite, which we often witness. For if thick beds of gneiss were deposited upon the granite, under the circumstances supposed by the hypothesis, it is easy to conceive how the internal heat should accumulate in the manner explained by Prof. Babbage, so as to melt the granite crust anew, and to extend the fusion into the lower beds of the gneiss; at least so as to produce an almost entire obliteration of the lines of stratification, and form numerous *niduses* of perfect granite in the gneiss. This hypothesis explains the passage of these two rocks into each other, better than the first hypothesis; because it supposes a higher temperature beneath and upon the earth's crust at the time of the formation of the gneiss.

Rem. The most important objections to this hypothesis are embraced in those which are urged against the doctrine of internal heat in general; and therefore, it will be necessary to state only the latter.

Objections to the Doctrine of Internal Heat.

Obj. 1. *It has been maintained that the high temperature of deep excavations may be explained by chemical changes going on in the rocks;* such as the decomposition of iron pyrites by mineral waters, the lights employed by the workmen, the heat of their bodies, and especially by the condensation of air at great depths.

Ans. In the experiments that have been made upon the temperature of mines, care has been taken to avoid all these sources of error except the last, (which are indeed sometimes very considerable,) and yet the general result is as has been stated; nor is there a single example on the other side to invalidate that result. As to the condensation of air in mines, Mr. Fox has shown that the air which ascends from their bottom is much warmer than when there; so that it carries away instead of producing heat. *Cordier's Essay on the Temperature of the Interior of the Earth. Edinburgh Journal of Science, April,* 1832.

Obj. 2. *The temperature of the Ocean.* Prof. Parrot, who urges this objection, recapitulated the results of the most accurate observations upon the temperature of the ocean: "1. That the temperature diminishes as the depth increases: 2. That it diminishes at first rapidly, then very slowly. From the surface to the depth of 2478 feet, it diminishes more than 41°F. and from that to 5490 feet, less than 2°." *American Journal of Science, Vol.* 26. *p.* 12. According to De La Beche, there are some exceptions to these conclusions, especially in high latitudes. In fresh water lakes, the same observer found that the temperature decreased till it had nearly reached 40°F, when it continued nearly the same to the greatest measured depth. *Manual of Geology, p.* 20. Facts of this sort Prof. Parrot considers as directly at variance with the idea of internal heat.

Ans. Taking the conclusions of Prof. Parrot as true, they are just what we might expect would be the temperature of the ocean, whether the earth had internal heat or not. For it appears that the strata of water arrange themselves according to their specific gravities. The warmest particles being the lightest, of course rise to the top; and the coldest sink to the bottom: just as we find to be the case in a vessel of water that is being heated over a fire. But when fresh water has descended to the temperature of 40°F. it begins to expand, and therefore water below that degree will not sink but rise. Yet experiments show that it rarely goes lower than that degree; and therefore, when the water has reached it, or nearly reached it, we might expect that the temperature at greater depths would be nearly the same. Salt water continues to contract until it reaches the freezing point, which varies from 32° to 4°, according to the amount of salt which it contains. Hence we might expect that the temperature of the sea, except perhaps in very cold latitudes, would decrease downwards until it reached a temperature below which it rarely descends; after which we should expect a uniform temperature to the greatest depths. A few observations, indeed, are on record, which can hardly be reconciled to the general principle that waters of lakes and oceans arrange themselves according to their specific gravities; yet such cases probably result from local causes of variation. Upon the whole it seems that the facts in respect to the ocean's temperature, neither prove nor disprove the doctrine of internal heat.

Rem. Some have supposed, that since the ocean has a depth of several miles, the water at its bottom ought to be in a state of ebullition, if the doctrine of internal heat be true. But there is no reason to suppose the earth's crust to be thinner there than on dry land; and hence no more heat will escape into the waters by radiating from the earth, than escapes into the air; which, as we have seen, according to Fourier, is a very small quantity: not sufficient to affect the temperature of water or air perceptibly.

Obj. 3. "If the central heat were as intense as is represented, there must be a circulation of currents, tending to equalize the temperature of the resulting fluid, and the solid crust itself would be melted."—" If the whole planet, for example, were composed of water covered with a spheroidal crust of ice fifty miles thick, and with an interior ocean having a central heat about 200 times that of the melting point of ice, &c :—if it must be conceded, in this case, that the whole spheroid would instantly be in a state of violent ebullition, that the ice instead of being strengthened annually by new internal layers, would soon melt and form part of an atmosphere of steam, on what principle can it be maintained that analogous effects would not follow in regard to the earth under the conditions assumed in the theory of central heat?" *Lyell's Prin. Geol. Vol.* 2. *p.* 440, *and* 449.

Ans. In the first place, it is not essential to the doctrine of central heat, that a temperature very much exceeding that requisite to melt rocks, (7000°F.) should exist in any part of the molten nucleus. It may even be admitted that the whole globe was cooled down very nearly to that point, before a crust began to form over it. For still, according to the conclusions of Fourier, it would require an immense period to cool the internal parts, so that they should lose their fluid incandescent state, after a crust of some 20 miles thick had been formed over them. In the second place, we have the case of currents of lava, which cool at their surface, so as to permit men to walk over them, while for years, and even decades of years, the lava beneath is in a molten state, and sometimes even in motion. And if a crust can thus readily be formed over lava, why might not one be formed over the whole globe, while its interior was in a melted state: and if a crust only a few feet in thickness, can so long preserve the internal mass of lava at an incandescent heat, why may not a crust upon the earth, many miles in thickness, preserve for thousands of years the nucleus of the earth in the same state? True, if we immerse a solid piece of metal in a melted mass of the same, the fragment will be melted; *because it cannot radiate the heat which passes into it:* but keep one side of the fragment exposed to a cold medium, as the crust of the earth is, and it will require very much stronger heat to melt the other side. If the crust of the globe were to be broken into fragments, and these plunged into fluid matter beneath, probably the whole would soon be melted, if the internal heat be strong enough. But so long as its outer surface is surrounded by a medium, whose temperature is at least 58° below zero, nothing but a heat inconceivably powerful, can make much impression on its interior surface. In the third place, a globe of water intensely heated at its centre, and covered by a crust of ice, is not a just illustration of a globe of earth in a similar condition, covered by a crust of rocks and soils. For between ice and water there is no intermediate or semi-fluid condition. As soon as the ice melts, there exists a perfect mobility among the particles; so that the hottest, because the lightest, would always be kept in contact with the surrounding crust of ice, and melt it continually more and more: especially as ice, being a perfect non-conductor of heat, would not permit any of it to pass through, and by radiation prevent the melting. On the other hand, between solid rock and perfectly fluid lava, there is every conceivable degree of spissitude; and of course every degree of mobility among the particles. Hence they could not in that semi-fluid stratum, arrange themselves in the order of their specific gravities; and therefore, the layer of greatest heat would not be in contact with the unmelted solid rock. True, the heat would be diffused outwards, but so long as the hardened crust could radiate the excess of temperature, the melting would not advance in that direction. This would take place only when the heat was so excessive, that the envelope could not throw it off into space.

Obj. 4. It is maintained, that if the earth was originally in a fluid or gaseous state, and subsequently condensed, the solidification would commence at the centre and proceed outwards. The solidification of a nucleus at the centre by pressure, would throw out much heat, by which a layer around the nucleus would be expanded, so as to become lighter, and to cause heavier particles to take its place. These would at length become solidified, and thus would this process gradually advance towards the circumference of the globe, until the whole was converted into a solid mass. This is the view of M. Poisson. *See Am. Journal of Science, Vol.* 34. *p.* 61.

Ans. If it be admitted that the order of solidification in a globe condensing from a fluid or a gaseous state, would be from the centre to the circumference, while that globe was surrounded by a medium of very high temperature, yet if the temperature were such as actually surrounds the earth, radiation must produce a crust over the surface; and when once a solid crust was formed, then the conclusions of Baron Fourier, already explained, would follow. Even though enormous pressure might make the central parts more dense than the crust, still this would so confine the heat that a high temperature might exist in the interior. In every case in which experiments have been made upon the cooling of intensely heated bodies, a crust forms over the surface, which much retards the refrigeration of the central parts. All known analogies, therefore, are opposed to this hypothesis.

Rem. M. Poisson resorts to a most extraordinary supposition to explain the observed increase of temperature as we descend into the earth. He assumes as true, the suggestion of the elder Herschel, that the solar system is in motion through space, and that the temperature of this space is so different in different parts, as to heat the earth to a great depth at one time, and then, while passing through the frigid regions, it is gradually giving off its heat. It is hardly necessary to say, that such a movement of the solar system as is here supposed has scarcely nothing but conjecture to prove it. But if it be admitted, we cannot imagine what evidence there is, that different portions of the space passed over should have more than a very slight difference of temperature. This is, therefore, an hypothesis based upon hypothesis. *Whewell's Hist. Induct. Sci. Vol.* 3. *p.* 564.

Hypothetical state of the Globe in the earliest Times.

Rem. The theory of central heat, as already explained, extends no further back in the world's history than to the time when the globe was in a state of fusion from heat: and the chemical theory, which ascribes subterranean heat to the oxidation of a metallic nucleus, does not necessarily describe the state of things in the beginning. But the mind naturally inquires, whichever of these theories is adopted, what was the state of things at the commencement, or at the earliest period of which we can obtain any glimpses. To gratify this curiosity the two following hypotheses have been suggested. It ought, however, to be remarked that though they be entirely groundless, the theories of central heat and of the oxidation of a metallic nucleus, may nevertheless be true.

First hypothesis. This is advanced by the advocates of original igneous fluidity, and supposes that previous to that time, the matter of the globe had been in a state so intensely heated, as to be entirely dissipated, or converted into vapor and gas. As the heat was gradually radiated into space, condensation would take place: and this process would evolve a vast amount of heat, by which the materials would be kept in a molten state, until at length a solid crust would be formed as already explained.

Arguments in favor of this Hypothesis.

1. The nature of comets shows that worlds may be in a gaseous state. These bodies appear to have "no more solidity or coherence than a cloud of dust, or a wreath of smoke,"—"through which the stars are visible with no perceptible diminution of their brightness." *Whewell's Bridgewater Treatise, p.* 152, 153. Sometimes, however, they appear more dense towards their centre, and well defined circular nuclei have been seen in a few. It has been thought, also, that some of them become more dense at their successive returns. Dr. Herschel regards them all as self luminous. Now in such facts do we not see a striking resemblance to the early condition of our globe, according to this hypothesis—to its condition before it had become so much condensed as to be a fluid incandescent mass.

2. The nebulæ appear to be similar in composition to comets: though not yet actually converted into comets. They prove that a vast amount of the matter of the universe actually exists in the state of vapor.

3. The sun, and probably the fixed stars, appear to be examples of immense globes so far condensed as to be in a fluid state by intense heat. This heat, perhaps, is still powerful enough to dissipate the more volatile materials, which form a vast zone around the sun's equator and produce the zodiacal light.

4. The process of refrigeration appears to be still farther advanced upon the moon: so much so, that it has ceased to be self-luminous. And yet its entire surface bears the marks of volcanic desolation: so that it is doubtful whether even yet it is in such a condition that beings like man could inhabit it. *Bakewell's Geology, p.* 384.

5. Some of the other planets appear to be in a transition state between habitable and uninhabitable worlds. Thus, a remarkable nebulosity surrounds the asteroid planets, Juno, Ceres, and Pallas. Jupiter is not improbably covered with water; and Saturn by a fluid lighter than water.

6. All these facts render it probable that other worlds are passing through the successive stages of refrigeration to which the hypothesis under consideration supposes the earth to have been subject. They afford us some glimpses of a far reaching law of nature on this subject.

Second hypothesis. This hypothesis supposes the globe to have been created a mass of combustibles and metals uncombined: to which were suddenly added water, the atmosphere, chlorine, iodine, and perhaps hydrogen. The chemical action that would ensue, would produce an intense ignition and combustion of the whole surface of the planet: a new and oxidized crust would be formed over it; that crust would be rent and dislocated, as we now find it to have been. But as the crust became thicker, water and other agents, which act energetically on the uncombined metals, would less frequently reach them, and at length the surface would become habitable. *Am. Jour. Science, Vol.* 14. *p.* 88.

Proof. It is not pretended that any facts directly corroborative of this hypothesis are known. But the facility with which it explains the changes that have taken place on the globe, is supposed to render it probable.

Prin. It is doubtful whether geologists will ever be able from their science to ascertain the state of things at the beginning, or the first condition of created matter. For though they may go far backward in tracing the changes which the earth has undergone, yet the condition of things becomes so different from the present the farther they penetrate the past, that the thread of analogy fails. Hence it is, that so many *theories of the earth* have not only been failures, but have brought ridicule upon the whole science of geology. When geologists shall be contented to trace effects to their proximate causes only, and leave the origin of things untouched, they will find their science resting on a firm foundation. *Whewell's Philosophy of the Inductive Sciences, Vol.* 2. *p.* 136. *London,* 1840.

Intensity of Action in the Causes of Geological Change, or the Doctrine of Catastrophes and of Uniformity.

First theory. Mr. Lyell contends that the causes of geological change now operating upon the globe, with no increase of intensity, that is, acting with no more energy than at present, are sufficient to account for all the revolutions which the crust of the earth has undergone. He admits of no irregularities or catastrophes greater than now take place; and supposes that the effects which transcend any single effect of existing causes, have been the result of repetitions, sometimes almost endless, of present agencies. In other words, he supposes that things have remained from the beginning subject to no greater changes than they experience at the present time. To prove these positions is the great object of his able work on the Principles of Geology.

Proof 1. It is agreed on all hands, that the nature of geological causes has been the same in all ages; although even as late as the time of Cuvier, he says that "none of the agents nature now employs were sufficient for the production of her ancient works."

2. An indefinite repetition of an agency on a limited scale, can produce the same effects as a paroxysmal effort of the same agency, however powerful: provided the former is able to produce any effect, as for instance, in the accumulation of detritus, the elevation of continents, the dislocation of strata, &c. Now it is unphilosophical to call in the aid of extraordinary agency, when its ordinary operation is sufficient to explain the phenomena.

3. Nearly every variety of rock found in the crust of the globe, has been shown to be in the course of formation by existing aqueous and igneous agencies: and if a few have not yet been detected in the process of formation, it is probably because they are produced in places inaccessible to observation.

Second theory. This theory admits that no causes of geological change, different in their nature from those now in action, have ever operated on the globe: in other words, that the geological processes now going on, are in all cases the antitypes of those which were formerly in operation: but it maintains that the existing causes operate now in many cases, with less intensity than formerly.

Proof 1. The spheroidal figure of the earth and other facts already detailed, seem to render almost certain the former fluidity of the globe. Now, whether that fluidity was aqueous or igneous, or both in part, it is certain that the agencies which produced it must have operated in earlier times with vastly greater intensity than at this day, and that their energy must have been constantly decreasing from that time to the present.

2. Still more direct is the evidence from the character of organic remains in high latitudes, of the prevalence of a temperature in early times hotter than tropical: too warm, indeed, to be explained by any supposed change of levels in the dry and. And if this be admitted, heat must have been more

powerful in its operation than at present; and this would increase the aqueous, atmospheric and organic agencies of those times.

3. No agency at present in operation, without a vast increase of energy, is adequate to the elevation, several thousand feet, of vast chains of mountains and continents; such as we know to have taken place in early times. A succession of elevations by earthquakes, repeated through an indefinite number of ages, the vertical movements being only a few feet at each recurrence, is a cause inadequate to the effect, if we admit that earthquakes have exhibited their maximum energy within historic times. Besides, it is difficult to conceive how a continent could be sustained several thousand feet high, unless melted matter be forced in beneath its crust. But earthquakes, and even the whole amount of volcanic power, if the doctrine of internal heat be rejected, could not supply any such prop. If we could suppose a succession of earthquakes, acting for thousands or millions of years along some anticlinal axis of great length, we have reason to suppose from their known operation, that sometimes they would elevate, and sometimes sink down the surface; so that the final resultant would be probably little change of level, and not an elevation like the Andes or the Himmalayah mountains.

4. In a majority of cases, the periods of disturbance on the globe appear to have been short compared with the periods of repose that have intervened: as is obvious from the fact that particular formations have the same strike and dip throughout their whole extent: unless some portions have been acted upon by more than one elevatory force: and then we find a sudden change of strike and dip in the formations above and below. Whereas, had any of the causes of elevation now in operation lifted up these formations by a repetition of their present comparatively minute effects, there ought to be a gradual decrease in the dip from the bottom of the formation upwards, and no sudden change of dip between any two consecutive formations, unless some strata are wanting. At the periods of these elevatory movements, therefore, the force must have been greater than any that is now exerted, to produce analogous effects.

5. The sudden and remarkable changes in the organic contents of the strata, as we pass from one formation to another, even when none of the regular strata are wanting, coincides exactly with the supposition of long periods of repose, succeeded by destructive catastrophes. Nor is the supposition that species of animals and plants have become gradually extinct, and have been replaced by new species, by a law of nature during periods

of repose, sustained by any facts that have occurred within the historic period: no example having been discovered of the creation of a new species by such a law; and only a few examples of the extinction of a species. *Whewell's Hist. Induc. Sci., Vol. 3. p.* 589.

6. We have no evidence that the most important of the older rocks, both stratified and unstratified, are produced by any causes now in operation. That they may be produced deep in the earth, where igneous causes are still in intense operation, is a plausible hypothesis, but unsustained by a single example of the production of mica slate, gneiss, granite, or syenite. The highly crystalline and in other respects peculiar character of these rocks, as well as their entire deficiency of traces of organic existence, when they were formed, point to a state of the globe different from the present, but different only because existing causes, especially heat, operated then with greater energy than at present.

7. Glacio-aqueous agency, since the deposition of the tertiary strata, requires for its explanation a greater intensity of action in existing geological agencies than is known at the present day. This point, however, has been so fully discussed in Section VII, that nothing more need be added here.

8. Upon the whole, with the exception of glacio-aqueous action, were we to confine our attention to the tertiary and alluvial strata, it might be possible to explain their phenomena by existing causes, operating with their present intensity. But when we examine the secondary and primary rocks, we are forced to the conclusion that this hypothesis is inadequate: and that we must admit a far greater intensity in geological agencies in early times than at present.

9. But the question here arises, how long a period shall we assume as a measure of the intensity of existing agencies? The most strenuous advocates of the doctrine of uniformity will admit of some oscillation in the intensity of these agencies; because a single year shows it. How then shall we determine how wide that oscillation may be? In order to obtain the average intensity, how can we say but that all geological cycles must be included? To make any particular portion of time the measure of all the rest, must be an arbitrary assumption. And therefore, we cannot ascertain what is the standard or the average of intensity: and until this can be done, is the subject considered under this head any thing more than a controversy about words? The alluvial period has been assumed in the above argument as the measure of all that have gone before it. But can any reason be given why this should be taken rather

than a longer one? and who knows how much greater intensity existing causes may yet exhibit, during the alluvial period, than they have done? Since we know that some catastrophes, however small they may be regarded, have occurred during the alluvial or historic period, all presumption against more powerful ones in future is taken away: and as to past changes, we must judge of the intensity of the producing agencies by the effects. *See an able view of this subject in Whewell's Philosophy of the Inductive Sciences, Vol. 2. p.* 123.

Character and Repletion of Metallic Veins.

Rem. The subject of metallic veins,—one of the most difficult in geology, although touched upon in several places in this work, has been mainly deferred to this place; because it could not be well understood without an acquaintance with nearly the whole of geology.

Descr. The metallic matter, called *ore*, rarely occupies the whole of the vein: but is disseminated more or less abundantly through the quartz, sulphate of baryta, wacke, granite, &c which constitutes the greater part of the vein, and is called the *gangue, matrix* or *veinstone.* Often the ore and the gangue form alternating layers. Sometimes there are cavities lined with crystals, which cavities are called *druses.*

Descr. Metallic like other veins vary very much in width, both in a vertical and a horizontal direction. They are of unknown depth; for scarcely ever have they been exhausted downward. The deepest mine that has been worked, is that at Truttenberg in Bohemia: which has been explored to the depth of 3000 feet.

Descr. In all cases metallic like other mineral veins, are filled with matter different from the rocks which they traverse. In some instances they are obviously of the same age with the containing rock, but in a majority of cases, they are fissures that have been subsequently filled. They exhibit almost every variety of dip and strike, and yet it has been thought that they very often affect an east and west direction, though frequently they run north and south, and their dip usually approaches the perpendicular. These veins often ramify and diminish until they finally disappear. Their width is very various; from a mere line, up to some hundreds of feet. The metallic veins of Cornwall vary from an inch to 30 feet in width. The contents are sometimes arranged in successive and often corresponding layers on each side.

Descr. The contents of metalliferous veins often vary in the same vein, in different rocks, through which it passes, both perpendicularly and in the direction of the vein. Its width also varies in the same manner.

Descr. Metallic veins are most numerous in primary and transition rocks. No vein is worked in Great Britain above the new red sandstone. Nor are any explored of much importance, above the carboniferous limestone. In the Pyrenees, however, hematitic and spathic iron occur in transition strata, in the lias, and the chalk. In the Cordilleras of Chili, also tertiary strata, which have become crystalline by the proximity of granite, are traversed by true metallic veins of iron, copper, arsenic, silver, and gold, which proceed from the underlying granite.

Descr. As a general fact, metallic veins are most productive near the junction of stratified and unstratified rocks. Their productiveness depends also on their direction, in some measure: an east and west direction being regarded as the most favorable in Cornwall; while the *cross courses*, or north and south veins, are usually unproductive.

Descr. Mr. Carne finds evidence in Cornwall of the existence of metallic veins of no less than six or eight different ages: a case analagous to the one exhibited on Plate 2. *Whewell's Hist. Induc. Sciences*, *Vol.* 3. *p.* 540.

Fig. 120.

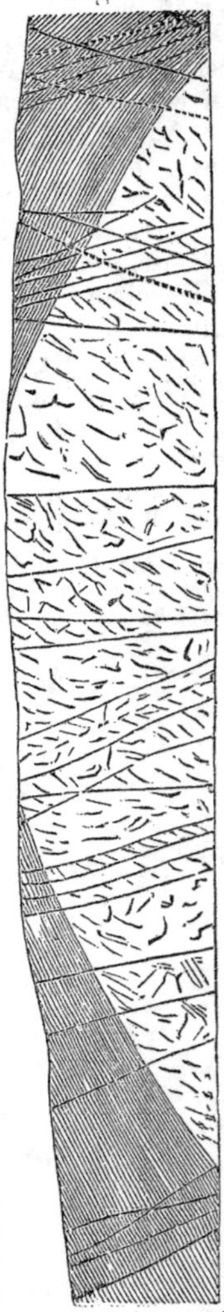

Fig. 120, is a section of tin and copper veins near Redruth in Cornwall. They generally pass from the killas, or slate, into the granite beneath. The section reaches to the depth of 1200 feet. The dotted lines represents the tin lodes, (veins) and the continuous lines, the copper lodes.

Theories to explain the Repletion of Veins in General.

1. Werner supposed that veins were fissures filled by aqueous infiltration from above. But it is probable that this hypothesis will apply to scarcely a single example of all the varieties of veins.

2. Hutton supposed that veins were filled by melted matter injected from beneath. And the facts that have been detailed in this work, make it almost certain, that a large part of the veins, filled by unstratified rock, were thus produced. Indeed, it is often practicable to trace these veins to the central mass from which they proceeded, and to follow them at the other extremity, as they thin off and are lost. It is almost equally certain that many metallic veins were thus produced.

3. Prof. Sedgwick supposes some veins to have been produced by chemical segregation from the rock in which they occur, while that was in a yielding state; just as the nodules of flint were segregated from chalk, or crystals of simple minerals from the rocks in which they are now found imbedded. That many veins were produced in this manner can hardly be doubted: for sometimes we find them passing by insensible gradation into the including rock, and thus showing that they are of contemporaneous origin, with the rock, while both were in a fluid state. In such cases chemical segregation is the only known principle by which the veins could have been formed.

4. Mr. Fox and M. Becquerel refer the origin of many metallic veins to electro-chemical agencies which are operating at the present day, to transfer the contents of veins even from the solid rocks, in which they are disseminated, into fissures in the same. The former of these gentlemen has shown conclusively, that the materials of metallic veins, arranged as they are in the earth, are capable of exerting a feeble electro-magnetic influence: that is, they constitute galvanic circuits, whereby numerous decompositions, and recompositions, and a transfer of elements to a considerable distance, may be effected. He was induced to commence experiments on this subject, by the analogy which he perceived between the arrangements of mineral veins and voltaic combinations. And he thinks if such an agency be admitted in the earth, it shows why metallic veins, having a nearly east and west direction, are richer in ore than others; since electro-magnetic currents would more readily pass in an east and west than in a north and south direction, in consequence of the magnetism of the earth. M. Becquerel has shown, that even insoluble metallic compounds may be produced by the slow and long continued reaction and transference of the elements of soluble compounds by galvanic action. He has also made an important practical application of these principles, which is said to be in successful operation in France: whereby the ores of silver, lead and copper, are reduced without the use of mercury. *Buckland's Bridgewater Treatise, 2d. Edition, p.* 552, *and* 615, *Vol.* 1. *and p.* 108, *Vol.* 2. This ingenious theory bids fair to solve many perplexing enigmas relating to metallic veins; and to prove that some of them may even now be in a course of formation.

5. M. Neckar and Dr. Buckland suggest, that some mineral veins may have been filled by the sublimation of their contents into fissures and cavities of the superincumbent rocks, by means of intensely heated mineral matter beneath. Thus, it has been shown that by heating galena in a tube, and causing its vapor to unite with that of water, a new deposition of that mineral was produced in the upper part of the tube; and in a similar manner boracic acid, which by itself does not sublime, may be carried upwards and deposited anew. *Buckland's Bridgewater Treatise, Vol.* 1. *p.* 551. *Phillips's Geology, p.* 273.

Conclusion. Probably it will be necessary to call in the aid of nearly all the preceding hypotheses to explain the complicated phenomena of mineral veins.

For accurate accounts of this difficult subject see *Phillips's Treatise on Geology, Vol.* 2, *Chapter VIII. Also De La Beche's Geological Report on Cornwall and Devon, Chapter X. Also Amsted's Geology, Vol.* 2, *p.* 244.

SECTION IX.

CONNECTION BETWEEN GEOLOGY AND NATURAL AND REVEALED RELIGION

1. *Illustrations of Natural Religion from Geology.*

Rem. The bearing of geology upon religion, has always excited a good deal of interest and discussion: and being in some respects peculiar and important, a treatise on geology, which omits this subject, must be considered as deficient.

Prin. Geology shows us that the existing system of things upon the globe had a beginning.

Proof. 1. Existing continents have been raised from the bottom of the sea, where most of their surface was formed by depositions. 2. With a few exceptions, the existing races of animals and plants must have been created since the deposition of all the rocks except the diluvial; since their remains do not occur in the older rocks. Hence it appears that not only the present races of organic beings, but the land which they inhabit, are of comparatively modern production.

Inf. 1. Hence it is inferred that the existing races of animals and plants must have resulted from the creative agency of the Supreme Being: for even if we admit that existing continents might have been brought into their present state by natural causes, the creation of an almost entirely new system of organic beings, could have resulted only from an exertion of an infinitely wise and powerful Being. Indeed, the bestowment of life must be regarded as the highest act of omnipotence.

Inf. 2. Hence the doctrine which maintains that the operations of nature have proceeded eternally as they now do, and that it is unnecessary to call in the agency of the Deity to explain natural phenomena, is shown to be erroneous.

Inf. 3. The preceding inferences being admitted, natural theology need not labor to disprove the eternity of matter; since its eternal duration might be admitted, without affecting any important doctrine. *See Chalmers' Works, Vol.* 1. *Chap. V. on Natural Theology;* where this subject is admirably treated.

Prin. Several different systems of organic life have appeared on the globe, adapted to its varying conditions, as to temperature, moisture, food, and other circumstances. In the opinion of many geologists, also, numerous changes took place on the globe previous to the creation of animals and plants; all of which tended to prepare it for their dwelling place.

Inf. 1. Hence it appears that the Deity has always exercised over the globe a superintending Providence; and whenever it was necessary, has interfered with the regular sequence of events.

Inf. 2. A presumption is also hence obtained, that the mat-

of the globe had a beginning: or at least, all presumption against its creation out of nothing, is taken away. For there must have been a commencement to a series of changes in which there is continued improvement, (such as the globe has actually experienced;) and it is a priori as probable, that at the beginning of these changes, matter was called into existence, as that at successive periods new races of animals and plants were created.

Prin. In all the conditions of the globe from the earliest times, and in the structure of all the organic beings that have successively peopled it, we find the same marks of wise and benevolent adaptation, as in existing races; and a perfect unity of design extending through every period of the world's history.

Proof. 1. The anatomical structure of animals and plants was very different at different epochs: but in all cases the change was fitted to adapt the species more perfectly to its peculiar condition. 2. To communicate the greatest aggregate amount of happiness, is a leading object in the arrangements of the present system of nature: and it is clear from geology, that this was the leading object in all previous systems. 3. The existence of carnivorous races among existing tribes of animals tends to increase the aggregate of enjoyment, first, by the happiness which those races themselves enjoy; secondly, by the great reduction of the suffering which disease and gradual decay would produce, were they not prevented by sudden death: and thirdly, by preventing any of the races from such an excessive multiplication as would exhaust their supply of food, and thus produce great suffering. Now we find that carnivorous races always existed on the globe; showing a perfect unity of design in this respect. Thus when the chambered shells, so abundant in the secondary rocks, and which were carnivorous, became extinct at the commencement of the tertiary epoch, numerous univalve molluscs were created, which were carnivorous: although till that time these races had been herbivorous.

Inf. From these statements we infer the absolute perfection, and especially the immutable wisdom of the Divine character. A minute examination of the works of creation as they now exist, discloses the infinite perfection of its Author, when they were brought into existence: and geology proves Him to have been unchangeably the same, through the vast periods of past duration, which that science shows to have elapsed since the original formation of the matter of our earth.

Rem. The whole of this subject is admirably developed in the late splendid Bridgewater Treatise, by Dr. Buckland.

Prin. Geology furnishes many peculiar proofs of the benevolence of the Deity. The following are the most striking.

1. *The formation of Soils by the decomposition of Rocks.*

Illus. The disintegration of rocks, which we every where witness, strikes the mind at first as an exhibition of decay indicating some defect of contrivance on the part of the Deity. But when we find that the soils resulting from this decomposition are exactly adapted to the growth of plants, and that these are essential to the existence of animals, we can no longer doubt but we have before us a bright exhibition of benevolent design.

2. *The Disturbances that have taken place in the Earth's Crust.*

Illus. To a person not familiar with geology, the elevation, disruption, contortion, and overturnings, exhibited by the rocks, present a scene of confusion and chaos rather than proofs of benevolent design. But suppose the strata had remained horizontal, as first deposited. Nearly all the beds of valuable rocks and minerals must have been hidden from human view, and rendered inaccessible. But the disturbances experienced by these strata have brought them within the reach of human industry. Design then is manifest in this apparent confusion.

3. *The formation of Valleys.*

Illus. In mountainous countries these have resulted mainly from the elevation and dislocation of the strata. They have, however, been greatly modified and rendered beautiful and arable, by means of atmospheric and aqueous agencies; and to these latter causes most of the valleys in level countries owe their origin. Now without valleys, the earth would be uninhabitable; because there could be no circulation of water, and stagnation and death would pervade all nature, even if we admit enough of inequality to redeem a part of the earth from the ocean.

4. *The distribution of Water.*

Illus. We might at first suppose, that in mountainous regions, all the water would soon be accumulated in the valleys. Whereas such are the nature and situation of the soil and rocks, that the ridges are usually as well watered as the valleys. The alternations of pervious with impervious strata form natural reservoirs of water in the earth, and those dislocations of the strata, termed faults, tend to render these reservoirs still more perfect, while the fact that springs occur in almost every part of the earth, show that enough communications exist to the surface to allow of the passage of sufficient water for the support of animals and vegetables. These springs, uniting into rivers, find their way into the ocean; where an equal quantity of water is evaporated, and brought back by clouds into the regions where this perpetual drain is going on. Thus a constant circulation is kept up; while the hydraulic arrangements of the earth's crust are such as to keep a constant supply in all those places where it is needed. Surely here is benevolent design; and design too brought about by apparent disorder and confusion.

5. *The distribution of Metallic Ores.*

Illus. If the earth has been in a state of fusion, we should expect that the metals, being generally heavier than other minerals, would have accumulated at the centre, and have disappeared from the earth's crust. But by means of sublimation, segregation, and other agencies, enough of these metals has been brought so near the surface as to be accessible to man. Yet they are not so abundant, nor so easily obtained, as not to demand patient industry and ingenuity, whose exercise is indispensable to human improvement and happiness. Again, the most important of these metallic ores,—iron, lead, copper, &c. are most abundantly distributed and most easily obtained.

6. *Glacio-aqueous Agency.*

Illus. The effect of those powerful agencies, until recently regarded as exclusively diluvial, that have swept over large portions of the earth's surface in past times, has been to wear down its more rocky and salient parts, to convert steep escarpments into gentle slopes, and to increase the quantity of soil, and spread it more extensively over the surface. Hence, though at first a desolating agency, its ultimate effect is most salutary.

7. *Volcanic Agency.*

Illus. It operates, in the first place, as a safety valve, to prevent those vast accumulations of heat which exist in the earth, from rending whole continents in pieces: in the second place, it aids in raising continents from the ocean and in the formation of valleys.

Obj. Why should not a benevolent Being, who is omnipotent, secure to his creatures the benefits which result from volcanic agency, without the attendant evils, such as the destruction of property and life?

Ans. This is a question that meets the student of natural theology at almost every step of his progress: for we find almost universally, that evils are incident to operations whose natural tendency and general effect are beneficial. Probably it is so, because a greater amount of good can thereby be secured in the end. But the existence of evil is one of those difficult subjects whose complete elucidation ought not to be expected in this world.

8. *The accumulation of extensive deposits of coal, rock salt, gypsum, marble, and other valuable minerals, for the use of man during the long periods that preceded his existence.*

Illus. While the earth was in a state unfit for the animals and plants now existing upon it, it was covered with a gigantic vegetation, whose relics became entombed, and were gradually converted into those beds of coal, which are now in the course of disinterment, and which are so important to human improvement and happiness. Then also, rock salt, gypsum, and marble were slowly preparing for the service of beings to be created centuries afterwards. Can there be a doubt but this is a beautiful example of the prospective benevolence of the Deity?

9. *The adaptation of the natures of different groups of animals to the varying condition of the globe.*

Illus. The Deity intended the world ultimately to become the residence of intellectual and moral beings: but for wise reasons he chose to bring it by slow processes of change into a fit condition for their residence. Yet his overflowing benevolence prompted Him to people the world, during this transition state, with animals whose natures were perfectly adapted to its condition. And as often as that condition changed, did he change its inhabitants and their constitution. He might have left it desolate during these mighty periods of preparation. But infinite benevolence would not permit.

The World in a fallen State.

Prin. In spite of these evidences of Divine benevolence, geology unites with all other sciences, and with experience, in showing the world to be in a fallen condition, and that this condition was foreseen and provided for, long before man's existence, so that he might find it a world well adapted to a state of probation.

Proof. 1. It appears that the laws and operations of nature have been the same, essentially, as at present in all ages. 2. That the same systems of sustenance, reproduction and death, have always prevailed.

Inf. 1. Hence it must always have been impossible, in this world, to have avoided severe suffering: ex. gr. pain and death.

Inf. 2. Hence it has never been such a world as perfect benevolence would have prepared for perfectly holy and happy beings; though benevolence has always so decidedly predominated in it, as to show it to be a world of probation and mercy, not of retribution.

Prin. Geology enlarges our conceptions of the plans of the Deity.

Exam. 1. The prevailing opinion, until recently, limits the duration of the globe to man's brief existence, which extends backward and forward only a few thousand years. But geology teaches us that this is only one of the units of a long series in its history. It develops a plan of the Deity respecting its preparation and use, grand in its outlines, and beautiful in its execution; reaching far back into past eternity, and looking forwards, perhaps indefinitely, into the future.

2. Each successive change in the condition of the earth thus far, appears to have been an improved condition: that is, better adapted for natures more and more perfect and complicated. In its earliest habitable state, its soil must have been scanty and sterile, and almost destitute of calcareous matter, except in the state of silicates, which plants decompose with difficulty. The surface also, was but little elevated above the waters: and of course the atmosphere must have been very damp; though the temperature was very high. Every subsequent change appears to have increased the quantity and fertility of the soil, the amount of the salts of lime and geine, and the dryness of the atmosphere. Should another change occur, similar to those through which it has already passed, we might expect the continents to be more fertile and capable of supporting a denser population.

3. It appears that one of the grand means by which the plans of the Deity in respect to the material world are accomplished, is constant change; partly mechanical, but chiefly chemical. In every part of our globe, on its surface, in its crust, and we have reason to suppose, even in its deep interior, these changes are in constant progress: and were they not, universal stagnation and death would be the result. We have reason to suspect also, that changes analogous to those which the earth has undergone, or is now undergoing, are taking place in other worlds; in the comets, the sun, the fixed stars, and the planets. In short, geology has given us a glimpse of a great principle of *instability*, by which the *stability* of the universe is secured; and at the same time, all these movements and revolutions in the forms of matter essential to the existence of organic nature, are produced. Formerly the examples of decay so common everywhere, were regarded as defects in nature: but they now appear to be an indication of wise and benevolent design:---a part of the vast plans of the Deity for securing the stability and happiness of the universe.

2 *Connection of Geology with Revealed Religion.*

Prin. Revelation does not attempt to give instruction in the principles of science: nor does it use the precise and accurate language of science: but the more indefinite language of common life. Nor does science attempt to teach the peculiar truths contained in revelation.

Inf. 1. Hence it is only where revelation incidentally touches upon the same points as science, that the two subjects can be brought into comparison.

Inf. 2. Hence there may be *apparent* discrepancy between the two subjects, when there is *real* agreement, on account of a difference in the language employed: ex. gr.: the Bible apparently contradicts astronomy, when it asserts that the earth is immoveable, and that the sun rises and sets: but that here is no real disagreement, is too obvious to require proof.

Inf. 3. Hence it is reasonable to expect, only that the principles of science, rightly understood, should not contradict the statements of revelation, rightly interpreted. Unexpected coincidences, however, may occur between the two subjects; and these will tend to strengthen our belief in the truth of both.

Inf. 4. Hence the points of apparent discrepancy ought to be more numerous than the points of agreement between science and revelation, in order to prove a real contradiction between them: for it is as difficult to explain an apparent agreement, where there is real discrepancy, as the reverse.

Points of Coincidence between Geology and Revelation.

1. They agree in representing our present continents as formerly covered by the ocean.

Proof. That they were thus submerged, is one of the best settled principles of geology; and that revelation teaches the same, appears from Genesis, 1: 1, 9.

2. They agree as to the agents employed to produce geological changes on the globe: viz. water and heat.

Proof. Water is the only agent directly named in Genesis. and the elevation of the land is imputed directly to the exertion of Omnipotence. But in *Psalm* 104: 2. 4 *to* 7, where this operation seems to be described, *the voice of God's thunder*, there represented as the agent, may reasonably be understood to refer to volcanic agency. This same agency is represented as having destroyed the cities of the plain, according to Dr. Henderson's translation of Job 22: 15 *to* 20. A future change in the earth, is also described as resulting from fire, 2*d. Peter*, 3: 10 *See Turner's Sacred History of the World, p.* 24, 25.

3. They agree in representing the work of creation as progressive, after the first production of the matter of the universe. *Genesis, First Chapter.*

4. They agree in the fact that man was among the latest of the animals created to inhabit the globe.

Rem. This is a very important point. For had the remains of man been found among the earliest organic relics, while the Bible represents him as the last animal created, it would have been difficult to see how the two records could be reconciled.

5. They agree in the fact, that the epoch when the existing races of animals and plants were placed upon the globe, was comparatively recent.

Proof. According to revelation, this epoch could not have been more than 6000 years ago; and although we cannot as yet connect geological and chronological time, there are facts which prove that the commencement of the present order of things, and of the existing races of animals and plants, cannot have been very remote. Their remains occur only in alluvial deposits. Now the quantity of alluvium at the mouths of rivers, although often advancing rapidly, is yet comparatively limited. The accumulation of fragments at the base of steep rocky precipices, is still in most cases going on: as is also the formation of peat. But had these processes commenced at an immeasurably remote period, they ought ere this to be completed. Wide oceans ought to be converted into alluvial plains, precipices should all be levelled, and peat swamps be so filled that the process of its formation would stop.

6. The facts of geology render the future destruction of the earth by fire, a not improbable event.

Proof. Nearly all geologists admit that the earth contains vast reservoirs of heat; and if these are brought into action by the fiat of the Almighty, *the elements might be melted and the earth and the things therein be burned up.* Or it is even easy to conceive how this internal heat, without miraculous interference, might, under certain circumstances, produce the same result.

Supposed Discrepancy between Geology and Revelation.

Descr. The supposed discrepancies between geology and revelation, relate first, to the age of the world, and secondly, to the period when death was introduced upon the globe.

Descr. Geologists suppose that the changes which have taken place on the globe, must have occupied immense periods of time; and that several successive systems of animals and plants inhabited the world previous to the creation of the existing races: whereas the Mosaic account, according to the common interpretation, represents the matter of the globe to have been produced out of nothing, only a few literal days previous to the creation of man; and that all the animals and plants that ever lived on the globe, were then brought into existence.

Rem. I am not aware that this statement has ever been formally adduced by any geological writer in opposition to revelation. But geologists having come to the conclusion that the earth, in some form, must have existed more than 6000 years, some Christian writers have inferred that this was opposed to the Mosaic account, and have attempted a defence of revelation. And hence has resulted the prevailing opinion, that geologists in general have been hostile to the Bible:—an opinion which may be refuted by an appeal to their writings.

Prin. In order to obviate this objection to revelation, it is only necessary to show, that one or more modes exist, of reconciling the apparent discrepancy, which it would be more reasonable to adopt, than to infer any real collision between the two records. Some of these modes of explanation will now be briefly described.

1. Some theological (but no geological) writers maintain, that the fossiliferous rocks were not the result of slow deposition and consolidation: but were created at once, with all their organic contents, just as we now find them.

Refutation. This is admitted to be possible; because God's power is infinite. But our only ground for judging as to the cause of any natural changes, is analogy:—and this is entirely opposed to the idea that rocks were thus produced: and every example that can be quoted of rocks in a course of formation, is in favor of their slow formation by second causes.

2. Some maintain that the fossiliferous rocks were deposited by the deluge of Noah.

Refutation. 1. That deluge must have been for the most part violent and tumultuous in its action on the globe: for the ocean must have flowed over the land in strong currents; and when it retired, urged on as it was by a wind, similar currents must have prevailed. But a large proportion of the rocks were evidently deposited in quiet waters. 2. If deposited by that deluge, the materials and entombed organic remains of the rocks ought to be confusedly mingled together; whereas in both these respects they are actually arranged with great regularity into groups. 3. The period occupied by the Noachian deluge was vastly too short for the deposition of rocks seven miles in thickness, and with a great number of entire and distinct changes in their nature and organic contents. 4. The organic remains in the rocks do not correspond to the animals and plants now living on the globe. But this deluge took place since the creation of the present races; and, therefore, by this hypothesis, they ought to be found in the rocks. Hence they were deposited before that event.

Rem. An apology is due to the geological reader, for introducing a formal refutation of an hypothesis, which, to him, appears so entirely absurd. The apology consists in the fact, that many intelligent men are still found maintaining this hypothesis.

3. Some suppose that the fossiliferous strata have been deposited in the interval of 1600 years between the creation of man and the deluge.

Refutation. 1. The time since the deluge has been twice as long as 1600 years; but the amount of alluvium deposited has not been one thousandth part as great as the whole fossiliferous rocks. Hence 1600 years is vastly too short a period for their deposition: since no reason can be given why the process of their formation was essentially more rapid before than after the deluge. 2. By this hypothesis the sea and land must have changed places at the deluge; in order to bring the fossiliferous rocks from the bottom of the sea. But geology renders it extremely probable that no interchange of this sort took place so recently. 3. On this hypothesis the organic remains ought to consist of species of the existing races, with man among the number: unless we suppose that several new and distinct creations of animals and plants have taken place on the globe since man was placed upon it, none of which are mentioned by Moses. 4. There is the strongest evidence that the primary as well as the fossiliferous rocks, have resulted from secondary agencies; that is, they have existed in some previous form, before assuming their present state. But these changes could not have taken place after the creation and multiplication of man; because to take away the primary rocks is to take away all *terra firma* on which he could have subsisted.

4. Some regard the six days of the creation, (called the *demiurgic days*) in the Mosaic account as not literal days of 24 hours, but periods of indefinite or unequal length, or as the representatives of indefinite periods.

Rem. Three varieties of opinion are embraced in this theory, as above stated. 1. The more common supposition is, that the term day is here to be understood figuratively, as embracing a long period of time: a mode of using the term, that is frequent in all languages. 2. Some, as Bishop Horsley and Professor Jameson, suggest that the revolution of the earth on its axis was at first "inconceivably slow," and that it did not acquire its present rate till the close of the fourth day; so that the first four days may have been of vast duration. *Philosophical Magazine, Vol.* 47. *p.* 243: *also Vol.* 46. *p.* 227. Still more recently this theory has been ably elucidated by Dr. Keith in his *Demonstration of the Truth of Christianity, p.* 127. *First American Edition,* 1839. 3. Others, as Hensler in Germany, and Professor Bush in this country, suppose that each of the six demiurgic days stands as the closing day, or representative, of an indefinite number of days, which make up six periods of unknown and perhaps unequal length: during which geological processes might have been going on. *American Biblical Repository, Vol.* 6, *p.* 236. *Bush's Questions and notes upon the Book of Genesis,* 2*d. ed.*

Arguments in favor of this Theory.

1. The word day is often used in Scripture to express a period of indefinite length, Ex. gr. Luke 17 : 24.—John 8: 56. Job 14: 6. &c. 2. The sun, moon, and stars were not created till the fourth day; so that the revolution of the earth on its axis in 24 hours did not probably exist before that time; and some other measure of time must have been adopted, which Moses tells us was light and darkness: and how often these succeeded one another is not revealed; and therefore is unknown. 3. The seventh day, or the sabbath, has not yet terminated; and will not, until God shall resume the work of creation; that is, it will continue from the beginning to the destruction of the world, and there is no reason why we ought not to regard the other demiurgic days as of at least equal length. 4. In order to reconcile the declarations of Scripture with the discoveries of astronomy, it is necessary to depart as much from the literal meaning as this interpretation demands. 5. This interpretation corresponds remarkably with the traditional cosmogonies of many heathen nations; as the ancient Etruscans, and the Hindoos, who describe the demiurgic days as immense periods. 6. This theory is thought by Professor Jameson and others to develope a striking coincidence between the epochs of creation as described by Moses and by geologists. *Bakewell's Geology, p.* 450.

Obj. 1. There is no evidence that the word day is used figuratively in the first chapter of Genesis, as it is in all other places in Scripture where it means an indefinite period, except perhaps Gen. 2: 4. On the contrary, the Mosaic description of the creation appears to be a very simple and perfectly literal history, adapted to the most uncultivated minds. 2. In the fourth commandment (Exodus 20: 9. 10. 11.) no one can doubt but the six days of labor and the Sabbath, spoken of in the 9th and 10th verses, are literal days. By what principle of interpreting language, can the same word in the next verse, where the creation is described, be understood to mean indefinite periods? See a parallel passage, Exodus. 31: 17. 3. It seems from Genesis 2: 5, compared with Gen. 1: 11, 12, that it had not rained on the earth till the third day. If the days were only of 24 hours, this would be very probable, but altogether absurd, if they were long periods. 4. Such a meaning is forced and unnatural; and therefore not to be adopted without a very urgent necessity. 5. By this theory, existing species of animals and plants ought to be found mixed with the extinct species in all the fossiliferous rocks: for Moses describes only one creation of the different races. Now the fact that they are not thus mixed, shows that they could not have been contemporaries. If then the Mosaic account includes the fossil species, it does not include the living species: and if it embraces the latter, it cannot comprehend the former. It is hence inferred, that the Mosaic account embraces only existing races, and if this be admitted, there is no necessity of supposing the demiurgic days to be longer than literal ones. 6. If this theory be admitted, instead of exhibiting coincidence between the Mosaic and the geological account of the epochs of creation, it produces between them manifest discrepancy. For Moses describes vegetables to have been created on the third day, but animals not until the fifth. Hence about one third of the fossiliferous rocks, reckoning upwards, or those deposited during the first three days, ought to contain only vegetables. Whereas animals are found as deep in the rocks as vegetables: nay, in the lowest group, nothing but animals has yet been found. 7. The conclusion from all these statements, is, that Moses does not describe the fossil but only the existing races of animals and plants; and if so, there is no necessity for an extension of his demiurgic days into long periods, in order to reconcile his account with geology.

Rem. For a fuller examination of the preceding theory, See Faber's *Treatise on the Patriarchal, Levitical, and Christian Dispensations: De Luc's Letters on the Physical History of the Earth: Bakewell's Geology*, 2*d. American Ed. p.* 439: *and the American Biblical Repository for October*, 1835. It ought perhaps to be remarked, that Faber, who had advocated this theory with great ability, has recently given it up. *Buckland's Bridgewater Treatise, Vol.* 1. *p.* 597. 2*d. Ed.* 1837.

5. Some suppose that the Mosaic account is a pictorial representation of the successive productions of the different parts of creation, having truth for its foundation, yet not to be regarded as literally and exactly true. The terms employed, however are to be understood in their literal sense.

Illus. Conceive of six separate pictures, showing the work in different stages of its progress. "And as the performance of the painter, (says Dr. Knapp,) though it must have natural truth for its foundation, must not be considered or judged of as a delineation of mathematical or scientific accuracy, so neither must this pictorial representation of the creation be regarded as literally and exactly true." *Knapp's Theology, Vol.* 1. *p.* 364.

Inf. Whether this interpretation of the Mosaic account be admissible, is a question of mere philology, and cannot be discussed in this place: but admitting its correctness, it affords a solution for the apparent discrepancy between geology and revelation: for when it is conceded that the earth may have existed a longer time than the usual interpretation of the Mosaic account allows, there is no reason why the time may not be indefinitely extended; which is all that geology requires.

6. The theory of interpretation which is now the most extensively adopted among geologists, supposes that Moses merely states that God created the world in the beginning, without fixing the date of that beginning; and that passing in silence an unknown period of its history, during which the extinct animals and plants found in the rocks might have lived and died, he describes only the present creation, which took place in six literal days, less than 6000 years ago.

Arguments in favor of this theory with objections against it. 1. It is maintained by some able writers, such as Dathe, Doederlin, &c. in Germany, Milton in England, and Prof. Bush in this country, that the language employed by Moses in the first chapter of Genesis, does not mean a creation of the world out of nothing; but only a renovation, or re-modelling of previously existing materials. *Penteteuchus a Dathio, p.* 8. *Doederlinii Theologia, p.* 485. *Bush's Questions and Notes on Genesis.* Such writers of course admit the existence of the globe during an indefinite period before the six demiurgic days. The arguments for their opinions may be found in their works above referred to. *See also American Biblical Repository, Oct.* 1835, *p.* 280. 2. The phrase, *in the beginning*, is certainly indefinite as to time, and therefore Moses in Genesis does not fix the time of the original creation, even if we admit that he does describe that event; and therefore, it is doing no violence to his language, to admit this long intervening period between the creation of the universe and of man. If it be said, that in the fourth commandment Moses does declare the creation of the world out of nothing to have been contemporaneous with the first demiurgic day, it may be replied, that when a writer describes an event more than once, his briefer description is to be explained by his more extended account: so that the fourth commandment is to be explained by the fuller description in Genesis, of the same event. 3. This view of the first chapter in Genesis has been adopted in its essence by many distinguished Christian writers, long before the existence of geology as a science: as for example, by Augustine, Theodoret, &c. in ancient times; and by Rosenmuller senior, Bishop Patrick, &c.

in modern times. 4. If such an interval be admitted, it is sufficient entirely to reconcile the scriptural and geological accounts: because, during that period, all the fossiliferous rocks except the alluvial, might have been formed. 5. Astronomy shows us that probably other worlds are now undergoing slowly the process of preparation for habitable globes, in a manner analogous to that which is supposed to have taken place in the materials of the earth, anterior to the demiurgic days. And thus we obtain a glimpse of a general principle in the universe. 6. If it be objected that according to Moses, the sun, moon, and stars were not created till the fourth day, it may be replied that a more just interpretation of his language shows his meaning to be, not that the heavenly bodies were created on the fourth day, but that they were then first appointed to serve their present offices; and that they might have been in existence through countless ages.

7. In his late able and most interesting work, *On the Relation between the Holy Scriptures and some parts of geological Science*, Dr. John Pye Smith, who is at the head of a Theological Institution at Homerton near London, and who is distinguished for his knowledge of theology, biblical philology and geology, has proposed such an addition to the interpretation of Genesis just explained, as in fact to form a new method of reconciling geology and revelation. His principal positions are the following. 1. The first verse of Genesis describes the creation of the matter of the whole universe, probably in the state of mere elements, at some indefinite epoch in past eternity. 2. The term earth, as used in the subsequent verses of Genesis describing the work of six days, was "designed to express the part of our world which God was adapting for the dwelling of man and the animals connected with him." 3. The narrative of the six days work is "a description in expressions adapted to the ideas and capacities of mankind in the earliest ages, of a series of operations, by which the Being of omnipotent wisdom and goodness adjusted and finished not the earth generally, but as the particular subject under consideration here, a PORTION of its surface for most glorious purposes. This portion of the earth I conceive to have been a large part of Asia lying between the Causican ridge, the Caspian Sea, and Tartary, on the north, the Persian and Indian Seas on the south, and the high mountain ridges which run at considerable distances, on the eastern and western flank." *P.* 285, *London Ed.* "3. This region was first by atmospheric and geological causes of previous operation under the will of the Almighty, brought into a condition of superficial ruin, or some kind of general disorder." Probably by volcanic agency it was submerged. covered with

fogs and clouds, and subsequently elevated and the atmosphere by the fourth day rendered pellucid.* 4. The sun, moon, and stars were not created on the fourth day: but then "made, constituted, or appointed, to be luminaries." 5. The Noachian deluge was limited to that part of the world occupied by the human race, and therefore we ought not to expect that any traces of it on the globe can now be distinguished from those of previous and analogous deluges.

Rem. It is impossible in this place to present even a summary of the powerful reasoning and accurate erudition by which Dr. Smith sustains the above positions in his Lectures. The evidence in favor of several of them has already been exhibited; and I shall merely state the two leading principles by which he supports what is peculiar in his system.

Proof 1. In the description of the Divine Character and of natural phenomena, the sacred writers use language accommodated to the knowledge that existed among the people whom they addressed, and conformed to their notions of the universe. Hence, when they wished to speak of the universe to the Jews, they called it the heavens and the earth. But when they spoke of the earth only, we are not to suppose that they used the term in an astronomical sense, but to designate that limited part of it which was inhabited. For this was all the idea which the mind of the Jew attached to it; since he knew nothing of the earth beyond those limits. Hence the six days work of creation may have been limited to a small part of the earth's actual surface. 2. This view corresponds to the fact that there appear to have been numerous centres of creation; both of animals and plants, instead of one spot from which all proceeded. 3. Also with the fact that a considerable number of species of animals and plants, which were created much earlier than man, as their remains in the tertiary strata show—still survive, and do not appear to have perished since their first creation. 4. Universal terms are often used (in Scripture) to signify only a very large amount in number or quantity: as for instance, *all the earth came to Egypt to buy from Joseph: for the famine was extreme in all the earth.* Hence the terms descriptive of the deluge may not have literally embraced the whole earth; but have included only the earth then inhabited.

8. Some contend, that even if all the methods of reconciling the two records that have been described, should be regarded by any as unsatisfactory, it would be premature, in the present state of geology and of sacred philology, to infer any real discrepancy between them: and especially to infer that the sacred historian is in error.

Proof 1. Because the rapid progress of geological discovery, and the not unfrequent changes of opinion among geologists on

* See note K.

important points, show us that possibly more light may yet come from that quarter. 2. Because the exegesis of the first chapter of Genesis cannot by any means be regarded as settled: in proof of which, it is only necessary to refer to the great diversity of opinion, on many parts of this chapter, yet to be found among very able commentators. Hence we may hope for new light from this quarter. 3. Other apparent discrepancies between science and revelation, even more striking than that above examined between geology and revelation, have disappeared when the subjects were better understood. For instance, the doctrine introduced by the astronomers 200 years ago, that the earth revolves on its axis, and that the heavenly bodies do not actually rise and set, seemed to the most acute and learned theologians of those times, to be in *point blank* opposition to the Bible; which declares, that *the sun ariseth and goeth down; and that God laid the foundation of the earth that it should not be removed forever.* They also *felt* the earth to be at rest; and *saw* the heavens in motion; so that this new doctrine of the Copernican system was opposed, not only to the Bible, but to the senses. Yet who now suspects any collision between astronomy and revelation? How premature then, to infer from a less striking apparent discrepancy between geology and the Bible, that any real opposition exists!

Second Supposed Discrepancy.

Descr. The general interpretation of the Bible has been, that until the fall of man, death did not exist in the world even among the inferior animals. For the Bible asserts that by *man came death,* (1 *Cor.* 15: 21.) *and by one man sin entered into the world and death by sin.* (*Rom.* 5: 12.) But geology teaches us that myriads of animals lived and died before the creation of man.

Solution of the difficulty. Admitting that geology does show that violent and painful death was in the world before the fall of man, the following suggestions furnish a plausible reconciliation of the supposed difficulty.

1. Not only geology, but zoology and comparative anatomy, teach us that death among the inferior animals did not result from the fall of man, but from the original constitution given them by their Creator. One large class of animals, the carnivorous, have organs expressly intended for destroying other classes for food. Nor will it avoid the difficulty to suppose, as some have done, (although obviously in the face of the plain meaning of the first chapter of Genesis,) that carnivorous animals were not created till the time of the deluge. For other

animals must have lived on vegetables, and in doing this, they must have destroyed a multitude of minute insects, of which several species inhabit almost every species of plant. Much more difficult would it have been, to avoid destroying millions of animalcula, which abound in many of the fluids which animals drink, and even in the air which they breathe. Still farther, throughout the whole range of organic nature, vegetable as well as animal, decay and dissolution are inevitable after a longer or a shorter period. In this respect, the human system is constituted just like all other organic natures. So that death appears to be a universal law of organic being, as it exists on earth. Moreover, without miraculous interference for protection, or an entire change of the present laws of nature, animals must have been exposed to occasional violent disorganization: as for instance, from the falling of heavy bodies upon them; or from the shock of projectiles; even though there were no tendency in their natures to dissolution. In short, death could not be excluded from the world, without an entire change in the constitution and course of nature; and such a change we have no reason to suppose, from the Mosaic account, took place when man fell. 2. But God could remove any race of animals, say man out of the world, and introduce them into another state, without violence, disease, or suffering; and make the change, in fact, like many changes in life, a pleasant one; free from those concomitants which now indeed constitute death. He has already removed a few from the world in this manner; as Enoch and Elijah, and the Bible informs us that those who remain alive at the second coming of Christ, will in a similar manner be *translated* and not die. (1 Cor. 15: 51, 52.) This probably would have been the happy lot of all mankind, if they had not sinned. That they could not have continued on earth indefinitely, is certain: provided the present laws of their multiplication were not suspended: because the world ere long would have been filled. 3. The threatening of death to Adam for disobedience, seems to imply a knowledge on his part, of what death was, that is, he had seen it among the inferior animals: for it would be a strange legislation, that imposed a penalty of which those under the law could form no idea. 4. The two most striking passages of scripture, respecting the introduction of death into the world, have been already quoted. In regard to that from Romans, 5: 12, *by one man sin entered into the world, and death by sin*, the conclusion of the sentence,—*and so death passed upon all men, for that all have sinned*,—shows that its meaning must be limited to the human race. For it says not that death passed upon all animals, but upon all men;

and because all had sinned, an act of which the inferior animals, destitute of moral natures, are not capable. In like manner, the passage from 1 Cor. 15 : 21, *For since by man came death,* is limited to the human race by the concluding part of the verse : *by man came also the resurrection of the dead.* For the object here is to draw a contrast between Adam and Christ, as to their influence upon the human family. If the inferior animals are included, then they must not only share in the resurrection, but be interested in the redemption by Christ. 5. Able writers on the Bible have adopted these views in regard to the nature and extent of death, long before geology was known as a science. To quote only Jeremy Taylor. "That death," says he, "which God threatened to Adam, and which passed upon his posterity, is not the going out of this world, but the manner of going. If he had stayed in innocence, he should have gone hence placidly and fairly, without vexatious and afflictive circumstances, he should not have died by sickness, misfortune, defect, or unwillingness, but when he fell he began to die," &c. *Holy Dying, p.* 295. *Amherst Ed.* 1831.

Second Answer to the Objection. The view that seems most satisfactorily to explain this subject, supposes that God, in view of the certainty of man's transgressions, adapted the world beforehand to a fallen creature, who must die. It could not be adapted both to mortal and immortal natures, and since sin and death are probably inseparable, it must be fitted to the character of the former. And as to the inferior animals, existence would be a blessing even though death awaited them; as we have endeavored elsewhere to show. Death, then, was introduced into the world as a prospective result of man's apostasy : and the present theory has this advantage, that it falls in with the common opinion, derived from the Bible, that all the misery, disorder, and suffering, of the present world, are the fruit of human transgressions : or, in the language of Scripture, *the whole creation groaneth and travaileth together in pain until now.*

General Inferences. The preceding statements show us, *first,* that the coincidences between geology and revelation, upon points where we might reasonably expect collision, if both the records were not essentially true, are much more numerous than the apparent discrepancies, and therefore the presumption is, that no real disagreement exists: *secondly,* it appears that there are several modes of reconciling the few apparent discrepancies, which, on general principles, it would be far more reasonable to adopt, than to infer any real discrepancy between the two records. *Thirdly,* it appears that geology ought to be regarded as a new means of illustrating, instead of opposing, revelation; since it leads us to understand certain passages, which had before been misinterpreted; just as astronomy did in respect to the heavenly bodies. *Finally*, it appears that the illustrations of natural religion from geology, are more numerous and important than from any other, and perhaps all other sciences.

SECTION X.

THE HISTORY OF GEOLOGY.

Prin. Geology is eminently an inductive science. Now it is only within the last half century, that sufficient facts had been collected to make any important correct inferences respecting the causes of geological change. Hence all the speculations of philosophers previous to that time, on this subject, must have been mere hypothesis; sometimes indeed displaying great ingenuity, and approximating closely to the truth, but more commonly, so extravagant as to be the butt of ridicule.

Prin. Geology is likewise dependent upon an advanced and accurate knowledge of chemistry, botany, and zoology: such a knowledge as has been attained only within the last half century. On this account also, all previous speculations on geology must have been crude and fanciful.

Inf. 1. Hence the earlier hypotheses on cosmogony, which have been so long the subject of ridicule, ought not to be connected with the science of geology, as they have long been, to its reproach.

Inf. 2. Hence the history of the early hypotheses, usually called Theories of the Earth, is of little importance in its bearing upon the science of geology; though highly amusing and instructive, as illustrating the struggles of the human mind after the truth. A brief sketch only will, therefore, be here given of these hypotheses.

Descr. One of the most prevalent opinions among ancient philosophers, and which constituted a fundamental principle in their cosmogonies, supposes the world to have been subject to successive destructions and renovations by fire and water. The Grecian philosophers, who derived their notions from the Egyptian, denominated those catastrophes the *Cataclysm*, or deluge; and the *Ecpyrosis*, or destruction by fire. The interval between these changes was variously estimated from 120.000 o 360.000 years.

Descr. Pythagoras entertained very accurate notions respecting the operation of existing causes of geological change on the globe; such as the changes of sea into dry land, and the reverse, the formation of deltas and other alluvial deposits: and the formation of islands by the action of oceanic currents. In fact, this philosopher approximated as nearly to the modern views of geologists on these subjects, as he did to those of modern astronomers respecting the heavenly bodies.

Descr. We find Strabo, the geographer, explaining the manner in which fossil marine shells were brought into their situation upon the dry land, in a manner that would do no discredit to a modern geologist. He supposes them deposited originally at the bottom of the ocean, whose bed was subsequently elevated by earthquakes and volcanic agency.

Descr. In modern times, after the dark ages, science began first to be revived among the Arabians. Even as early as the tenth century, some of their writers, as Avicenna and Omar, produced some works on mineralogy and geology, which were not without considerable merit.

Descr. Among Christian nations, geological facts first began to excite attention in Italy, in the early part of the sixteenth century. Two questions were stated respecting organic remains: first, whether they ever belonged to living animals and plants: secondly, if the affirmative of this question be admitted, whether the petrifaction and situation of these remains could be explained by the deluge of Noah.

Descr. These questions occupied the learned world for nearly 300 years. At the commencement of the controversy in Italy, in 1517, Fracastoro maintained, in the true spirit of the geology of the present day, that fossil shells all once belonged to living animals, and that the Noachian deluge was too transient an event to explain the phenomena of their fossilization. But Mattioli regarded them as the result of the operation of a certain *materia pinguis*, or "fatty matter," fermented by heat. Fallopio, Professor of Anatomy, supposes that they acquire their forms in some cases, by "the tumultuous movements of terrestrial exhalations;" and that the tusks of elephants were mere earthy concretions. Mercati conceived that their peculiar configuration was derived from the influence of the heavenly bodies; while Olivi regarded them as mere "sports of nature." Felix Plater, Professor of Anatomy at Basil, in 1517, referred the bones of an elephant, found at Lucerne, to a giant at least 19 feet high: and in England similar bones were regarded as those of the fallen angels!

Descr. At the beginning of the 18th century, numerous theologians in England, France, Germany, and Italy, engaged eagerly in the controversy respecting organic remains. The point which they discussed with the greatest zeal, was the connection of fossils with the deluge of Noah. That these were all deposited by that event, was for more than a century the prevailing doctrine, which was maintained with great assurance; and a denial of it regarded as nearly equivalent to a denial of the whole Bible.

Descr. The question also, whether fossils ever had an animated existence, was discussed in England till near the close of the 17th century. In 1677, Dr. Plot attributed their origin to "a plastic virtue latent in the earth." Scheuchzer in Italy, however, in ridicule of this opinion, published a work entitled, *Querulæ Piscium;* or the *Complaints of the Fishes;* in which those animals are made to remonstrate with great earnestness that they are denied an animated existence.

Descr. Such discussions, however, tended to lead men to the collection of facts; and in 1678, we find Lester publishing an accurate account of British shells, to which were added the fossil species, under the name, however, of *turbinated* and *bivalve stones*. He also first proposed the construction of regular geological maps.

Descr. In 1680 the distinguished mathematician Leibnitz published his "Protogæa;" in which he developed a theory of the formation of the earth, and of subsequent changes in its crust, almost exactly like that which is now so widely adopted among geologists under the name of the igneous theory.

Descr. In the posthumous works of Robert Hooke, an English physician, published in 1668, are many views much in advance of his time, respecting geological changes and phenomena; especially respecting earthquakes and organic remains; which he maintains could not have been produced by the Noachian deluge.

Descr. The famous naturalist, Ray, who published in 1692, had similar views with Hooke: but he made improvements upon his predecessor; especially in describing the effects of aqueous agencies in modifying the earth's surface.

Descr. In most of the theories of the earth, however, that appeared in England in the latter part of the 17th and the first part of the 18th century, a strong determination is manifested to connect geology with theology. This gave rise to what has been called the Physico-Theological School of writers. In 1690, Burnet published a visionary work, entitled "The Sacred Theory of the Earth; containing an account of the Original of the Earth, and all the general changes which it hath already undergone, or is to undergo, till the consummation of all things." This being written in an elegant style, attracted no little atten tion. About the same time, Woodward, a Professor in Medicine, published a theory of the earth, in which he maintained that "the whole terrestrial globe was taken to pieces and dissolved at the flood, and that the strata settled down from this promiscuous mass, as an earthy sediment, from a fluid." In 1724, his disciple, Hutchinson, came out with the first part of his "Moses's Principia;" in which he attacked the theory of his master, as well as Newton's theory of gravitation: and he and his numerous followers maintained, that the Scriptures, when rightly understood, contain a perfect system of natural philosophy. This was the dogma which chiefly distinguished the Hutchinsonian, or Physico-Theological school; and its pernicious influence on the cause of religion and learning has scarcely yet ceased.

Descr. The Italian geologists, who were contemporaries of these English cosmogonists, were far more rational in their views. They devoted themselves with considerable success to an examination of geological phenomena, and rejected the visionary notions of the English. The writings of Vallisneri, Moro, Generelli, Donati, Targiona, &c. discover a good deal of industry in observation and acuteness in reasoning.

Descr. In 1749, Buffon, the French naturalist, produced an elegantly written hypothesis upon the formation of the earth, based chiefly upon the views of Leibnitz already explained. Some of his views gave offence to the Faculty of Theology at Paris; and he was compelled, like Galileo, to retract opinions, which at this day are maintained by geologists, with no suspicion that they contradict the Scriptures.

Descr. About the middle of the 18th century, some valuable works were produced by Lehman, a German mineralogist, by the botanist Gesner of Zurich, and by Mitchell an Englishman, on earthquakes. About the same time, appeared the work of Catcott on the Deluge: and although of the physico-theological school, he gave some valuable facts relating to diluvial currents.

Descr. Numerous other writers on geology, towards the close of the 18th century, might be named. But Pallas and Saussure were the most distinguished. The former examined the mountains of Siberia, and pointed out the order in which the rocks there occur. The latter spent most of his time in studying the nature of the Alps, and provided valuable materials for his successors.

Descr. In Germany, France, and Hungary, schools have long existed for giving instruction in the art of mining. In 1775, Werner was appointed professor in one of these institutions, at Freyburg, in Saxony; and in his lectures he attempted certain generalizations in regard to the position and origin of rocks, that were very extensively adopted, and for a long time excited much controversy. He supposes that all rocks, the unstratified as well as the stratified, were deposited by water; and that originally every formation was universal on the globe: and that veins were filled by matter introduced from above, in aqueous solution. On account of his referring all deposits to the agency of water, his views were denominated the Neptunian Theory.

Descr. Nearly at the same time, a Scotch geologist by the name of Hutton, published a Theory of the Earth, opposed in most respects to the doctrines of Werner. He supposes that the rocks which form our present continents were derived from the ruins of former continents; which were abraded and carried into the sea by the agency of running water; just as the same agency is now spreading over the bottom of the ocean, deposits of mud, sand, and gravel. Afterwards the unstratified rocks, in a melted state, were protruded through these deposits, by which they were consolidated, rendered more or less crystalline, and elevated into their present condition. Many of the fissures also, were filled with metallic and other matter, injected from beneath. When our present continents are nearly all worn down, Hutton supposes this process of consolidation and elevation may be repeated. Indeed, in these changes he sees "no traces of a beginning, no prospect of an end." Professor Playfair, however, in his illustration of this theory, endeavors to show that Hutton did not mean by such language that the world is eternal: but only that geology, like astronomy, does not disclose to us the time when this series of changes commenced.

Descr. These rival hypotheses excited a great deal of discussion, both on the European continent and in England, for a great number of years. The final result is, that the theory of Werner has been almost universally abandoned; especially so far as the unstratified rocks are concerned; while that of Hutton, denominated also the Plutonian theory, has, in its essential principles, been adopted by most geologists of the present day.

Descr. During these discussions between the Neptunists and the Plutonists, a humble individual in England, Mr. William Smith, was accomplishing more for geology than all its learned professors. He explored the whole of England on foot: and in 1790, published a "Tabular View of the British Strata," in which he classified the secondary rocks; and although not acquainted with Werner's arrangement, he proposed essentially the same order of superposition among the rocks as that geologist. He also ascertained that strata, somewhat remote from one another, could be identified by their organic remains. By extraordinary perseverance, he was able, in 1815, to publish a geological map of the whole of England.

Descr. One important effect of excessive theorizing, was to produce an almost universal scepticism in unprejudiced philosophical minds in respect to all geological hypotheses; and to make them feel the importance of amassing facts. Out of these feelings grew the London Geological Society, which was founded in 1807; and which has contributed more than any other public institution, to advance the science. This Society selected a sentence for their motto, from the Novum Organum of Lord Bacon, which invites "those to join them as the true sons of science, who have a desire, and a determination, not so much to adhere to things already discovered, and to use them, as to push forward to farther discoveries; and to conquer nature, not by disputing an adversary, but by labor; and who, finally, do not indulge in beautiful and probable speculation, but endeavor to attain certainty in their knowledge."

Descr. The example thus set in England has been followed in nearly every part of the civilized world; and geological societies, both local and national, have been formed in so great numbers, that even their names cannot here be given. The London Geological Society has already produced thirteen quarto volumes of transactions. Many of the ablest philosophers of the age have devoted themselves to geological researches: and have applied the principles of induction with great success; so that probably no other science has made so rapid advances within the last half century, as geology. Very numerous geological collections have been formed in almost every part of the civilized world: accurate geological maps have been constructed of nearly all Europe and the United States; and in 1843 the Geological Society of France published a general geological map of the whole globe; showing the leading formations over nearly the whole of its surface. This map, however, is in part theoretical.

Descr. The rapid progress of the collateral branches of science, and their successful application to palæontology, are to be regarded as among the most important means by which geology has been thus rapidly advanced. The most important application of this sort, was that of comparative anatomy, to determine the character of organic remains, by the Baron Cuvier, in his great work entitled *Ossemens Fossiles, &c.* in seven quarto volumes; first published in 1812. Numerous successful applications have also been made of the discoveries in the various departments of zoology and botany, to the same object: the result of which has been, the great works of Goldfuss, Sow erby, Bronn, &c. on petrified molluscs, zoophytes, &c.; that of Agassiz on fossil fishes; those of Adolphe Brongniart and of Lindley and Hutton, on fossil plants, and those of Owen on British fossil reptiles and mammalia, on the Dinornis, and other papers too numerous to mention. The splendid work of Cuvier and Alexander Brongniart, "on the mineral geography and organic remains of the neighborhood of Paris," published in 1811, constituted an era in geology; because it directed the attention of geologists to the tertiary strata, from which they have since gathered so rich a harvest.

Descr. The present state of facts and theories in geology may be learnt from the preceding pages. Very many of the principles, which are now regarded as belonging to the science, may be considered as settled, as much as they are in most physical sciences: though some of them may doubtless experience slight modifications. A few points of importance yet remain in a great measure unsettled. Perhaps on no one is there more diversity of opinion, than concerning diluvial action. Indeed, the whole history of opinions on this subject is very instructive. When the subject was first discussed, as much as 300 years ago, it was assumed as a most unquestionable fact, that whatever marks of a deluge any part of the earth's surface exhibited, or even of the former presence of the ocean on the land, all must be referred to the deluge of Noah. Nay, it was soon maintained that the whole solid frame-work of the globe was dissolved and re-deposited by the diluvial waters. One after another have these extravagancies of hypothesis been given up, and nearly all geologists have come to the conclusion, though without denying the occurrence of the Noachian deluge, that no certain marks of that event are now to be discovered on the globe. Nay, the question now is, whether there is any evidence of the occurrence of a general flood at any epoch. Not a few believe that no such evidence exists; while those who admit of a general deluge, for the most part, regard it as having taken place anterior to man's existence on the globe. It is now generally admitted also, that ice, in the form of glaciers and icebergs, has been a most important event in producing the phenomena of drift. After centuries of discussion, it is also beginning to be found out, that the facts are yet very imperfectly known; and to an examination of these, geologists are now devoting great attention.

Descr. In American mineralogy and geology, almost literally nothing had been done at the commencement of the nineteenth century. A few individuals, indeed, among whom may be mentioned Dr. Seybert of Philadelphia, Dr. S. L. Mitchill of New York, and Dr. B. Waterhouse of Harvard University, had commenced rather earlier than this, to make collections, and to call the public mind to these subjects. But the return of Mr. B. D. Perkins and of Dr. A. Bruce from Europe, in 1802, and 1803, with beautiful collections, and of Col. Gibbs, in 1805, with his splendid cabinet, began to awaken more interest; so that ere long courses of lectures upon them were introduced into several of our colleges. In 1807, William Maclure returned from Europe, and commenced, single handed, the Herculean task of exploring the geology of the United States; and after several years of labor, during which he crossed the Allegany range of mountains at different places not less than fifty times, he actually produced a geological map of this whole country; which, though it gives only the Wernerian classes of rocks, forms a most valuable outline, and is a monument of great industry and perseverance In 1810, Dr. Bruce commenced his Mineralogical Journal, the first scientific journal ever undertaken in this country. But though ably commenced, it reached only the fourth number, in consequence of the sickness and death of its editor. In 1818, Professor Silliman, who had long been among the most distinguished and successful pioneers and teachers of these sciences, commenced the American Journal of Science; a work which has been continued, often at great personal and pecuniary sacrifices on the part of its editor, till it has now completed its 52d volume: and which has been a most efficient means of promoting, not only mineralogy and geology, but all the physical sciences in the United States. In this connection, the valuable Elementary Treatise on Mineralogy and Geology by Professor Cleveland, another of the early and distinguished cultivators of those sciences in this country, ought to be mentioned. It was published in 1816; and for a long time continued to be almost the only American work on these subjects of any importance. Since it was out of print, the able works of Professor Shepard and Mr. Dana on Mineralogy have taken its place.

Descr. In 1818, the American Geological Society was organized. But it has never accomplished much for the science; not having published any separate volume of transactions. The Academy of Natural Sciences at Philadelphia, the Lyceum of Natural History at New York, and the American Academy of Arts and Sciences at Boston, have from time to time given valuable papers in their transactions on mineralogy and geology; while several societies of more recent organization, but limited to particular districts of country, have been still more devoted to these subjects. The want of a general organization for the whole country, has long been felt to be a most powerful obstacle to the progress of Geology: and in April, 1840, those gentlemen who are engaged in the state surveys, met in Philadelphia, and formed themselves into the *Association of American Geologists*. Their second annual meeting was held in the same city on the first Monday of April, 1841: its third meeting in Boston, in 1842: its fourth in Albany, in 1843: its fifth in Washington, in 1844: its sixth in New Haven, in 1845: its seventh in New York, in 1846: and its eighth is appointed in Boston, in September, 1847. Its first volume of Transactions was published in 1843; and it has accumulated materials enough for two or three volumes more.

Descr. An important feature in the history of American geology, is the numerous geological surveys that have been executed, or are still in progress, under the patronage and direction of the State authorities, as well as the United States Government. The leading object of these surveys, is to develop those mineral resources of the country, that are of economical value. But with a commendable liberality, the Legislatures have encouraged accurate researches into the scientific geology, and sometimes also into the botany and zoology of the several states. So numerous have these surveys become, that I may not be able to give in all cases their entire history up to the present time; but since they constitute an important era in our geology, I shall give all the useful facts within my reach.

Descr. The first survey authorized by the government of a state, was that of North Carolina; which was committed to Professor Olmsted, who made a valuable report, in two parts, the first in 1824, and the second in 1825; in which the economical geology of a considerable part of the state was given in a pamphlet of 141 pages. In 1842, Professor E. Mitchell published a work on Geology, embracing an account of that of North Carolina, with a geological map of the state.

Descr. The year following, South Carolina gave a similar commission to Professor Vanuxem, whose report was published only in the newspapers.

Descr. The survey of Massachusetts was proposed by Gov. Lincoln in 1830, and commenced the same year by the author of this work. The first Report on the Economical Geology of the State was made in 1822, in a pamphlet of 70 pages. In 1833 a full report on the whole subject was made, in a volume of 702 pages, with an atlas of plates and a geological map: of which a second edition was directed to be printed in 1834. In 1837, on recommendation of Governor Everett, a farther examination of some parts of the geology of the state was ordered: and in

1838 a report of this re-examination, embracing only the economical geology, was made in a pamphlet of 139 pages. The final report was published in two quarto volumes of 300 and 544 pages, with 56 plates and 282 wood cuts, in 1841. The report of 1833, embraced also the zoology and botany of the state; and when the re-examination was ordered, these departments were committed to Professors Dewey and Emmons, Rev. W. O. B. Peabody, Drs. Harris, Gould, and Storer, and George B. Emerson, Esq., all of whom have made extended reports.

Descr. Tennessee was only two or three years later than Massachusetts in commencing a geological survey, under the superintendence of Professor Troost. In 1845 he had made eight annual reports, the three last of which were pamphlets of 32, 37, and 75 pages; with a geological map of the whole state in that of 1839.

Descr. In 1834, Professor Ducatel was directed to make a Reconnoissance of the state of Maryland; and his report the next year was a pamphlet of 38 pages. This survey embraces the topography; of which Professor Alexander has charge. Seven annual reports in addition to the reconnoissance have been made by Prof. Ducatel, of about 50 pages each, with maps and sections.

Descr. G. W. Featherstonhaugh, in 1834, received a commission from the government of the United States, to examine geologically "the territory of Arkansas and the adjacent public lands." His first report of 97 pages, with an extensive geological section, embracing 1600 miles, appeared in 1835, and his second of 161 pages, with four sections, in 1836.

Descr. The survey of New Jersey was ordered in 1835, and a report made by Professor Henry D. Rogers in 1836, in a pamphlet of 188 pages, with extensive sections; the survey having been ordered the preceding year. His final report of 361 pages was published in 1840, accompanied by a geological map of the state.

Descr. The survey of New York was commenced in 1836. It embraced zoology and botany as well as geology; and was put into the hands of four principal geologists, viz. Profs. Mather, Vanuxem, and Emmons, with Mr. Conrad, who were to employ each an assistant; one botanist, viz. Prof. Torrey; one zoologist, viz. Dr. J. E. De Kay; and one chemist, viz. Professor L. C. Beck. Subsequently Mr. Conrad was appointed palæontologist to the survey; and Mr. J. Hall took his place as geologist. These gentlemen made their first report in 1837, in a pamphlet of 212 pages, their second report in 1838, of 384 pages; their third report in 1839, of 351 pages; their fourth report in 1840, of 484 pages; and their fifth report in 1841, of 184 pages. The survey is now completed, and at least ten quarto volumes of the final report, abundantly and splendidly illustrated, have been published, viz. one volume each by Mather, Vanuxem, Emmons, and Hall, upon the geology; one volume by Beck, upon the mineralogy; one by Torrey, upon the botany; and four by De Kay, upon the zoology. Several other volumes, with a geological map, will be published before the completion of this noble work.

Descr. The first report, which was a reconnoissance of the geology of Virginia, was made by Prof. Wm. B. Rogers in 1835, of 36 quarto pages: his second report, in 1836, of 30 pages; his third in 1837, of 54 pages; his fourth in 1838, of 32 quarto pages; and his fifth in 1839, of 161 pages, and the sixth of 132 pages. Prof. Rogers is aided in this work by five assistants: two of whom are occupied in the chemical laboratory.

Descr. A survey was directed of the state of Maine in 1836, under the care of Dr. Charles T. Jackson. He made his first report in 1837, of 128 pages: his second in 1838 of 168 pages; and his third in 1839, of 340 pages. The work is now suspended, at least for a season: though it is hoped that it will not be finally abandoned. Dr. Jackson had also a commission from Massachusetts and Maine, to examine the unsettled lands owned by them jointly, and he has made two reports on this subject in 1837 and 1838.

Descr. The survey of Connecticut was commenced in 1835; and committed to Dr. J. G. Percival for the geology, and to Prof. C. U. Shepard for the mineralogy. The latter gentleman made a report, in 1837, of 188 pages. Dr. Percival reported in 1842, in an octavo volume of 495 pages, with a geological map.

Descr. The survey of Pennsylvania was begun in 1836, by Prof. Henry D. Rogers; who the same year made a report of 22 pages: and in 1837, another of 93 pages: also one in 1838, of 119 pages; another in 1839, of 252 pages; and a fifth in 1840, of 179 pages. Prof. Rogers has five assistants, one of whom is devoted to the laboratory, and four sub-assistants. The surveys of Pennsylvania and Virginia have been completed, but the final reports have not yet been published.

Descr. The first Geological Report on the state of Ohio, was submitted to the government in 1839 of 134 pages. This survey is under the direction of Prof. W. W. Mather, as principal geologist, aided by Dr. S. P. Hildreth, J. P. Kirtland, and

Locke, and Professor Briggs, J. W. Foster, Esq. and Col. J. W. Whittlesey. The latter gentleman has charge of the topographical department, and Drs. Kirtland and Locke of the zoology and botany. Their second report in 1838 contains 286 pages, with numerous drawings. This survey is for the present suspended.

Descr. In Delaware, the work was commenced in 1837, by James C. Booth, Esq., who has made two annual reports, in 1838 and 1839, in a pamphlet of 26 pages, and his final report of 188 pages, was published in 1842.

Descr. In Michigan, the survey was committed to Douglass Houghton, Esq., with two sub-assistants in geology, one assistant in zoology, one botanist and zoologist, and one topographer and draughtsman. In 1838, the first report appeared of 37 pages: in 1839, the second report of 123 pages; in 1840, the third report of 124 pages, and in 1841, the fourth of 184 pages.

Descr. The survey of Indiana was commenced in 1837, by Dr. D. D. Owen; who in 1838, produced his first report of 34 pages; and in 1839, his second report, of 54 pages.

Descr. In 1839, the legislature of Rhode Island commissioned Dr. Charles T. Jackson to make a geological and agricultural survey of that state, and his report of 322 pages with a geological map has been published.

In 1840, the same gentleman was appointed to survey New Hampshire; his first annual report of 164 pages was published in 1841, and his final report in 1845, in a quarto volume of 376 pages, with illustrations.

Descr. In Georgia, Mr. John R. Cotting was commissioned in 1836; and he informs me that half the state has been surveyed; that three section lines, from 300 to 430 miles long, have been explored; and that it is intended to publish a volume of 600 pages, shortly.

Descr. In 1838, the legislature of Kentucky directed a Geological Reconnoissance of the State to be made, preparatory to a more accurate survey. This was executed by Prof. W. W. Mather, who reported at the close of the year in a pamphlet of 40 pages. But I believe the work has not yet been prosecuted farther: although the developments made in Prof. Mather's report, of the mineral resources of the state, are such that there can hardly be a doubt but the survey will be continued, by a people as intelligent as those of Kentucky.

Descr. Professor W. M. Carpenter is now engaged (1842) by direction of the Legislature, in making an examination preliminary to a general geological survey of the State of Louisiana.

In 1839, Dr. D. D. Owen was commissioned by the government of the United States, to examine the territory of Iowa: or at least, its most promising mineral districts. The work was prosecuted by Mr. Owen with as many as 120 assistants; and a report of 161 pages was published in 1840.

Descr. In 1845, Professor Charles B. Adams was appointed by the governor to make a survey of Vermont. His first report of 92 pages, was published in the autumn of that year, his second in 1846 of 267 pages, and he is prosecuting the work with energy.

Results. Thus it appears, that within the last 25 years, surveys have been commenced in 22 states and two territories of this Union; embracing an area of nearly 700.000 square miles. For several years, not less than 65 geologists and assistant geologists have been constantly employed in this work. Geological maps of Massachusetts, Rhode Island, New Jersey, Tennessee, Connecticut, New York, and North Carolina, have already been published; and they are nearly prepared in many other states, so that we are now able to construct a complete geological map of the United States. The collections of specimens that have been made and will be deposited in the capitals of the different states and of the United States will be of great extent and value, and the vast number of chemical analyses that have been made of different substances will prove of the utmost value. *See my Anniversary Address before the Association of American Geologists in Philadelphia, in April,* 1841, *where the facts are given in greater detail.*

Descr. It ought not to be omitted in this connection, that at an earlier date than any of the state surveys were undertaken, the late Hon. Stephen Van Rensselaer, with great liberality and munificence, directed a geological survey to be made of the region bordering upon the Erie Canal; and that Mr. Amos Eaton, who was employed, made a report of 163 pages in 1824; with an extended section from the Atlantic to Lake Erie.

Descr. Within a few years past, the British provinces of Nova Scotia and New Brunswick have been geologically examined by Dr. Gessner, who has published reports of his labors.

Descr. In 1845, the government of Canada appointed Mr. Logan to make a geological survey of that province, and he is now engaged in that work. He has published at least one annual report.

Descr. As early as 1822, a geological survey of France was commenced by order of the Government, under the direction of MM. Brochant de Villiers, De Beaumont, and Dufrénoy. This work has been prosecuted with great ability for 20 years, and recently a large geological map of the whole kingdom has appeared. *Whewell's History of the Inductive Sciences, Vol. 3. p.* 534. *London,* 1837.

Descr. In his Anniversary Address before the London Geological Society in February, 1836, *Phil. Mag. Vol.* 8. 3*d Series, p.* 568, Mr. Lyell says, "early in the Spring of last year, an application was made by the Master General and Board of Ordnance, to Dr. Buckland and Mr. Sedgwick, as Professors of Geology in the Universities of Oxford and Cambridge; and to myself, as President of the Geological Society, to offer our opinion as to the expediency of combining a geological examination of the English counties with the geographical surveys now in progress." The plan was adopted; and H. L. De la Beche commissioned to make the Survey. In 1839 he made a report on the Geology of Cornwall, Devon, and West Somerset, in a volume of 684 pages, with 12 plates.

Inf. Those whose recollection enables them to compare the state of geological science 30 years ago with its present condition, and the almost universal interest now taken in it, with the almost entire absence of all interest or knowledge on the subject then, will hardly venture to predict what will be its condition 30 years hence.

SECTION XI.

GEOGRAPHICAL GEOLOGY.

Def. Under the term Geographical Geology, I embrace simply a summary description of the geology of the different countries of the globe, in a geographical order. The term might, also, properly include the geography of the globe at different geological periods: for it is easy to know what was the geography of the world at these successive periods; and a map of it might be made answering to those times. Thus, during the tertiary period, all those portions of the globe now colored as such, on a geological map, might be represented as water; and we should see what was then dry land. Convert all the secondary into water, and what remains would show the character of our continents during that period; and so of the older fossiliferous formations. This subject, however, cannot be treated fully in this work. *See Lyell's Principles of Geo-*

logy, and Boue's Memoires Geologiques, &c., Tome Premier Paris, 1832.

Descr. M. Boue has lately published a geological map of the whole globe under the auspices of the Geological Society of France; on which the great classes of rocks are colored over a large part of the earth's surface: not all, however, as the result of actual observation; but partly from theoretical considerations. This map, somewhat modified by later observations, together with a geological map of North America, I had intended to annex to this edition of the present work. But this would have so increased the price of it, that it has been thought best to publish those maps in a small separate work, as a companion to the present one. In that work I shall exhibit a brief view of the geology of the globe, so far as it is known: and therefore, I have not devoted so much space to geographical geology in this section; since I trust all who study the elementary geology, will possess themselves of the smaller work, which will be published soon. A few glances at those maps, will give more knowledge of the geology of the globe, than the fullest detailed descriptions.

EUROPE.

England.

Rem. The rocks of no country on the globe have been studied so carefully and successfully as those of England. Indeed, the descriptions of rocks in general, given in this treatise, is based upon those of Great Britain: and, therefore, little more will here be needed than to describe briefly their distribution in that country.

Descr. The primary rocks are confined chiefly to the west or mountainous part of England, viz. Wales, Cumberland, Cornwall, and Devon. In proceeding easterly, the country becomes hilly and then level, and the rocks generally become newer and newer, until we reach the alluvium of the eastern coast. The unstratified rocks are mostly confined to the western districts, as are the Cambrian and the Silurian Groups. The estuary of the Thames contains the principal deposit of the tertiary strata.

Descr. Probably coal is the most important mineral found in England; and that country is remarkably well supplied with it. Conybeare and Phillips describe five great coal districts: 1. That north of the Trent: 2. Central coal district: 3. Western coal district: 4. Middle western, or Shropshire: 5. South western. Some statements, however, have already been made in respect to the coal deposits of England in Section III. The annual amount of coal dug and consumed in England and Ireland is about 15.500.000 tons.

Descr. Extensive deposits of iron often accompany the coal of England; as well as the limestone necessary for smelting the ore; and the firestones for the furnaces; while the coal supplies the fuel. In 1839, the amount of iron manufactured from ore in England, was 1.312.000 tons. *Mining Review, Dec.* 1839, *p.* 182.

Descr. England contains several mining districts where other metals are found; the chief of which are, copper, tin, lead and antimony. In Cornwall and Devon, the mines of these metals are in primary or transition rocks. Rich lead mines with copper and zinc, are also found in the mountain limestone of England and Wales. The amount of oxide of tin raised in 1839, was 4868 tons. *Mining Review for Feby.* 1840, *p.* 14. The amount of lead in 1835, was 46.112 tons, and the amount of copper in 1837 was 11.209 tons.

Descr. In the new red sandstone at Northwich, is a valuable deposit of rock salt, the two beds being 60 feet in thickness and a mile and a half long. The salt springs at Droitwich produce 700.000 bushels of salt yearly, and the whole amount in all England, is 15.000.000 bushels. In the London clay, the septaria produces the valuable Roman Cement; which is carried all over the world. Many other valuable minerals exist in the rocks of England, which cannot be here specified.

Scotland.

Descr. Scotland is much more mountainous than England, and is composed generally of primary and older secondary rocks. All the varieties of the former occur here, and are fully developed. They consist of granite, syenite, porphyry, trap and serpentine; which are unstratified, and of gneiss, mica slate, quartz rock, and clay slate, which are stratified. The secondary rocks are graywacke, old and new red sandstone, mountain limestone, coal formation, lias and oolite, which are covered with diluvium and alluvium.

Descr. The amount of iron smelted in Scotland in 1839, was 200.000 tons. Copper, lead and silver, are also obtained there. The annual value of lead is about 603.000, and of the silver, $44.400. Several important coal fields exists, which furnish a large supply. The Mid Lothian coal fields produce annually 390.000 tons; and they are calculated to contain 2250 millions of tons: which would supply the whole of Great Britain, which annually consumes 30 millions of tons, for 75 years. *Mining Review, Oct.* 1839, *p.* 149. The Brora coal is said to be in oolite: but its quality is poor.

Descr. The Western Islands of Scotland are remarkable for unstratified rocks, which produce much wild and interesting scenery. *See Macculloch's Western Islands, in three Volumes. London,* 1810.

Ireland.

Descr. All the primary and older secondary rocks both stratified and unstratified, occur in Ireland on an extensive scale. Chalk also is found beneath the trap. Perhaps the most striking feature of Irish Geology is the vast deposit of trap in the northern part of the island. It occupies an area of 800 square miles, and is 545 feet in average thickness. Where it extends to the coast, it exhibits the famous columns, which constitute the Giant's Causeway and Fingal's Cave: the latter being upon the island of Staffa. The chalk on which this trap rests, is often changed into granular limestone. Beneath the chalk lie green sand, lias, and variegated marls.

Descr. Ireland contains several coal fields. Nearly all that in the province of Munster is anthracite; which is used to some extent for burning lime. The other basins produce bituminous coal. The peat bogs in that country occupy, it is said, more than one tenth of the island, and are from 15 to 25 feet thick. Iron is also wrought to some extent; as are copper and lead. The single mine of Allihies produces more than 2000 tons of copper annually; and those of Coonbane and Tigrony, in 1811, produced 1046 tons. Native gold also occurs in considerable quantity.

Denmark.

Descr. This low and flat country is composed almost entirely of the deposits of weald clay, and chalk, covered by tertiary and drift; and these in their turn often by alluvium.

Descr. Iceland and the Faroe Islands, which are under the jurisdiction of Denmark, are composed almost entirely of ignigenous rocks of two eras: the first, that of dolerite or greenstone; and the second, that of modern volcanos. Some of the most terrible eruptions of modern times have occurred in Iceland. The amygdaloidal rocks of these islands are celebrated for the splendid zeolites which they afford.

Sweden, Norway, and Lapland.

Descr. These extensive countries form the ancient Scandinavia. It is rough and mountainous, and the rocks are mostly primary, with the older secondary; though chalk is said to occur; and in Sweden some recent tértiary strata. The mountains are mostly gneiss with mica and talcose slates, and in Norway are extensive deposits of porphyry and syenite. But the diluvium or bowlder formation that is spread over this country, is perhaps the most remarkable in the whole series. This, however, has been already described in a preceding section. I would only remark here, that diluvial phenomena in Scandinavia appear to resemble those in New England more nearly than in any European country. And there is a good deal of resemblance in the rocks and minerals; especially between those of Norway and New England.

Descr. The mines of Sweden have long been well known. Those of gold and silver yield much less now than formerly. Those of Kongsberg have been considered the richest in Europe. A mass of native silver, was once found here weighing 600 pounds. The copper mines are very prolific: that of Fahlun alone yielding annually 510 tons. Cobalt is also wrought in several places. From the sulphurets of iron and copper, sulphate of iron or copperas is manufactured in large quantity. In Sweden and Norway, 120.000 tons of iron were manufactured in 1839. The immense beds of iron are usually connected with the gneiss. Extensive quarries of granite, marble, and porphyry, the latter the finest in Europe, are also worked.

Holland and Belgium.

Descr. These countries are proverbial for the flatness of their almost entire surface; although some parts are hilly. But much of the land lies even below the level of the sea, and is defended by dykes. As might be expected, no primary rocks occur: nor any unstratified rocks. Clay slate is the oldest; and most of these secondary strata are superimposed upon this, with one or two tertiary basins, and a coat of diluvium.

Descr. Both anthracite and bituminous coal are found in Belgium; the latter in great abundance. In 1837 her 250 coal mines yielded 1.600.000 tons: and the iron mines in 1839, 80.000. *Mining Review for July*, 1838, *p.* 109*; and for January*, 1840, *p.* 6.

France.

Descr. This extensive country, embracing an area of more

than 200.000 square miles, is generally level: though in its central parts mountains rise 5000 or 6000 feet; and the highest mountains of Europe, the Alps, are on its borders. Nearly all the stratified and unstratified rocks, from the oldest to the most recent, are found here. The carboniferous strata are far less extensive than in England: but the new red sandstone, and the jura or oolitic limestones form extensive tracts. Chalk also occurs in vast abundance: and there are not less than six extensive basins of tertiary deposits. But perhaps the most remarkable of the formations is the district of extinct volcanos in Auvergne; which has long been classic ground for the geologist. The rocks are trachyte, basalt, and tufa.

Descr. In 1837, the one hundred and ninety-eight coal mines of France produced 1.150.000 tons of coal; and her iron mines, 600.000 tons of iron in 1839. *Mining Review, July,* 1838, *and January,* 1840. Her lead mines yield annually about 1525 tons, and her silver mines about 2800 pounds troy. Besides these, there are mines of copper, antimony, manganese, alum, and vitriol. Cobalt, arsenic, nickel, and tin, also occur in small quantity. Quarries of marble, slate, gypsum, flint, and buhrstone, are likewise extensively wrought.

Spain.

Descr. The surface of Spain is strikingly irregular: some of the mountains, as the Pyrenees, rising more than 11.000 feet above the ocean. The central axis of these mountains is usually primary, or older secondary rocks; both stratified and unstratified. The middle secondary strata, however, sometimes rise to a great elevation, as in the Pyrenees. The more recent secondary, even to the chalk, as well as tertiary deposits, occur. Around Olot, in Catalonia, is a region of extinct volcanos, occupying 20 square leagues, according to Maclure, who first pointed it out. *Lyell's Elements of Geology, Vol.* 2. *p.* 265.

Descr. Lead occurs in great quantity in Spain; one district yielded in 1828 no less than 30.000 tons. This deposit is in transition limestone. There are also rich quicksilver mines in clay slate. An extensive deposit of rock salt exists at Cardona, in the cretaceous rocks. The iron smelted in 1839, was only 18.000 tons, although the ores are said to be very abundant. Extensive deposits of coal are found in the province of Asturias. The sands of the river in this country have long been known as yielding gold.

Portugal.

Descr. The geological structure of Portugal is similar to that of Spain: the older stratified and unstratified rocks constituting the mountains. Gold occurs in the sands of a few of the rivers, and is still sought, but with small success. There are also mines of coal, of graphite, of iron, and of quicksilver; and formerly tin was explored.

Italy.

Descr. Italy contains the loftiest mountains in Europe, viz. a part of the Alps, including Mount Blanc, St. Bernard, and the Apennine chain. Between these mountains are most beautiful plains. The Apennines are made up in a good measure of limestone, belonging to the newest of the primary, or the oldest of the secondary rocks: though in some parts are slate, euphotide, and gabbro. The sub-Apennine hills, at the base of the Apennines, are composed of tertiary strata, often containing species of shells such as now exist in the Mediterranean. And there is evidence that the Apennine chain has been elevated 3000 or 4000 feet since the deposition of these tertiary strata. In Calabria, granite, gneiss, and mica slate exist. Primary rocks occur also in the Alps and on the island of Sicily: though most of the rocks on this island are secondary and tertiary. Most of the English newer secondary rocks occur here. In this island, also, is volcanic Etna, nearly 10.000 feet high. On the main land, Vesuvius is the only acting volcano, nearly 4000 feet in height: though near Naples is a solfatara, which continually gives off gases and vapor. Brocchi estimates the number of craters of extinct volcanos in the vicinity of Naples, at 27. The Lipari Islands, as also Procida and Ischia, are all of volcanic origin; and among the former is the perpetually active volcano, Stromboli. In Corsica and Sardinia, the prevailing rocks are primary and the older secondary. Jura limestone, however, is found there; and in Sardinia, are tertiary strata covered by drift with volcanic rocks. Malta and Goro are chiefly composed of soft limestone. Along the shores of the Mediterranean are deposits of alluvium: and in the fissures and caverns of the rocks, is a calcareous deposit, called bone breccia, made up of the fragments of the bones of extinct animals, cemented by carbonate of lime.

Descr. Italy has long been celebrated for its marbles, such as the Carara, the Pentelic, Lumachella, &c. Coal occurs there, but is not abundant. In the province of Coserza, are extensive beds of rock salt. Silver, lead, copper, and iron, have been wrought: the latter especially. In 1839, Italy, with the islands of Elba and Sardinia, produced 50.000 tons of iron. The iron mines of Elba have been wrought from the earliest times. Sicily has long supplied nearly all Europe with sulphur, dug from the tertiary clays.

Switzerland.

Descr. The geology of this country of mountains has been explored by many of the ablest geologists of Europe, commencing with Saussure. And although its structure is complicated, it has afforded many fine illustrations of geological principles. The central parts of the Alps are primary: or gneiss, mica slate, talcose slate, hornblende slate, and limestone. In the gneiss, talc often replaces the mica; and the rock is then called protogine: of which Mount Blanc is formed. On the flanks of the mountains are deposits of new red sandstone, lias, oolite, green sand, chalk, and tertiary strata; over which are accumulations of diluvium. The tertiary rocks are from 2000 to 4000 feet above the ocean, and mark the height to which these mountains have been raised since those strata were deposited. All the older unstratified rocks, as granite, syenite, porphyry, and greenstone, are found in the Alps.

Germany.

Descr. The south western, and eastern borders of Germany, consist of lofty mountains of primitive rocks: which occur also in several places more central, as in the Hartz mountains. Secondary rocks occur in many places; though relatively less abundant than in Great Britain. Yet nearly all the fossiliferous rocks of Great Britain are found in Germany, and in the same relative position. A part of the extensive plains of northern Germany is composed of tertiary rocks, covered with diluvial detritus from Scandinavia. As many as four other tertiary basins occur in Germany. Indeed, nearly every stratified rock that has been found in any part of the world exists in Germany.

Descr. Several coal deposits are found in this country, which annually produce about 1.000.000 tons of coal. Salt and gypsum, are also found in the new red sandstone. and of the former the annual produce is 157.500 tons. The amount of iron manufactured in 1839, in Germany, was about 300 000 tons: of which Austria produced 100.000; Prussia the same and the Hartz mountains, 70.000. The annual produce of the gold mines of Germany, is 182 marks: of silver, 123.000 marks: of copper, 1950 tons: of lead, 9560 tons: of tin, 400 tons: of mercury and cinnabar, 799 tons: of cobalt, 825 tons: of calamine, 4140 tons: of arsenic, 530 tons: of bismuth, 75 tons: of antimony, 120 tons: of manganese, 90 tons. Indeed, this country is the most remarkable one in Europe for mining operations, and for the scientific skill with which they are conducted.

Hungary with Transylvania, Sclavonia, Croatia, and the Bannat.

Descr. This extensive region has the Carpathian mountains on its north and eastern border, and the Julian Alps and other mountains on the south. The country in the interior is in general hilly, except several plains of great extent. Primary rocks occur in several places; but they are not relatively very abundant. Nearly all the secondary and tertiary stratified deposits hitherto described, occur here; and above them diluvium and alluvium. The common unstratified rocks are also found here; and in addition we have trachyte in abundance, with trachytic porphyries, tufas, and conglomerates; which are of more recent data than the tertiary strata: and one solfatara at least exists in the trachytic regions of Transylvania.

Descr. Hungary produces more gold and silver than any country in Europe. Hassel states the annual amount of the former to be 1050 pounds; and of the latter 41.600 pounds. But this is greater than the quantity now obtained. The iron annually smelted is about 10.000 tons: of copper, 19.000 tons: of lead, 1225 tons: and large quantities of coal and salt are obtained.

Poland and Russia in Europe.

Descr. Poland and Russia are for the most part a vast plain, bounded on the east by the Uralian mountains, and on the south by the Carpathian and Silesian chains. These mountains are chiefly primary and older secondary; as are also Finland and Russian Lapland. But the tertiary and alluvial strata predominate in most parts of the country. Nearly all the stratified secondary rocks occur; such as old and new red sandstone with the coal formation intervening; lias and other limestones, and green sand and chalk. The tertiary strata correspond rather to those in Hungary and France, than to those of England and Italy. Over the tertiary is a deposit of clay, sand, and bowlders, from 30 to 100 feet thick, which has been brought from the north; and this deposit contains the bones of many extinct terrestrial animals. Unstratified rocks are uncommon: but trachyte occurs in the Caucasian chain.

Descr. In Poland occurs in the tertiary strata, the most extensive deposit of salt in Europe. Several other deposits of this mineral are found in the new red sandstone with gypsum, in Russia. Some coal mines have been opened: but they are not extensively wrought. The iron mines of Russia and Poland yielded in 1839, about 158.000 tons. Copper is obtained also, on the western side of the Uralian mountains.

Greece and Turkey in Europe.

Descr. The geology of these countries is very similar to that of Hungary, but it has not been minutely described.

Descr. The following Table gives a more recent view than the numbers already presented, of the annual produce and value of the mining operations in Europe; so far as metals are concerned. *Ansted's Geology, Vol. 2. p. 303.*

	Tin.	Copper.	Mercury.	Zinc.	Lead.	Iron.	Silver.	Gold.	Value in Pounds Sterling
	Tons.	Tons.	Tons.	Tons.	Tons.	Tons.	Pounds.	Pounds.	
British Isles,	3600	12.000		2000	46.000	1.350.000	6000		17.600.000
Russia and Poland,		3300		4000	600	300.000	38.500	12.000	5.400.000
France,		80			400	430.000	3314		5.280.000
Austria,	20	3500	250	75	4500	70.500	42.500	3250	2.680.000
German Confederacy,	300	2000	650		8000	70.000	5350	60	2.580.000
Spain,		25	1750	85	20.500	15.000			2.160.000
Sweden and Norway,	65	1240		300	40	85.000	10.350	4	2.160.000
Prussia,		540		500	6000	70.000	10.000		1.196.000
Belgium,				1750	325	162.000	350		1.160.000
Tuscany, Elba, &c.						24.000			600.000
Piedmont, Switzerland, Savoy, &c.,					350	22.750	1250	13	440.000
Denmark,		700				12.000			360.000
Total,	4165	21.385	2650	8710	86.765	2.611.250	111.614	15.327	41.616.000

ASIA.

Turkey in Asia and Arabia.

Descr. Although many insulated facts are known respecting the rocks of these extensive and mountainous regions, ye no connected view of their geology can be made out. It is certain, however, that rocks of every age, stratified and unstratified, are found there. A few statements of interest can be made respecting the geology of Syria, Palestine, and Arabia.

Descr. In Syria and Palestine, limestone, probably of the newer secondary class, is the most abundant rock. It forms the greater part of the mountains of Lebanon, Anti-Lebanon,

Carmel, the mountains of Galilee, and the ridges stretching south from the Dead Sea. Much of it is yellowish white and compact, not a little resembling the stone used in lithography. At least this is the character of the limestone on which Jerusalem is built; and of which the ancient temple of Solomon and the houses of the city were constructed; and it was the material from which the tesselated pavement of the streets was formed.*

Descr. Mount Lebanon has been recently described by M. Botta, as consisting of three groups of calcareous rocks. The upper group consists of limestone of variable hardness, alternating with marls: the middle group embraces siliceous beds and nodules, with fossil echini and fishes; and the lowest group is mostly sandstone, with beds of silico-calcareous matter, iron ore and lignite. The whole formation he refers to the chalk. *Traite Elementaire de Geologie, Par M. Rozet, p.* 524.

Descr. Mount Hor and Wady Mousa, in the ancient Edom, are composed of variegated or new red sandstone; and the excavations of Petra, the ancient capital of Edom, are in this rock; which extends to Mount Sinai, and still farther south. On the borders of the Dead Sea marly strata occur, and a remarkable ridge of rock salt, from 100 to 150 feet high, called Usdum. *Robinson and Smith's Biblical Researches in Palestine, &c. Vol.* 2. *p.* 482.

Descr. Some of the highest peaks around the Red Sea, are said to be composed of granite rock. Mount Sinai is wholly granite, or perhaps syenite: as are also the mountains on each side of the Arabian Gulf. Porphyry and greenstone exist also in the same mountains. Trap rocks occur at Akaba, the eastern extremity of the Red Sea: and according to Burckhardt, ancient volcanic craters may be seen in the same region. According to Von Boch, the valley of the Jordan, from Mount Libanus to the Red Sea, is a fissure, through which volcanic agency has been active, and the character of the Dead Sea, as well as the thermal springs on its margin, the existence of volcanic rocks in the same region, the rock salt and great amount of bitumen, and the columnar and amygdaloidal rocks existing near the Jordan, render it highly probable that this igneous agency has been exerted at some former period. *American Biblical Repository, Vol.* 3, *Second Series, pp.* 24, 324.

Descr. Barometrical observations had made it almost certain that the Dead Sea is sunk many hundred feet below the level

* A recent analysis of this limestone gave me in 100 parts,—

Earthy Residuum, ...	1.00
Carbonate of Magnesia, . . .	0.83
Carbonate of Lime, ...	98.17
	100.00

of the Mediterranean: and very recently (1841) the point has been settled by British officers, who have surveyed the region between the two seas trigonometrically. They find the Dead Sea to be 1337 feet below the Mediterranean.

Rem. I have received an extensive series of specimens illustrating the geology of Syria and Palestine, especially that of Mount Lebanon, from Rev. Story Hebard, American missionary in Syria, who had paid a good deal of attention to geology before leaving this country. They confirm the preceding statements essentially. The limestones of Mount Lebanon are often pulverulent, and even chalky; though some of them are compact. They often contain siliceous nodules; sometimes as large as a man's head, which, on breaking open, present splendid geodes of crystals of quartz, or sometimes of mammillary chalcedony. Solid nodules of hornstone, passing into chalcedony, are also common. The limestone contains fine petrifactions of one or two species of Tellina, and Venus, a Terebra, a Trochus, a Dolium, an Ostrea, an Echinoderm, and fishes on a white marl slate somewhat like that of Monte Bolca, and with homocercal tails. The sandstones are highly ferruginous, and resemble those in this country that are referred to the lower part of the cretaceous group. In this sandstone occur beds of asphaltum, especially at Carmel on Mount Lebanon, and at Hermon on Anti-Libanus, at the former place in ferruginous sandstone. Lignite is also found on Lebanon. Near the tops of this mountain occurs a brittle bituminous shale, and on the borders of the Dead Sea and the Sea of Tiberias, a black highly bituminous limestone. Genuine greenstone is found with the limestone on the east side of Anti-Libanus; and a vesicular basalt with olivine. The same rock exists at Hauran, southeast of Damascus. Genuine basaltic vesicular lava occurs in an ancient crater five miles west of Safet; and a still more porous variety on the north shore of the Dead Sea. The ruins of Jericho furnish a similar rock. (*See the Specimens in the Cabinet of Amherst College.*)

Descr. "The compact limestone rocks which bound the Nile in the whole of Upper Egypt," says Mr. Weaver in his *Observations on Ehrenberg's Discoveries in Mag. Nat. Hist. for June*, 1841, *p.* 305, "and extend far into the Sahara, or Desert, as well as the West Asiatic compact limestones in the north of Arabia, are in the mass composed of the coral animalcules (Polythalamia of Ehrenberg, or the Foraminifera of D'Orbigny) of the European chalk. This affords a new insight into the ancient history of the formation of Lybia, from Syene to the Atlas, and of Arabia, from Sinai to Lebanon: thus opening a large field to organic distribution."

Rem. Professor Bailey, of West Point, having seen the above paragraph, requested me (March, 1842) to send him specimens from my collection above referred to, of limestones from Syria and Palestine. Having complied with the request, Professor Bailey has kindly communicated to me the following results, which confirm in an interesting manner the statements of Ehrenberg. The localities mentioned by Ehrenberg are "Hamam Favaün and Tor in the Sinaian portion of Arabia," "constituting hilly masses in Upper Egypt," and "continued eastward into the Great Desert plain tending towards Palestine." The specimens examined by Prof. Bailey are mostly from places much farther north.

No. 1. From the interior of the Pyramid of Cheops: Polythalamia distinctly present, but rare.

No. 2. West side of Anti-Libanus: Polythalamia abundant. Fig. 122.

No. 30. Anti-Libanus: clay colored limestone: Polythalamia abundant.
No. 3. Near Damascus: Polythalamia abundant. Fig. 121.
No. 7. Mount of Olives: Polythalamia abundant. Fig. 122.
No. 8. Beyroot: Polythalamia abundant. Fig. 122.

Prof. Bailey has added sketches of Polythalamia from the English chalk, and from a cretaceous formation at a mission station on the Upper Mississippi; Fig. 121. The scale shown on Fig. 122 is the same for all. It is $\frac{25}{100}$ths of a millimetre, magnified equally with the sketches.

Inf. It is impossible to compare the sketches on Figs. 121, 122, without being convinced that they all were derived from the same formation; that is, from the chalk formation. The evidence would hardly be more satisfactory if the organic remains were those of mastodons or megatheroids, having as close a resemblance as is shown in these figures. What an interesting example of the constancy of nature in her minutest works, and of the triumph of science over difficulties apparently insuperable!

Fig. 121.

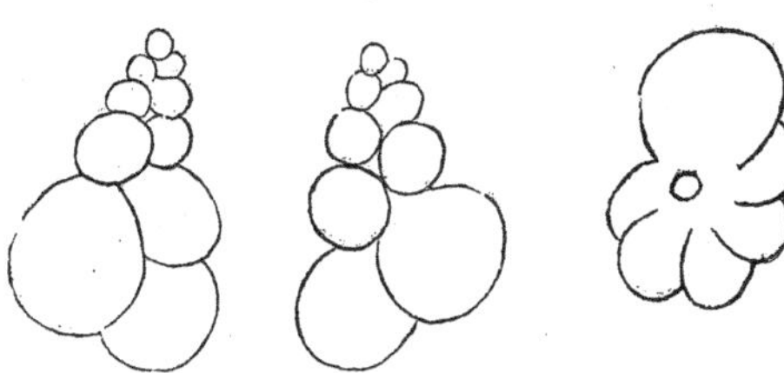

From Damascus.

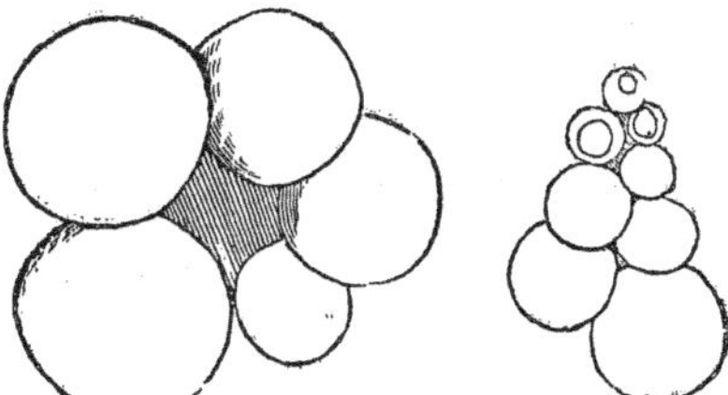

From English Chalk.

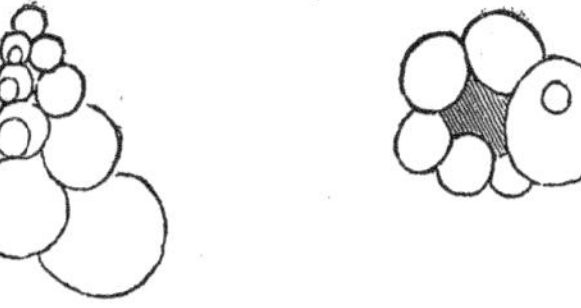

From a Missionary Station on the Upper Mississippi.

Fig. 122.

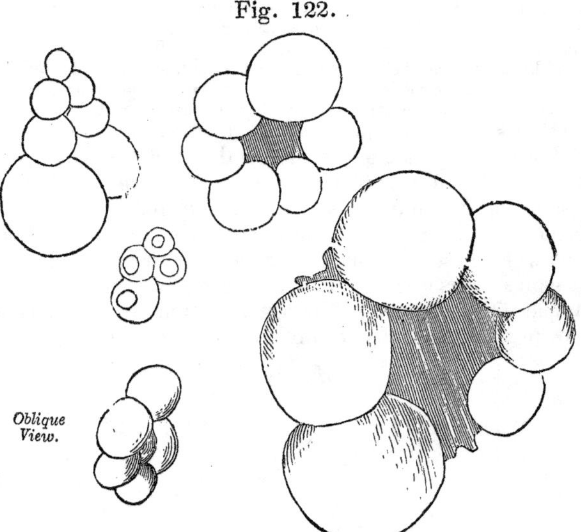

From the Mount of Olives.

From West Side of Anti-Libanus.

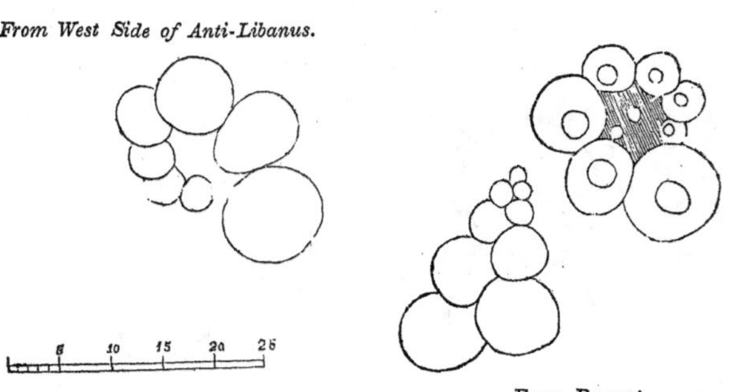

From Beyroot.

Descr. In the western part of Asia Minor to the east of Smyrna, and in the vicinity of the ancient Sardis and Philadelphia, is a region interesting to the geologist, called the *burnt country*, from the striking marks it bears of former volcanic action. Here are seen numerous streams and beds of lava, and 30 distinct volcanos, whose craters are covered by soil and vegetation. The deep fissures cut through the lava by running water, show that the period of action in these volcanos must have been very remote. Three of them, however, are much more recent: yet even these have been quiet for at least 3000 years. *Dr. Smith's Scripture and Geology, 2d Ed. p.* 149.

Persia, including Cabul, Affghanistan, Baloochistan, &c., as far as the River Indus.

Descr. The little that is known of the geology of this vast region, consists of a few insulated facts. On the north, commences the largest inland sea in the world, viz. the Caspian; which is 600 miles long, and 300 broad, and salt like the ocean. In the northwest part of Persia is the lake Ooroomiah, 300 miles in circumference, whose waters hold in solution one fifth of their weight in salt, and contain so much sulphuretted hydrogen as to tarnish silver rapidly, after having been brought to this country.* The sea of Aral, in Tartary, east of the Caspian, is also salt. These facts, particularly those in respect to Ooroomiah, indicate volcanic agency in those regions; and the remarkable springs and deposition of asphaltum, in the region of ancient Assyria, as well as the sulphur in the soil, lead to a similar conclusion. An extinct volcanic crater is said to exist in the vicinity of Mount Ararat; and that mountain itself is made up of ancient lava. A crater also exists in the Kourdish mountains. Near Tiflis, in Georgia, are hot springs; and a boiling spring on the river Akhoor. In the Caucasian mountains are extensive groups of basaltic columns. Near lake Ooroomiah occurs the beautiful white or yellowish calcareous alabaster of Tabreez, which is deposited by thermal waters in great abundance. Near the ruins of the city of Tact-i-Solomon is another enormous deposit of calc sinter. Copper, lead, and silver are mined near Tokat, where 50 furnaces are in operation. Iron, copper, and silver are also wrought near Samsoon. *Sir Robert Ker Porter's Travels from* 1817 *to* 1820. *See Am. Journal Science, Vol.* 37. *p.* 347.

Descr. From specimens and their labels sent by Rev. Justin Perkins and J. L. Merrick, American missionaries in Persia, I learn the following facts.

---

* I have recently analyzed this water, and found 500 grains of it to contain—

Of solid matter, ... 102.1 grs.

which was composed of

Chloride of Calcium, . .	0.74 grs.
Chloride of Magnesium, .	5.76
Chloride of Sodium, . . .	90.58
Sulphate of Soda, . . .	5.67
	102.75

Excess above the solid contents 65

Bromine was detected, but its quantity not ascertained.

1. That on the vast plain northwest of Mount Ararat, are moraines like those in this country. 2. That near Ooroomiah are extensive deposits of very fine gypseous alabaster. 3. That gray sandstone and conglomerate with gypsum are abundant in the same region. 4. That quartz rock exists there, as well as in some of the islands of the lake, and in several other parts of Persia. 5. That limestone is abundant around Ooroomiah. 6. That red sandstone and conglomerate (probably the new red sandstone,) constitute a mountain just back of Tabreez, some part of which appears as if it had been exposed to very powerful heat; and that similar rocks occur in other parts of Persia. 7. That beautiful rock salt is dug from a mountain near Tabreez, as well as on the plain of Khoy with gypsum, and at various places along the river Aras in Georgia. 8. That rich ore of copper, composed of native copper, and the green and blue carbonates, exists in large quantity 30 miles northeast of Tabreez, from which the Persian government have manufactured cannon. 9. That the marble in the ruins of Persepolis resembles lias.

Tartary.

Descr. The interior portion of Asia is crossed by four grand systems of mountains from east to west; viz. on the south, the Himmalayah, next northerly the Kuenlun, then the Thian-chan, and the Altaian. The first is in India; the second separates Tartary from Thibet, and the last from Siberia; while the Thian-chan occupies its central parts. The Altaian chain presents its magnificent displays of primary and transition rocks; over which are secondary and tertiary deposits. In these rocks are numerous metallic deposits. The quantity of gold annually derived from thence, is 1140 pounds avoirdupois, and of silver, 41.992 pounds.

Descr. Active and extinct volcanos occur in central Asia; occupying a space of 2500 square leagues. The most remarkable volcanic mountains are Pechan, Houtcheon, Ouroumptsi, Kobok, and Aral-toube. These are between 900 and 1200 miles from the sea.

Siberia.

Descr. The geology of the Uralian mountains, which separate Siberia from Europe, is similar to the Altaian range in its general features.

Descr. The Uralian mountains, especially on the Siberian side, have long been celebrated for their mineral treasures. In 1828, about 53 tons of gold were thence obtained. As long ago as 1782, this mountain yielded 66.000 tons of iron, and 31.500 tons of copper. The Altaian chain of mountains, also, yields an immense quantity of silver, as well as much gold and copper. In 22 years, 12.348 pounds of gold, and 324.000 pounds of silver, were dug from the single mountain of Schlangenberg. In 1828, the entire produce of silver in Siberia was 182 tons. In the Uralian mountains, within a few years, platinum has been found in such quantity that it is now used as a coin in Russia.

Descr. In 1843 the eastern parts of Siberia yielded £2.250.000 of gold, and the total produce of the whole Russian Empire is nearly £3.000.000. There is said to be a tract in Siberia, larger than France, every part of which is more or less auriferous; and even the subjacent rocks, when pounded up, yield a certain percentage of gold. That country will prove a second Brazil. *Murchison's Geol. Russia, Vol.* 1. *p.* 648, *Lond.* 1845.

Descr. Other interesting minerals are found in Siberia: among which are the diamond, topaz, emerald, beryl, onyx, lapis lazuli, beautiful crystals of quartz, rubellite, aventurine, carnelian, chalcedony, agates, &c.

Hindostan, Thibet, and Further India.

Descr. The two first of these countries are separated by the Himmalayah mountains, the highest on the globe, and their geology becomes, therefore, of great interest. On approaching this chain from the plains on the south, we first meet with a sandstone belonging to the newer secondary series; next in succession argillaceous slate, mica slate, talcose slate, quartz rock, hornblende slate, and limestone. The highest part of the range is composed chiefly of gneiss, traversed by granite. The mica slate is traversed by porphyry. At the base of these mountains, also, occur tertiary strata, in which the bones of the mastodon have been found in Burmah; and between the river Sutlej and the Ganges, bones of the elephant, mastodon, hippopotamus, rhinoceros, elk, horse, deer, crocodiles, gavials, sivatherium, and the monkey. Diluvial deposits are said also to occur in these mountains. In Middle India, the vast plains are composed mostly of clays, sand, and gravel, with organic remains of animals and fossil wood. Good coal also occurs here in a formation resting on granite. Peninsular India is composed in a great measure of unstratified rocks; such as granite, syenite, and trap: though nearly all the stratified primary ones are present also. Extensive deposits of secondary and tertiary strata exist there likewise; as well as diluvium and alluvium. The geology of Further India is but little known.

Descr. Southeastern Asia has long been known as a region of gems. The diamond occurs in several places in Hindostan, both in conglomerate and the alluvium. Corundum, from its coarsest to its finest state, is found there. Pegu is well known to produce the most beautiful variety of the oriental ruby, which is a sapphire. Topaz, zircon, tourmaline, garnets, carnelian, jasper, agates, amethyst, catseye, chrysolite, &c. are common. The annual value of carnelian formerly exported from India, was **$50.000.** Burmah is celebrated for the vast amount of petroleum collected in wells: which amounts to 95.781 tons per annum. India also contains gold, tin, iron, lead, and zinc: though not enough of these metals is obtained for home consumption. Rock salt is also found. From Thibet is brought a large quantity of borax. This is obtained from a làke, which contains also common salt. The Thibetans are said to work mines of copper and mercury.

Ceylon.

Descr. The geology of this large island, lying near the coast of southern India, is quite simple. It is made up almost entirely of primary rock, most of it stratified. The prevailing rock is gneiss: which is associated with dolomite, and subordinate masses of quartz rock. Granite, syenite, and greenstone, also exist there.

Descr. Iron is almost the only valuable metal found in Ceylon. But the gems are numerous. Amethyst, rose quartz, catseye, prase, topaz, schorl, garnet, pyrope, cinnamon stone, zircon, spinelle, sapphire, and corundum, have been found; most of them in connection with graphite in gneiss. *Transactions of the Geological Society, Vol.* 5. *Part* 2. *p.* 311.

China.

Descr. Little is known of the geology of China. We only know that granite, syenite, porphyry, sandstone, &c., occur there. But its minerals are more known. Gold and silver, mercury and copper, lead, tin, and arsenic, are among those which are wrought. Rubies, corundum, topaz, lapis lazuli, jasper, agate, jade, porcelain clay, and marble, are enumerated among its valuable minerals.

Japan.

Descr. It is known that these islands contain several volcanos, but their other geological features are little understood. Gold and silver abound, and copper and mercury occur. Sulphur exists in great quantities: and coal is found, as well as amber, porcelain clay, marble, &c.

East Indian Archipelago.

Descr. In Sumatra are four volcanos, nearly 12.000 feet high: while granite, trap, limestone, and other primitive rocks exist there, as well as tertiary clays. In Java, are several ranges of volcanic mountains; and at their bases are deposits, perhaps tertiary, of limestone, clay, and marl, with rock salt. Banca is said to be composed of gneiss and mica slate with granite. In Borneo, primary formations are abundant, forming the axes of the principal mountain chains; while secondary, tertiary, and alluvial deposits cccupy the lower regions. Volcanos also exist here. The Philippine islands are likewise volcanic: so are also the Moluccas: of which Celebes contains primary rocks.

Descr. The mineral treasures of this Archipelago are abundant. Gold occurs in nearly all the islands: and the amount annually collected, is estimated at 2.922.300. Tin is abundant especially in the island of Banca; where in 1817, about 2083 tons were smelted. Copper also occurs. Sulphur is obtained in great abundance and purity in Java. Borneo is celebrated for its diamonds: one of which in the possession of a prince of Matan, is valued at $1.194.350.

Australia, or New Holland.

Descr. The details of the geology of particular portions of this vast island, or rather continent, have not been given: yet numerous facts on the subject are known. The coast is for the most part rocky and mountainous; and the rocks which have been recognized, are granite, mica slate, talcose slate, quartz rock, ancient sandstones and limestones, the coal formation, red marl, with salt and oolite: also porphyry, greenstone, clinkstone, amygdaloid, and serpentine. There are some traces of former volcanos. Topaz and agate are the most interesting minerals.

Descr. Interesting collections of fossil bones from the limestone caverns of New Holland have been carried to England, and referred to at least 14 species of animals. These animals are all similar to those found living in Australia; such as the kangaroo, dasyurus, hypsiprimnus, wombat, a rodent, a saurian, and an elephant. These bones are much gnawed like those in the English caverns, and were probably carried into the caverns by carnivorous races.

Polynesia.

Descr. The predominant formations in this vast group of islands are the volcanic and the coral reefs. Those islands formed by the latter, are low and level: but the others rise to a great height. Mouna Roa in the Sandwich islands is more than 16.000 feet high; and Mouna Kea, 18.400 feet. These islands, with a surface of 4000 square miles, are all volcanic: Kirauea being the most remarkable volcano on the globe. The Society islands are composed mostly of igneous rocks of greater age; as basalt, which frequently contains most of the zeolitic minerals. The Friendly islands are mostly volcanic; as are also the Marquesas, the Gallipagos, the New Hebrides, and Gambier's islands. Juan Fernandes is composed wholly of basaltic greenstone. The South Shetland and Orkney islands are made up of primary rocks, along with those of more recent igneous origin. The coral islands are Elizabeth Island, Whitsunday, Queen Charlotte, Lagoon, Egremont, &c. &c. The shores of nearly all the volcanic islands are lined with coral.

AFRICA.

Egypt.

Descr. The southern part of this country is primary; consisting of granite, syenite, and other primary rocks. The felspar in the granite and syenite is red, and the rock has been used for many of the monuments of Egypt, as Pompey's Pillar, Cleopatra's Needle, &c. Syenite took its name from one of these localities, near Syene. North of Syene succeeds a region of sandstone, which appears to belong to the more recent of the secondary rocks. This is succeeded by the limestone tract, which reaches from Thebes several days journey south. Between Thebes and the Mediterranean, the country is wholly alluvial. On the west side of the Nile are the most remarkable examples of dunes that are known.

Nubia and Abyssinia.

Descr. Of the latter country nothing of importance is known respecting its geology, though some gold has been found, and large quantities of salt are obtained from a large plain; perhaps the site of a former salt lake. In Nubia, travellers mention granite, syenite, porphyry, sandstone, and limestone; and formerly mines of gold were wrought there.

Barbary or Northern Africa.

Descr. The greater part of northern Africa is a level sandy plain. But the chain of mount Atlas extends nearly across the whole of Barbary from west to east, and rises to the height of 13.000 feet. Its higher parts are composed of granite, gneiss, mica slate and clay slate. The northern part of the chain is composed, in a great measure, of calcareous rocks, resembling the lias, though sometimes approaching the transition limestones. Secondary sandstones also occur, and above these, tertiary strata. Among the secondary and tertiary formations, trap rocks are included. Salt springs and gypsum are found.

Western Africa.

Descr. This embraces the west coast of Africa, from the river Senegal in 18° north latitude, to the river of Benguela in 18° south latitude. Little is known of its geology. The hills around Sierra Leone are granitic. The district through which the river Zaire flows, is composed of gneiss, mica slate, clay slate, primary limestone, granite, syenite, and queenstone. Not a little gold (estimated in the beginning of the last century from 200.000 to 300.000 pounds sterling annually,) is obtained from the interior, especially along the Gold coast, and at the head of the Senegal and Gambia rivers. In Angola are very extensive deposits of salt, and some iron and copper.

Southern Africa.

Descr. The southern part of Africa is crossed by three ranges of mountains, running parallel to the coast; the height increasing towards the interior, till the innermost range rises 10.000 feet. North of these mountains are extensive table lands. The rocks composing these mountains are gneiss, clay slate, graywacke, quartz rock, and sandstone. Sandstone is most abundant; and this forms most of the rock in the table land beyond the mountains, as far north as 30° of latitude. Its strata are usually horizontal; and they give a tabular shape to the summits of the mountains. Table Mountain at the cape of Good Hope, is formed in this manner. Through these rocks are found protruding masses of granite, as well as basalt, pitch stone, and red iron ore. The relative situation of all these rocks is the same as in Europe.

Eastern and Central Africa, and Sahara, or the Great Desert.

Descr. Nothing appears to be known of the geology of the eastern side of Africa. Of Central Africa, which is traversed by the mountains of the Moon, much more is known. The more elevated parts consist of gneiss, mica slate, quartz rock, hornblende rock, clay slate, and limestone; traversed by granite, greenstone, and other trap rocks. At Goree are fine examples of basaltic columns. Extensive beds of rock salt are quarried in Soudan, and salt and natron lakes occur. It is also from this country, than most of the gold carried to the western and the eastern coast is obtained. The Great Sahara or Desert, is chiefly covered with quartzose and calcareous sand; and is the most desolate region on the globe. Occasionally, however, the rocks rise through the sand, and are found in the eastern part to consist of limestone and sandstone, containing rock salt and gypsum, and traversed by trap rocks. Where the Sahara reaches the coast, the rocks are chiefly basalt.

African Islands.

Descr. A large proportion of the islands around Africa are volcanic, and are composed of lava or basalt. Of this description are the Azores, the Canaries, where the volcanic peak of Teneriffe rises to the height of 12.000 feet: also the Cape Verd islands, St. Helena, Bourbon, &c.

AMERICA.

Descr. The longest chain of mountains on the globe, and with one exception the highest, traverses the whole length of this continent near its western side. In South America it is called the Andes: on the isthmus of Panama it sinks into a comparatively low ridge: but rises into the table land of Mexico and Guatimala, 6000 feet high; and some conical peaks shoot up much higher. Beyond Mexico this chain is prolonged in the Rocky mountains; which rise in some places about 12.000 feet. This chain continues nearly or quite to the Northern Ocean. Parallel to the Atlantic, the Appalachian, or Allegany mountains traverse the U. States, rarely rising more than 6000 feet and usually not more than 2000 or 3000 feet. Detached branches from these mountains extend irregularly into Canada, Labrador, and around Hudson's Bay. The mountains which form the West Indies, may be regarded as the southerly prolongation of the Alleganies. Beyond the delta of Orinoco, they appear again in numerous ridges in Guiana and Brazil, extending to the La Plata.

Descr. America is distinguished by vast plains as well as grand mountains. One of these lies along the Atlantic, of variable width, through the whole of the United States, extending to the Appalachian ridge, and occupying a wide extent through the whole of South America. A similar plain, but much narrower, occupies the western side of the continent. But the most extensive plain is the vast region lying in North America between the Appalachian chain and the Rocky mountains. Another, almost equally vast, occurs in the heart of South America. In its northern part is a vast expanse along the Orinoco: and in the southern part, are the immense Pampas of La Plata.

South America.

Descr. The basis of the vast chain of the Andes is composed essentially of gneiss and granite: but these are covered by an immense deposit of ancient volcanic rocks; such as pyroxenic porphyries, diorites, basalt, trachytes, with wacke-like conglomerates. The elevated table lands in the vicinity of this range, are covered in a measure by fossiliferous limestone, which occurs from 9000 to 14.000 feet above the ocean: and by new red sandstone, embracing ores of copper and gypsum

The lower table land is covered by diluvial detritus, embracing gold. The mountains in the central parts of America, lying east of the principal Cordilleras, are in a measure composed of various slates, with quartz rock and sandstone: the central axis being usually primary, and the slopes the older secondary. The plains between the different chains are extensively covered by tertiary strata. These strata, with white marl analogous to the lower chalk, and a sand like the green sand, are well developed on the shores of Patagonia. Indeed, the chalk formation has an immense development in South America, and rises in some places nearly to the height of 13.000 feet. *De Buch's Petrifactions receuillies in Amerique, &c. p.* 18. In Brazil, granite frequently abounds in the lower parts of the country, as well as in the mountains, where it is associated with nearly every other stratified and unstratified rock of the primary group. In short, nearly every stratified and unstratified rock on the globe, from the oldest to the newest, from granite to lava, and from gneiss to alluvium, are developed in South America on a magnificent scale, and occupy the same relative position as in other parts of the globe. *Humboldt on the Superposition of Rocks: Rozet's Geologic, p.* 525.

Descr. South America has long been celebrated for its mines of gold, silver, platinum, and diamonds. In Chili the annual produce of the gold and silver mines is about $8.500.000. More than a hundred copper mines exist, and are much more profitable than those of gold and silver. Mines of quicksilver, lead, tin, iron, and antimony also exist there, as well as deposits of nitre, rock salt, and coal. But mining operations are at present in a wretched state; owing to wars and political convulsions.

Descr. Brazil is most celebrated for its gems, especially its diamonds. They are mostly explored in the beds of rivers by washing the soil. Upon the first discovery of these mines, they sent forth a thousand ounces of diamonds; which affected the market powerfully. At present the annual produce is about 22.000 carats; or 183 ounces. The Brazilian topazes and emeralds are very fine, as well as the chrysoberyl, amethyst, and quartz crystals. Gold is also sought after with success in the auriferous sands. Copper ores are likewise abundant; and deposits of common salt and nitre occur.

Descr. Colombia affords some diamonds; also emeralds, sapphires, hyacinths, precious garnets, turquoises, and amethysts. Great expectations have also been excited respecting gold and silver; but from 1810 to 1820, only $2.000.000 were annually coined from the native metal: but much more might be obtained with proper management. Copper and quicksilver mines are wrought, as well as tin and rock salt; the latter yielding a revenue of $150,000.

Descr. Peru, including Bolivia, or Upper Peru, is perhaps the most remarkable region on the globe for the precious metals. Gold, silver, and mercury, are the most important. The silver mountain of Potosi alone, yielded in 225 years, no less than $1.647.901.018; and yet but a small part of it has been excavated: the whole mountain being a mass of ore: and it is 18 miles in circumference. One mine of quicksilver is 70 feet thick, and formerly yielded an immense amount. The other mines of gold and silver are very numerous; but their produce is much less than formerly, on account of the imperfect manner in which they are wrought. There are also mines of copper, lead, tin, and rock salt. Most of the mines are situated in the eastern range of the Andes, which consists of mica slate, syenite, porphyry, new red sandstone, and oolite.

West Indies.

Descr. These islands appear like the fragments of a former continent; and most of them contain mountains, which, in Cuba, Hayti, and Jamaica, rise to the height of 8000 or 10.000 feet. The highest mountains in Cuba are mica slate, and through the secondary formations of the lower regions, project gneiss, granite, and syenite. Veins of gold, silver, and copper, also occur here, and coal exists in a vein,—which is a very rare occurrence. *Philosophical Magazine, Vol.* 10, *p.* 760. In the south part of Hayti is a mountain of granite. The highest part of Jamaica is composed chiefly of graywacke with trap rocks. Upon these lie red sandstone, marl, and limestone, with trap and porphyry; the whole covered by diluvium and alluvium. The eastern part of the Caribbean group, as Tobago, Barbadoes, Antigua, Bermuda, &c. are principally composed of limestone, probably tertiary. Antigua has been described minutely by Dr. Nugent, *Transactions of the Geological Society, Vol.* 5, and more recently by Prof. Hovey. *American Journal of Science, Vol.* 35, *p.* 75. It consists of graywacke, (*indurated clay* of Prof. Hovey,) recent calcareous deposits, and trap. Lying above the graywacke, is a singular siliceous deposit embracing an immense number of silicified trees of every size, up to 20 inches in diameter; along with vast numbers of shells. These form most splendid agates, and admit of a high polish. (*See fine specimens in Prof. Hovey's Cabinet bequeathed to Amherst College.*)

Descr. Prof. Hovey has also given an interesting account

of the geology of St. Croix. *Am. Journal of Science, Vol.* 35. *p.* 64. He finds there the same stratified formations as occur in nearly all the West India islands; viz. graywacke, and tertiary and alluvial limestone; but unstratified rocks are wanting The tertiary strata consist mostly of limestone, more or less in durated, and abounding in shells and corals; many of which are identical with those now living on the ocean. A similar de posit is now forming upon the shores: and it was in this recent deposit that the human skeletons, now in the British museum and garden of plants, were found. (Numerous specimens of the rocks and fossils of St. Croix and Antigua may be seen in Prof. Hovey's collection above referred to.)

Descr. The western parts of the Caribbean islands are mostly volcanic: such as Grenada, St. Vincent, St. Lucia, Martinique, Dominica, Guadaloupe, Montserrat, Nevis, St. Christopher's, and Jamaica. On some of these islands no eruption has occurred within the historic period: but craters are visible, and trachytic and basaltic rocks are common.

Descr. Trinidad is a continuation of the continent, and consists mostly of primary rocks. The famous Pitch Lake, three miles in circumference, has already been described.

Guatimala and Mexico.

Rem. The geology of Guatimala is very similar to that of Colombia, on the south, and of Mexico on the north: being in fact the same as that of the Andes generally; and, therefore, in this brief sketch nothing farther need be added, except to say that this country contains mines of silver and of sulphur.

Descr. Mexico consists mainly of a vast and very elevated table land; being in fact a flattening down of the Andes on the south, and the Rocky mountains on the north. This vast plain 1500 miles over, is occasionally diversified by an elevated insulated peak, which is often a volcano, recent or extinct. The basis of the country seems to be primary rocks; such as gneiss and granite; but its upper portion is covered with porphyry and trachyte. Secondary sandstone and limestone also occur. A line of volcanos, of which there are five principal vents, traverses Mexico from east to west, about in the latitude of the capital.

Descr. The rocks of Mexico are rich in gold and silver, particularly the latter. They occur in veins traversing clay slate and talcose slate, transition limestone, graywacke, and porphyry. New Spain yields annually 1.541.015 pounds troy, of silver: or two thirds of the silver which is obtained on the whole globe; and ten times as much as is produced by all the mines of Europe. The number of mines is 3000. The quantity of gold annually obtained is only 4315 pounds troy. Tin, copper, lead, mercury, and iron, are mined in Mexico to some extent, and the ores are abundant. Beds of rock salt occur on the Rio Colorado.

Westerly and Northerly Regions of America.

Descr. The regions embraced under this name are those lying west of the United States and north of the Canadas The Rocky mountains, which are a prolongation of the Cordilleras of South America, are the principal chain, and are composed of primary rocks with extensive secondary deposits upon their sides, among which are clay slate, graywacke, and limestone. On Mackenzie river, brown coal occurs in tertiary clay. The coast of the Arctic Ocean, from Cape Lyon to Copper Mine river, is composed of clay slate covered by trap rocks: and most of this northern region is strewed over by bowlders of limestone, granite, porphyry, greenstone, &c.

Descr. Melville island, in latitude 74° 26′ north, is composed of sandstone, probably of the coal formation, with remains of coal plants of a tropical character. Around Hudson's Bay, are found all the great classes of rocks, except the tertiary; and almost every other variety found in the other parts of North America.

Descr. No part of the extensive regions above described appears to be truly volcanic, except a strip between the Rocky mountains and the Pacific Ocean. Whether any recent volcanos exist there, may be doubtful: but genuine lava is brought from thence, as well as trachyte, and other volcanic productions; and immense deposits of basalt, massive and columnar, are found along the principal rivers. Volcanic mounds, cracked at the top, and surrounded by fissures, are numerous over the whole region. This is probably a continuation northward of the great volcanic chain of the Cordilleras of South America and Mexico.

Descr. The mineral treasures of this region have been but imperfectly explored. Near the head of Salmon river, on the west side of the Rocky mountains, Rev. Mr. Parker saw a bed of pure rock salt, among strata which he says resemble the description of those at Wieliczka, in Poland. He used the salt and found it good. The great salt lake in the same region farther south, contains so much salt that it is deposited in large quantities on the shore. Epsom salts occur also, on both sides of the Rocky mountains, in thick incrustations in the bottom of ponds, which have been evaporated in the summer. *Explor ing Tour beyond the Rocky mountains, p.* 328, *second Editior Ithaca,* 1840. Beds of rock salt occur in other places besides those mentioned above, west of the Rocky mountains and of great thickness. *Prof. H. D. Rogers's Report to the Britis Association at the Fourth Annual Meeting.*

United States, and British America.

Rem. A more accurate idea can be given of the geology of this vast region, by considering the United States and the British possessions as a single extended district.

Descr. In proceeding southeasterly from the coast of Labrador, we meet ere long with ranges of mountains, at first not high, having a southwest and northeast direction. On the south shore of the St. Lawrence they become very elevated, forming in fact the northeastern termination of the Allegany range. In the northern part of Maine they rise to the height of 5300 feet; and in New Hampshire, (the White hills) 6234 feet. The western side of this range forms the eastern side of Lakes Champlain and George, and the valley of the Hudson river. The range occupies all New England, and extends southwesterly into the middle and southern states, crossing the Hudson at the Highlands: and in Pennsylvania, Maryland, Virginia, North Carolina, South Carolina, and Tennessee, forming several parallel and lofty ranges of mountains; four of which are sometimes reckoned; distinguished as well by their physical features, as by their geology. The most easterly, Prof. Rogers denominates the *Eastern system* of mountains: the next in order, the *Blue Ridge system:* the third the *Appalachian*, and the last the *Allegany system.* They rise in Black mountain in North Carolina, to the height of 6476 feet, and in Roan mountain, 6234 feet.

Descr. On the north shore of the St. Lawrence, are two ranges of mountains running parallel to the river; one at the distance of 15 or 20 miles, and the other 200 miles distant. On the west side of Lake Champlain, is a remarkable development of these mountains, in Essex county in New York, where they rise to the height of 5400 feet. From Essex county these mountains stretch to Kingston, in Upper Canada, where they become blended with another low range, which has been traced along the northern shores of Ontario and Huron, and to the west of Superior, and probably extends nearly or quite to the Rocky mountains. These latter mountains form the western boundary of the region under consideration.

Descr. The vast region of comparatively level country between the Rocky and Allegany mountains, forms the valley of the Mississippi, which extends from the centre of Alabama, probably to the Arctic Sea; being not less than 2000 miles long and 1200 miles broad, and embracing more than 2.500.000 square miles. At its southern extremity, this valley unites with the Atlantic slope. This slope is several hundred miles wide at its southern extremity, but becomes gradually narrower until it terminates at New York; with the exception of Long Island, Nantucket, and Martha's Vineyard.

Descr. The way is now prepared for explaining the simple and grand outlines of the geology of this wide region. All the mountainous region east of Hudson river, even to Labrador, with some few exceptions to be noticed subsequently, is composed of primary rocks: such as gneiss, mica slate, talcose slate, quartz rock, and limestone; intersected and upheaved by granite, syenite, porphyry, and greenstone. South of the Hudson, this primary range is continued, as the Highlands, through New Jersey and Pennsylvania, where it terminates. Its southeastern border passes near New York City, including Staten Island, and extending to Perth Amboy. Here it is covered by red sandstone, until near Trenton it reappears and forms a second band of primary rocks, parallel to that just described as terminating in Pennsylvania, and separated from it by a trough of red sandstone. This most easterly belt, from 80 to 100 miles wide, ranges through Virginia, North and South Carolina, and Georgia, as far as Alabama river, where it passes beneath the alluvium of the Mississippi valley. On the southeast side, these primary rocks are covered by the cretaceous and tertiary strata of the Atlantic slope.

Descr. The mountains of Essex county, on the west of Lake Champlain, are composed chiefly of Labrador felspar and hypersthene rock: similar to the rocks in the northern part of Lower Canada, and the coast of Labrador: and a similar rock is found north of the lakes in Upper Canada: and occupying a large space south of Lake Superior in the western peninsula of Michigan; so that probably they may form a connected range almost to the Mississippi. Nearly all the rocks in Upper Canada appear to be primary.

Descr. It appears then, that the great central basin of the United States is surrounded by primary rocks, except on the north and south. The rocks which rest on the primary over most of this area consist of the varieties of those great groups formerly denominated graywacke, old red sandstone, and the carboniferous or coal formation. The first is now called the Upper and Lower Silurian System; and the second, the Devonian System. In this country, however, the Silurian System has been called the New York System, and the Apalachian Palæozoic System. The subdivisions of these great groups will be found upon the Table of various Classifications, affixed to the first section of this work. But the extent which they occupy, can be learnt only by a Geological Map; such as the author of this work means soon to publish as a companion to the Elementary Geology.

Descr. It is a remarkable fact, that while in Europe gypsum and rock salt are usually found in the new red sandstone group lying above the coal measures, in this country gypsum and brine springs, so far as ascertained, are universally situated in the rocks beneath the coal formation.

Descr. Very recently, for the first time in the United States, a bed of rock salt has been discovered, 18 miles from Abingdon in the southwest part of Virginia. It was struck in digging a well for salt water. Soil and rock were penetrated 50 or 60 feet, then a bed of gypsum 160 feet thick, and next the salt, which was found to be 60 or 70 feet thick. The rocks embracing the gypsum and the salt are sandstone and marly clay. *Am. Jour. Sci., Vol.* 41. *p.* 215, *July*, 1841.

Descr. It seems now to be well ascertained that the carboniferous or mountain limestone is developed on a vast scale in our western states: that it extends westward from Pennsylvania, at least to Fort Leavenworth on the Missouri river, and northward to the Falls of St. Anthony on the Mississippi.

Descr. In many parts of the vast deposit of transition rocks that have been described, a regular coal formation rests upon them; and it is a singular fact, that the coal along the southeastern part of the great valley of the Mississippi, is usually destitute of bitumen; and as we recede from the primary strata, it becomes more and more bituminous.

Descr. Some of these coal deposits, already described in Section III, are probably the largest in the world. The Appalachian Coal Field, lying mainly in Pennsylvania and Virginia, is 720 miles long, and 180 miles broad at its widest place; and consequently must contain as much as 100.000 square miles. Another field in Kentucky, Ohio, and Illinois, is 350 miles long and 175 broad; containing, of course, something like 26.000 square miles. A third, in Michigan, is about 150 miles long and 80 broad; containing 12.000 square miles. Besides these deposits, very extensive ones occur in New Brunswick and Nova Scotia, (covering at least 12.000 square miles;) and small ones in Massachusetts and Rhode Island, and in the eastern part of Virginia: all of them containing at least 150.000 square miles! That is, this vast area is probably all underlaid by beds of coal, varying in thickness from an inch or two, up to 50 or 60 feet; and in number, from one to 50! Who can form an adequate idea of the immense amount of coal which they contain! If in Great Britain the inhabitants already begin to look forward, not without some anxiety, to a period when their coal fields will be exhausted, we need not surely in this country be haunted by any such fears.

Descr. It has been intimated that there are some exceptions to the general statement that the whole country between Hudson river and Labrador is primary. Thus, along the St. Lawrence is a deposit of Lower Silurian rocks. The Upper Silurian occupies a considerable part of New Brunswick and Nova Scotia; and some of the eastern part of Maine: connected with trap and gypsum, and salt springs. In New Brunswick and Nova Scotia, as already observed, we have large deposits of carboniferous rocks. In Rhode Island and the eastern part of Massachusetts, are interrupted patches of a rock of the lower secondary series, containing anthracite. In the valley of Connecticut river also, strata of red sandstone occur, with protruding greenstone.

Descr. In going south of the Hudson river, the same red sandstone just described, as occurring in the valley of Connecticut river, appears in New Jersey; where it forms a wide belt southeast of the Highlands. Thence it passes through Pennsylvania, from Bucks to York counties: thence into Frederick county in Maryland; thence into Virginia; thence into North Carolina; and probably still farther south. Throughout the whole extent of this deposit, from Nova Scotia to Virginia, ores of copper, bituminous shale and limestones, and protruding masses of greenstone, are associated with it. In Virginia the deposit appeared to be eminently calcareous; and one of its lowest beds is the well known brecciated Potomac marble.

Descr. I have for many years been in the habit of regarding this rock as the equivalent of the European new red sandstone formation; and have so named it in my reports on the geology of Massachusetts. The geologists of our country, have generally adopted this opinion. And the more I examine, the stronger evidence I find, that it is the new red sandstone. There is not room in this work, to state fully the grounds of this opinion. But the main argument is as follows. Profs. W. B. and H. D. Rogers seem to have shown, that this rock in New Jersey, Pennsylvania and Virginia, lies above the coal formation. Now in Massachusetts, Connecticut, and New Jersey, fossil fishes have been found in it with heterocercal tails, of the genus Palæoniscus. Hence, according to Agassiz, this rock must be older than the oolite: because fish with such tails do not occur as high as the oolite. If, therefore, this formation lies above the coal formation, and below the oolite, it must be the new red sandstone.

Descr. Prof. W. B. Rogers has found that the sandstone containing beds of workable coal near Richmond in Virginia is of the age of the European oolite. This is proved by the vegetable remains occurring in it. But no where else in this country has this formation been identified. Very extensive strata occur, however, which have been referred by our geologists, on account of their organic remains, to the cretaceous group. They are entirely wanting in chalk, which gives the name to the European series; and consist in this country, of sands, sandstones, marls, and limestone. In New Jersey, Delaware, and Maryland, they are characterized by the green sand, which is so valuable as a fertilizer. Farther south they consist almost wholly of marls and limestones. From these strata Dr. Morton has enumerated 108 species of zoophyta, mollusca, and echinodermata: only one of which is common to this country and Europe. But there is such a resemblance between the genera, and those of the chalk, that it is supposed the rocks must belong to the cretaceous period. Several interesting genera of saurians, tortoises and fishes, also occur in these rocks; among which are the plesiosaurus, ichthyosaurus, mososaurus and the batrachiosaurus. Among the fishes are several species of sharks; some of which are common to the chalk of England.

Descr. The cretaceous rocks of this country probably commence as far east as the islands of Nantucket and Martha's Vineyard, which they may underlie, as well as Long Island; although hidden by drift and tertiary strata. They occupy a wide belt of country from New Jersey to Alabama; and much surface, also, in Mississippi, Louisiana, Tennessee, Arkansas, and according to Mr. Nicollet, extend from Council Bluffs on the Missouri, 1000 miles along that river, to the mouth of the Yellow Stone.

Descr. The principal tertiary deposits of the United States occur along the Atlantic coast of the southern states, to the southeast of the cretaceous rocks just described. These tertiary groups, however, commence as far east as the island of Martha's Vineyard, on the coast of Massachusetts; where they are well developed, and seem to belong to the miocene or eocene period. They next appear in the southeast part of New Jersey; from whence they extend almost continuously through the eastern portions of Delaware, Maryland, Virginia, and North Carolina; and in interrupted patches through South Carolina, Georgia,

Alabama, Mississippi, and Louisiana. Mr. Conrad and others suppose that they are able to distinguish in this vast area of our tertiary formations the three groups which are recognized in Europe; viz. the newer pliocene, the miocene, and the eocene. The post tertiary also, with shells of a highly arctic character, has been recognized in New York and Canada.

Descr. The drift of the United States has already been sufficiently described in Section VI.

Descr. The most extensive alluvial deposit in this country is the Delta of the Mississippi. But similar deposits exist along the banks and at the mouths of all our larger streams.

Descr. Either in alluvium or drift, the following extinct species of the higher orders of mammalia have been found. *Elephas primigenius* occurs in Kentucky, South Carolina, New Jersey, Maryland, and Mississippi: *mastodon maximus* in Kentucky, New York, Virginia, New Jersey, Connecticut, Mississippi, Ohio, Indiana, and most of the western states: *megatherium* on Skiddaway island, coast of Georgia; *megalonyx* in Virginia and Kentucky; several species of *ox* in Kentucky, and Mississippi; *cervus americanus*, or fossil elk, in Kentucky, and New Jersey: *walrus* in Virginia. Mr. Cooper estimates that the specimens already carried away from the Big Bone Lick in Kentucky, must have belonged to 100 skeletons of mastodon, 20 of the elephant, one of the megalonyx, 3 of the ox, and 2 of the elk. He is of opinion that these remains occur in drift. But at some other localities it is not improbable they may have become extinct still more recently.

Descr. No evidence of volcanic agency in this country east of the Rocky mountains, since the protrusion of greenstone, has yet been found. Pumice does indeed float down the Missouri; but it probably comes from the Rocky mountains. Thermal springs also occur on the western borders of Massachusetts; and near Seneca Falls in New York: and Dr. Daubeny considers the mineral springs at Ballston and Saratoga as thermal; likewise numerous and abundant springs west of the Blue Ridge in Virginia, whose maximum temperature is 102° F: in Buncombe county, North Carolina, with a temperature of 125°: in Arkansas near the river Washita, 200 miles west of the Mississippi, where are not less than 70 springs with a temperature from 118° to 148°. *Daubeny's Sketch of the Geology of North America, Oxford*, 1839, *p.* 69.

Descr. The mineral resources of the United States and the British possessions have as yet only begun to be developed The most important mineral in an economical point of view, is coal; but the extent of our coal fields and mining operations has already been given on pages 63 and 343. Details respecting the brine springs of the western states and Upper Canada have likewise been presented on page 196. The coast of Massachusetts and Maine furnishes a vast supply of beautiful granite. Vast deposits of the most beautiful porphyry also occur in Massachusetts, and of serpentine and steatite in various states; as well as inexhaustible supplies of primary and transition marbles.

Descr. An extensive deposit of gold exists in the primary range of strata extending from the river Coosa in Alabama, to the Rappahannock river in Virginia. The metal occurs in veins of porous quartz, traversing gneiss, mica slate, and especially talcose slate. It is found also in the detrital deposit covering the rocks. Silver has sometimes been found in connection with the gold, as well as lead and copper. Even as far north as Somerset in Vermont, gold occurs in the quartz traversing talco-micaceous slate: and probably it exists in the space intermediate between this place and Virginia. The value of gold sent to the mint from the gold region of the United States from 1823 to 1836, was 4.377.500 dollars; and it was thought that this was not more than one half of the actual product of the mines.

Descr. In the northwest part of the United States occurs one of the most remarkable deposits of lead in the world. The rock containing it is the carboniferous or mountain limestone; the same that is so rich in lead in Great Britain. The extreme length of this lead region, from east to west, is 87 miles, and its greatest breadth, 54 miles; covering about 80 townships, or 2880 square miles. It lies chiefly in Wisconsin; embracing a strip of eight townships in Iowa, and ten townships in Illinois. In 1839, it produced 30 million pounds of lead; and is probably capable of yielding 150 million pounds annually; or more than is obtained in all Europe. Lead exists also in considerable quantity in other parts of the United States; as in Hampshire county, Massachusetts, in Rossie, New York, &c. *Dr. Owen's Report on the Mineral Lands of the United States,* 1840.

Descr. The same region at the west that yields the lead, is also rich in copper ore, which has not yet been so extensively worked as the lead, for want of a market. At the Mineral Point Mines, however, upwards of a million and a half pounds of copper ore have been raised. *Owen's Report, p.* 40. Great quantities of zinc occur also in the same region; although neglected:—and the same metal is usually associated with galena throughout the country.

Descr. Another remarkable deposit of copper exists in the

western part of Michigan, on the south shore of Lake Superior. South of this district, masses of metallic copper have been found in the drift over an area of several thousand square miles A single mass on the Ontonagon river, contains about four tons of copper, which has lately been carried to Washington. Dr. Houghton has recently ascertained that these fragments are derived from veins of native copper, with the gray and red oxides, carbonate, silicate, and other ores, traversing greenstone, amygdaloid, and an overlying conglomerate, very probably a member of the new red sandstone series. These veins will probably be very rich and productive. *Dr. Houghton's Report on the Geology of Michigan, for* 1841.

Descr. Within a few years, many companies have been formed in the eastern cities for prosecuting mining operations in the copper region of Lake Superior. Some of the veins have been wrought to a considerable depth, and found to be very rich in copper; and in some instances native silver is associated with the copper. Some of the masses of native copper found there, exceed by far, in size, anything of the kind ever before discovered. It is said that recently it is found that a similar mineral region exists on the north side of Lake Superior in Canada. Upon the whole, while it is probable that a large part of the adventurers and speculators in these operations will lose their time and money, a few will prove eminently successful.

Descr. Copper, both native and as ore, is common in the trap and new red sandstone in Massachusetts, Connecticut, New Jersey, Virginia, &c.; but it has not yet been extensively explored. Tin has been found in small quantity in Massachusetts, Connecticut, and New Hampshire; but not in workable quantity.

Descr. The deposits of specular iron ore in Missouri, are among the largest on the globe. They are connected with porphyry, and are separated from the metalliferous transition limestone of that region, by a deposit of granite with trap dykes, six miles in width. Pilot Knob, which rises 500 feet, is partly, and the Iron mountain, 300 feet high, and 2 miles in circumference, is entirely composed of this ore. Vast deposits of iron ore exist also in the northern parts of New York; and sufficient for present use is found in almost every part of the country. The primary regions of New England, New Jersey, and Pennsylvania, abound especially in the magnetic oxide. In 1840, the quantity of pig and bar iron manufactured in the United States was 84.136 tons. In Nova Scotia are valuable mines of iron.

Descr. The gems or precious stones of this country have as yet excited but little attention, because the expense of cutting them here is so much greater than in Europe. Hence we cannot say, in many cases, whether our localities will furnish elegant specimens or not. It may be well, however, to mention the most important. Prof. Shepard has lately discovered the diamond in North Carolina. *Amer. Jour. Sci., Vol. 2. p. 253, Second Series.* The blue sapphire occurs in Orange county, New York, in New Jersey, North Carolina, and Connecticut: the chrysoberyl at Haddam in Connecticut, and in Saratoga county, New York; the spinel, especially the blue variety, in Orange county, New York; as well as in Massachusetts; the topaz at Monroe in Connecticut, in great quantity, but coarse; the beryl in a variety of places; but pure enough for jewelry, at Royalston in Massachusetts, and at Haddam in Connecticut, (at the former place a rare variety can hardly be distinguished from the chrysolite); the zircon or hyacinth in the northern part of Vermont, in New York, in Canada, and North Carolina: the garnet in almost every part of the primary region of our country; and the pyrope especially, of great beauty, at Sturbridge in Massachusetts: the cinnamon stone, a variety of garnet, at Carlisle in the same state: the green and red tourmaline, of great beauty, at Paris in Maine; also at Chesterfield in Massachusetts; limpid quartz at Little Falls and Fairfield in New York, and a multitude of other localities: the smoky quartz in a variety of places; as at Canton in Connecticut, Williamsburgh and Brookfield in Massachusetts: the amethyst in Nova Scotia is very fine, and at Bristol in Rhode Island, and many other places: jasper in a great variety of places: hornstone and chalcedony of various sorts, forming agates, in the trap ranges of the eastern states, along the shores of Lake Superior, &c.; chrysoprase at Newfane in Vermont: iolite at Haddam in Connecticut, and Brimfield in Massachusetts: adularia of various colors in Brimfield, Southbridge, &c. in Massachusetts: labradorite in Essex county, New York, and hypersthene at the same place: kyanite at Chesterfield in Massachusetts: fluor spar in the northern part of New York; Satin spar at Newbury, &c. precious serpentine at the same place: red oxide of titanium at Middletown in Connecticut, and Windsor and Barre in Massachusetts.

General Inferences.

Inf. 1. It appears that the axes of all the principal chains of mountains on the globe, are composed of primary rocks, stratified and unstratified; while the secondary series lie upon their flanks, at a lower level; and the tertiary strata at a still lower level.

Inf. 2. Hence a similar process of the elevation of continents at successive epochs, has been going on in all parts of the world.

Inf. 3. Hence there is every reason to believe that continents once above the waters, have sunk beneath them, as those now above the waters were gradually raised: for since the quantity of matter in the globe has always remained the same, its diameter cannot be enlarged permanently; and therefore as one part rose, other parts must sink.

Inf. 4. Hence the geology of any district that embraces all the principal groups of rocks, affords us a type of the geology of the globe. This is what we should expect from the uniformity and constancy of nature's operations; and facts show that such is the case.

Inf. 5. Hence we have no reason to expect, that new discoveries in unexplored parts of the earth, will essentially change the important principles of geology. Slight modifications of those principles are all that can reasonably be expected from future researches.

INDEX.

H

I.

Q

R

www.ingramcontent.com/pod-product-compliance
Lightning Source LLC
LaVergne TN
LVHW010146110826
845151LV00002B/436

* 9 7 8 1 4 2 5 5 3 8 2 7 9 *